Second Edition

Transportation Tunnels

S. Ponnuswamy
Former Additional General Manager
Southern Railway, Chennai
and Guest Faculty, IIT Madras

D. Johnson Victor
Former Professor of Civil Engineering
Indian Institute of Technology, Madras

CRC Press is an imprint of the
Taylor & Francis Group, an **informa** business
A BALKEMA BOOK

Published by: CRC Press/Balkema
P.O. Box 11320, 2301 EH Leiden,The Netherlands
e-mail: Pub.NL@taylorandfrancis.com
www.crcpress.com – www.taylorandfrancis.com

First issued in paperback 2020

CRC Press/Balkema is an imprint of the Taylor & Francis Group, an informa business

ISBN 13: 978-0-367-57483-3 (pbk)
ISBN 13: 978-1-138-02981-1 (hbk)

Visit the Taylor & Francis Web site at
http://www.taylorandfrancis.com

and the CRC Press Web site at
http://www.crcpress.com

Foreword

With the ever increasing need to strengthen the transport infrastructure in terms of expressway, metro railway, and long distance high-speed railway networks, tunnelling has assumed a special importance over the past four decades. The construction of the 50 km long Eurotunnel across the English Channel has given a further impetus to the planners to conceive more and more challenging tunnelling schemes all over the world in order to improve the transport infrastructure. Planning, design and construction of transportation tunnels are specialised disciplines of the Engineering Science. Tunnelling has also gained a special significance in India in the recent past due to the immediate need to strengthen the metropolitan transport systems in the country. Successful implementation of large scale tunnelling in the Konkan Railway Project has further added to the confidence of engineers. The opening up of infrastructure projects to private investments coupled with adoption of the BOT concept for execution of such projects will further give a tremendous boost to "Tunnelling" in this subcontinent.

To my knowledge presently there is no exhaustive book on "Tunnelling" with reference to conditions obtaining in India. This book will, therefore, fulfil the much needed requirement of students and professionals in the field of tunnelling. The book has effectively covered important aspects of planning, design and construction of tunnels. Geological aspects including the details of soil and site investigations required to be carried out before undertaking tunnelling works have been lucidly brought out. Surveying and setting out which are very important aspects in execution of tunnelling works and secondary supporting system has been explained in sufficient details. Text in most of the chapters is supplemented by illustrative sketches and worked-out examples which will be of immense help to the students and designers. Apart from covering the conventional tunnelling methods in hard and soft rock formations, special attention has been given to the subject of Metro Tunnels by including a separate chapter on this topic. This will be of great help to the planners and engineers of Metro-Rail systems.

In spite of the difficult nature of the subject matter, the authors have succeeded in presenting the topic lucidly and at the same time maintaining a proper sequence. Due to this the book has become easily understandable. At the end of the book the authors have given a list of references, which will be of immense use to serious students of this discipline. This book will become an important reference book in the libraries of tunnelling engineers and will also be used as a much needed text book on the subject of tunnelling in the Engineering College.

28th March 1995

E. Sreedharan
Chairman
Konkan Railway Corporation Ltd.
New Bombay

Preface to Second Edition

Almost twenty years have passed since the First Edition was done. In this period, considerable changes have taken place in tunnelling technology, especially in terms of mechanization of work, introduction of road transport vehicles with better turning circle and loading capabilities for mucking and ventilation. Indian Railways have been and are implementing a large number of tunnels, especially in the Jammu-Kashmir area and the North- east. These works have been very challenging, since they pass through mixed and more of soft grounds. Along with Drill-Blast methodology mostly adopted in earlier tunnels they have been using NATM in a large scale. Designs are being refined using observational data during construction, calling for instrumentation on a larger scale.

With a number of cities going in for metro rail, large lengths of Subway construction have been and are being implemented. Major part of tunnelling for subways are being done with Tunnel Boring Machines, accompanied with soil stabilization with grouting, and monitoring using instrumentation.

Such developments have increased the demand for personnel with know-how on tunnelling. Even serving maintenance engineers have to be equipped with better knowledge on tunnel inspection and maintenance. Consequently, more technical institutions are offering courses on tunnelling. Hence it was felt desirable to bring out a new edition of the book, and the authors started planning and did preliminary work and started putting together material for this second edition. During the process, unfortunately, Dr. Johnson Victor passed away and the other author has continued the work in line with the original plans. With the publisher enthusiastically welcoming the idea, the revision work was followed up, to make the main author's wish a reality.

Two additional chapters on 'Instrumentation in Tunnels' and "Inspection and Maintenance of Tunnels' have been added to make it broad based. Some case studies from recent works have been added for the benefit of younger practicing engineers.

As in the case of the earlier edition, most of the additional information has been compiled from available literature, journals and internet; and also from personnel working in the field. A lot of details have been compiled from the literature sourced from CMRL, KRCL and IPWE(I) and IRICEN, USBRL, NF Railway and other web sites and ICJ and other journals, which is gratefully acknowledged. Help has been forthcoming from a number of authors' erstwhile colleagues and a number of experts in the field. Dr. G. Narayanan, who has had considerable hands-on experience on railway tunnels reviewed a major part of the book and gave a number of suggestions for addition and modifications. Mr. S. Suryanarayanan who had worked on a number of tunnels in the DBK project reviewed the case studies and also the chapter on Instrumentation. Mr. S. Subramanian, another railway expert, has reviewed the chapter on Maintenance, and enriched this with additional information on mobile inspection platforms. M/s R. Ramanathan, L. Prakash, S.D. Limaye, V. Somasundaram, V.K. Singh, Dr. Esther Malini and a number of serving engineers have helped with information on subway tunnels and tunnelling problems. The help received from all of them is gratefully acknowledged. Heartfelt thanks are due to Mr. S. Dinesh for help in preparation of a number of figures added and Ms. Sri Vidhya for providing secretarial help.

Thanking the publisher for the excellent work done in bringing out this edition in a short time, this edition is gratefully dedicated to the memory of the co-author, Dr. Johnson Victor.

March, 2016 **S. Ponnuswamy**

Preface to First Edition

Tunnelling has assumed increased importance in recent years because of the need to expand the transportation infrastructure in the country to cope with the anticipated magnitude of the economic activities. Major development projects such as the Konkan Railway project, rapid transit schemes in metropolitan cities, expressways connecting major cities, and plans for strengthening National Highways call for numerous tunnelling works of varied magnitude, requirements and complexity. At the present time, there is a lack of suitable text books to train civil engineers in the art and science of tunnelling, with emphasis on conditions obtaining in India. The authors have attempted to fill this need. It is hoped that the curriculum for Civil Engineering courses would be updated to include adequate coverage of tunnel engineering.

This book has been based on the observations at construction sites and the study of published and unpublished literature on the subject by the authors over many years, and especially on their experience in teaching a graduate course on Transportation Structures at the Indian Institute of Technology, Madras. The book is primarily addressed to final year undergraduate and graduate students in Civil Engineering, and is expected to serve as reference material to the practicing engineer. A specialised book of such coverage and complexity has to necessarily benefit from inferences from the relevant IS codes and available published literature and many unpublished quoting the reference. The authors record their gratitude to the authors of the above sources. Readers requiring additional information on particular topic are encouraged to study the concerned source material.

The authors are deeply indebted to Mr. E. Sreedharan, Chairman, Konkan Railway Corporation Limited for kindly writing the Foreword to this book. Mr. Sreedharan's contribution to the field of tunnel engineering in India has been a source of inspiration to may an Indian engineer.

The authors would appreciate suggestions from the readers for the improvement of the text. Sincere thanks are due the publishers for their effective cooperation.

Madras
June 1995

S. PONNUSWAMY
D. JOHNSON VICTOR

Contents

CHAPTER

1

Introduction

1.1 GENERAL

Tunnels are artificial passages built underground to facilitate transportation or conveyance of people, materials, water, sewage, other fluids and gas in pipes, electric power etc., across obstructions like hills, rivers and other obstructions like buildings, industrial structures and other communication lines like major roads and rail tracks. An alternative definition of tunnels refers to underground structures which apart from serving the above noted purposes are built using special underground excavation methods without disturbing the surface. Even underground garages and power houses are treated as tunnels.

Tunnels have been built from time immemorial for various purposes, such as defence, assault/ escape and normal traffic across fortifications and water bodies. The earliest known tunnel was constructed about 4000 years ago by Queen Semiramis in ancient Babylon under the Euphrates River to connect her palace and the temple of love[1]. The tunnel was 1 km long and was of section 3.6 m × 4.5 m. It was constructed using the "cut-and cover" method with brickwork in bituminous mortar and vaulted roof. In Indian history, references are available to tunnels connecting forts and points of escape and private passages for the royalty from the palaces to river fronts and temples. (One such has been reported to have existed near River Godavari at Rajahmundry).

Current-day vehicular tunnels may be built for highways or railways and may be unidirectional or dual-directional. Often tunnels reduce distances. For example, the Banihal (Jawahar Road) tunnel joining the Kashmir valley with the rest of the country has reduced the road distance by 18 km, besides facilitating year-round communication. The world's second largest tunnel is the under Sea, Channel Tunnel linking Great Britain and France by rail. Considered an engineering marvel of the twentieth century, the tunnel is 50.5 km long and lies 50 m below the seabed for most of its length.

1.2 CLASSIFICATION

Tunnels can be broadly divided into two categories: (a) transportation tunnels and (b) conveyance tunnels. Some define (a) as traffic tunnels and define transportation tunnels to include tunnels used for conveyance of water to Hydro electric Power plants, water supply tunnels, sewage tunnels and tunnels used in industrial plants for conveyance of materials like those housing conveyors etc..

Transportation tunnels can be further classified as: (a) railway tunnels, (b) highway tunnels, (c) pedestrian tunnels, (d) navigation tunnels and (e) subway tunnels.

Conveyance tunnels serve to convey liquids and may include: (i) hydroelectric power station tunnels, (ii) water supply tunnels, (iii) tunnels for the intake and conduit of public utilities, (v) sewer tunnels and (v) tunnels in industrial plants, such as conveyor-belt tunnels.

We can also include under (i) above tunnels which have been driven for purposes of diversion of water during construction of dams. The earliest example of this use in India is the Periyar tunnel, which has been used as a permanent means for diverting water from the western slopes of the Western Ghats to the East. A recent major example of such conveyance tunnel is that used for temporary diversion in connection with the Bhakra dam construction.

1.3 TRAFFIC/ VEHICULAR TUNNELS

1.3.1 Major Transportation Tunnels.

Railway and highway tunnels are similar in nature and normally refer to surface-to-surface route tunnels, i.e., those provided for the purpose of crossing hills and mountains as distinct from subway tunnels (also known as tubes) used for underground railway in cities. Typical examples of a few remarkable railway and highway tunnels[2] are given in Table 1.1.

Table 1.1 Selected Railway and Highway Tunnels

Sl No.	Name	Country	Year	Length, km
1.	Seikan	Japan	1988	54.1
2.	Channel	UK-France	1993	50.5
3.	Simplon I & II	Switzerland-Italy	1906 & 1922	19.8
4.	Kanmom	Japan	1974	18.6
5.	Apennine	Italy	1934	18.5
6.	St. Gotthard	Switzerland	1882	15.0
7.	Lotschberg	Switzerland	1913	14.5
8.	Cascade	USA	1929	12.6
9.	Moffat	USA	1928	10.0
10.	Pir Panjal	India	2013	10.9
11.	Karbude	India	1995	6.5
Highway Tunnels				
12.	St. Gotthard	Switzerland	1980	16.2
13.	Arlberg	Austria	1978	14.0
14.	Frejus	France - Italy	1979	12.8
15.	Mont Blanc	France - Italy	1965	11.7
16.	Enassan	Japan	1977	8.4
17.	Transbay	USA	1973	5.8
18.	Kanmon	Japan	1958	3.4
19.	Mersey	UK	1934	3.2
20.	Holland	USA	1927	2.6
21.	Jammu-Srinagar (Banihal)	India	1961	2.6

Tunnels are warranted in the context of transportation for the following purposes:

(a) to avoid a long circuitous route around a mountain or spur;

(b) to avoid slips of slides of open cuts in soft strata;

(c) to avoid steep gradients in hilly terrain;

(d) to avoid crossing precipitous ridges or high peaks or zones likely to be under snow for a major part of the year;

(e) to avoid acquisition of valuable property or to avoid interfering or damaging a heritage structure.

However, tunnels are attendant with some disadvantages, such as:

(a) high initial cost;

(b) long construction period;

(c) specialised work, needing special equipment and highly skilled labour.

From the point of view of economics, a tunnel is preferred when depth of cutting through hard solid exceeds 18 to 20 meters.

Amongst transportation tunnels, railway tunnels are more numerous. Most have been constructed under water also, e.g., tunnel of New Tokaido line connecting two islands across the sea channel and the Channel Tunnel connecting France and England.

Subaqueous tunnels require fairly heavy approach gradient, take longer to construct and involve more personal risk to the workers. Maintenance costs are also higher, especially in seismic zones.

1.3.2 Subway Tunnels

Urban underground railways are mostly in the form of tunnels, otherwise known as tubes (after the shape used in boring). The earliest constructed was that provided for the London Tube, the first section of which was commissioned in 1863. The special requirement of underground and subaqueous railway tunnels, which distinguish them from the other railway tunnels, are:

(a) Increased safety requirements **due to high density** and high speed of traffic and likely disastrous consequences of any derailment/accident;
(b) Careful water sealing;
(c) High standard of cleanliness
(d) Ventilation with or without air conditioning, with duplicate power supply facilities;
(e) Full length needs to be provided good illumination and communication facilities for help in inspection/maintenance operations and for emergency evacuation of commuters.
(f) Availability of fire emergency facilities over full length and adherence to National Fire Hazard regulations.

1.3.3 Highway Tunnels

These are similar to railway tunnels except that due to the steeper permissible ruling gradient they may be shorter and require fewer spiral alignments. In cross-section they are now comparatively less in height and wider.

The additional factors to be considered in design and construction of highway tunnels are:

(i) Size: They have to be wider to accommodate the number of lanes of roadway to be carried. Hence, their width-height ratio is more than that of railway tunnels.

(ii) Shape: In view of the greater width required and also the need to carry additional services, a circular shape more often fits in better, with services carried through ducts provided in the lower half of circle.
(iii) Geometry: The curvature has to take into account the higher speeds of vehicles (which cannot be externally controlled) and also the need for a good view of each other by opposing traffic lane users.
(iv) Ventilation: Artificial ventilation (by induced draft through ducts) becomes a 'must' in view of harmful fumes and gases emitted by cars, buses and trucks.
(v) Lighting: Artificial lighting also is necessary for proper viewing inside by various types of users.
(vi) Drainage: Since the road surface and pavements have to be kept dry and non-slippery, no dripping from roof or sides can be allowed. Lining has to be waterproof and effective side drains to lead out seepage and other water become necessary.
(vii) Lining: Even where structurally not needed, lining is necessary for purposes of aesthetics, better lighting (reflection) and to control seepage. Properly cambered road surface, footpath and drains have to be provided at the invert level also.

1.3.4 Navigation Tunnels

Water transport is a form used by the man since the early days using rivers and streams for the purpose. Canals had been primarily built for purpose of irrigation and as early as 16th century, they have been used for inland transportation, one of the earliest examples being a canal built parallel to River Exe in Devon (UK) for the purpose in 1564. Industrial development gave a boost to this form of transport, before the advent of railways in the eighteenth century. Taking them across mountains through tunnels was a natural corollary and Navigation tunnels came into being. Rapid development of canals for inland transport and tunnels where required grew rapidly. For quite some time, they competed with the railways also. Rapid growth of railways and roads, with advent of motorised vehicles capable of providing quicker transport and door to door service, edged canal form of transport from reckoning.

Navigation tunnels are similar to highway tunnels. They have to be comparatively wider to allow for maneuvering of boats and provide sufficient space between boats plying in opposite directions. The waterway has to be provided with raised walkways on either side for movement of people. The waterway portion has to be impervious to prevent loss of water. They will have gentle or no gradient to suit the flow of water.

1.4 CONVEYANCE TUNNELS

The conveyance tunnels cover those provided for water supply, sewerage and those provided for housing pipe lines for fluids or conveyor belts, as in case of mines, power houses etc.. They also include tunnels provided in industrial plants for local conveyance of materials and products in any of the forms mentioned above. This classification includes also tunnels provided for diversion of water permanently for feeding power houses or transferring from one valley to another for irrigation and other purposes.

1.5 RAILWAY TUNNELS

The earliest railway tunnels to be built for steam powered Railways seem to be the ones built in Derbyshire in United Kingdom in 1830, most of them being short ones. The longest one to be built then was the 'Wymington'tunnel, 1690 m long built in 1859. The most notable early railway tunnels known world over are the Simplon Tunnels I and II in Switzerland connecting Italy with Germany, the first one built in 1906 and the second one in 1922. They are single line tunnels. They have been supplemented recently by the World's longest rail tunnel (as in 2016) known as Gotthard Base Tunnel, comprising two parallel circular (8.83 m - 9.50 m dia.) sections. The work on them is reported to be just complete, and they are under trial. They are likely to be commissioned in June 2016. It is 57 km long running parallel to each other with interconnections at intervals. Two multi-purpose stations in between have been (each in a single encompassing tunnel). They are designed for operation of passenger trains which can be run at 250 kmph and freight trains hauling 3500 t each. They are expected to save over an hour in running time of high speed trains for passengers. Longest under water rail tunnel as on date if the 53.8 km long Seikan Tunnel in Japan connecting Honshu and Hokkaido islands. Of this length, 23.5 km is under sea bed, considered deepest in the world. The other notable under sea tunnel is the 50.5 km long Channel Tunnel (Chunnel) connecting France and United Kingdom, designed for high speed train operation also. It has the longest under-sea length of 37.9 km. Though talked of since 1802, its construction was started in 1988 and completed in 1994. It consists of two circular sections of 7.6 m dia each, with a smaller 3.0 m dia service tunnel running parallel in between, with multiple purpose of providing ventilation and emergency services etc.. It also served as a pilot tunnel to know the kind of soil to be bored through, in advance, helping the drilling of main tunnels. It is a joint venture project involving private finance, loans and equity, a BOOT project covered by a 60 year concession. The longest tunnel in USA is the replacement Cascade Tunnel in Washington State. The first one built on this section in 1900 was a single line one 4.23 km long. In replacement, this 12.54 km long single track was built in 1929.

For more details on this topic, see Chapter 6.

1.6 SCOPE OF THIS BOOK

This book briefly covers the theoretical aspects of preliminary studies and basic requirements, geological investigations, design requirements and practical aspects of location and setting out, various operations including boring/blasting, mucking, lining, drainage and ventilation of tunnels in general and transportation tunnels in particular.

1.7 DEFINITIONS

In discussions of the various operations involved in tunnelling, one comes across a number of technical terms. Some of these terms are defined below for ready reference[5,6].

Adit: A tunnel or open cut driven from the surface to the main tunnel for providing access or addition to the number of working faces of the main tunnel.

Benching: Operation of excavation in the lower portion of the tunnel section after the top heading has been driven.

Blocking: Filling gap between the excavated rock surface and the ribs to transfer the (external and) rock load to the ribs.

Bracing: Structural frame connection provided between ribs/posts to prevent the latter from buckling or shifting (alternatively, this purpose may be served by lags fixed to the frame).

Cover: Cover on a tunnel in any direction is the distance from the tunnel profile to the outermost ground surface in that direction. If the thickness of the overburden is large, (more than three times the diameter of the tunnel) its equivalent, as determined in terms of the density of rock may also be treated as cover.

Cut hole: Group of holes fired first in a round of blasting to provide additional free faces for the succeeding shots. (Definition applies only to drilling patterns.)

Detonator: A catalyst used for activating a high explosive charge. Catalyst itself is activated by a safety fuse or by electricity.

Drift: A horizontal tunnel (mini-tunnel of a small section and length) driven as a part of stage working or for exploratory purpose from an underground face or from the surface for exploration purposes.

Drilling pattern: A layout arrangement showing location, direction and depth of the holes drilled into the face of a tunnel.

Easer (holes): A ring of holes drilled around cut holes and fired soon after cut holes.

Explosive: Any mixture of chemical compounds which is capable of producing an explosion by its own energy. (It can be black powder, dynamite, nitroglycerine compounds, and any fulminate or explosive substance having explosive power equal to or greater than black powder.)

Heading: Generally applies to the face of the tunnel where actual tunnelling operations are in progress. If it is prefixed by the word 'top' or 'bottom', it denotes that part of section of the tunnel excavated first/ in advance.

Jumbo: A mobile platform with a number of decks used at the heading of large tunnels generally for drilling holes. It is also used for scaling, erection of roof supports like rock anchors, and for primary lining by guniting, shotcreting etc.

Laggings: Structural elements (planks, steel sheets, precast RC slabs) spanning between the main supporting ribs used for supporting sides or overburden.

Mucking: Includes all operations covering grabbing, loading and removal of blasted stones/material after blasting.

Overbreak: That portion of the profile which is excavated beyond the prescribed boundary line of the intended profile.

Payline or B-line: Refers to an assumed 'profile line' set beyond the desired profile line or A-line. It denotes the mean line upto which payment for excavation and concrete lining is to be made, whether the actual (accepted) excavation falls inside or outside it.

Primer cartridge: The explosive cartridge into which the detonator is inserted.

Profile line or A-line: Line of profile as per approved design, taking into consideration minimum clearances over moving dimensions and suiting the desired geometrical shape.

Rib, rib and post, or rib post and invert strut: Various components of the support system

Rock load: Height of mass of rock which exerts pressure on the support (and lining). This is computed taking into consideration nature of rock and size/ shape of tunnel.

Scaling: Operation of removal of all loose rocks and bits from the basted surface after blasting is completed.

Tunnel support: Structure erected inside the tunnel to support the strata above and around the excavated cavity, till the (permanent support) lining is placed. They include: (a) supports which are left in place and/ or embedded permanently or (b) temporary supports which are erected during excavation and removed before or during erection of either the permanent lining or providing permanent supports.

Wall plates: Longitudinal members provided generally at springing level to serve as sills for ribs above and transmit the load from the ribs through blocks or posts to the base.

Soft strata: Strata requiring supports to be installed within a very short period of excavation, but which at the same time cannot be easily excavated by hand tools. They include soft rocks (usually sedimentary or metamorphic) which are joined and faulted.

Soils: Disintegrated rocks or other loose strata requiring support immediately after and/or during underground excavation and which can be excavated by hand tools.

Stemming: Inert material like clay used for packing the shot hole over the last explosive charge upto its outer end.

Stoping: Overhead excavation operation, by drilling from below on a tunnel face (reversing work by excavating bench before doing top heading).

Trimmer (hole): Holes drilled on the periphery of an excavation and fired for achieving the intended final outline of excavation.

1.8 REFERENCES

1. Szechy, Karlowi, The Art of Tunnelling, Academia Kinda, Budapest, Hungary, 1970
2. Fauchtinger, M.E.; Hockbrucke oder Tunnel, Strausse und Autobahn, 1956
3. Ministry of Railways, 'Important Tunnels in India', Research, Design and Standards Organisation, Lucknow.

4. Agarwal, M.M. and Miglani, K.K. (2014) 'Global Experience of Design, Construction and Maintenance with Special Reference to Indian Railways', National Technical Seminar, on Management of P.Way Works through need based Outsourcing and Design, Construction and Maintenance of Railway Tunnels. Jaipur, Institution Permanent way Engineers (India), New Delhi
5. IS: 5878,- 1970. Construction of Tunnels, Part I, Bureau of Indian Standards.
6. Pequignot, C.A., 1963, 'Tunnels and Tunnelling', Hutchinson Scientific and Technical, London, 555 P.

CHAPTER

2

Route Selection and Preliminary Investigations

2.1 ROUTE SELECTION

2.1.1 Economic Consideration

Decision-making on any transportation project has to follow a detailed Techno -economic analysis. This is particularly significant in respect of tunnels and bridges which are the most expensive parts of civil engineering structures on a roadway or railway project. A length of road/track through a tunnel may cost up to even 10 times the cost of the road/track on plain land and 4 to 6 times that of the same in a cutting open to sky in hilly areas. Hence tunnels can be justified only by the compensating savings in terms of distance, time of travel and operating cost for the volume of traffic to be handled. Saving in length of road/ line will have a direct effect on the savings in length of haul and indirectly on operating costs in order to offset high construction costs.

The comparative costs of transportation tunnels may be computed as below[1]:

The construction cost C_t of the line involving a tunnel and the construction cost C_o of the alternative without a tunnel are given in equations (2.1) and (2.2):

$$C_t = L_t \cdot C_t + L_a \cdot C_o \tag{2.1}$$

$$C_o = L_o \cdot C_o \tag{2.2}$$

Where L_t, L_a and L_o are length of the tunnel, length of line open to sky on the approaches leading to and from the tunnel and length of alternative line without tunnel respectively in metres: C_t and C_0 are the total capital costs per unit length (linear metre or km) of tunnel and open line respectively.

Let O_t, be the total annual operating cost (including cost of maintenance of tunnel/open cut) for all vehicles using the tunnel route and O_0 the corresponding total costs of operation of vehicles on the alternative without a tunnel. Then the excess capital invested in the tunnel (using simple arithmetic and no interest charges) will be paid back in t years as given in equation (2.3)

$$t = \frac{C_t \cdot C_0}{O_t \cdot O_0} \tag{2.3}$$

The construction of tunnel will be justified when t is less than the commonly accepted amortization period. Under Indian conditions, the amortization period may be taken as 25 to 30 years with interest charges also taken into consideration.

Comparison may be effected not only based on the period of refund, but also on the basis of minimum annual costs U_{min} of construction (capacity demand) and of operation (Pequignot, 1963). The basic relationship for the same can be written in the following form:

$$U_{min} = g \cdot B + U, \tag{2.4}$$

where g is the standard efficiency or capacity demand coefficient; B the first investment cost and U the annual operation cost.

Similarly, the limit of depth of cutting at the entrance section for short tunnels through mountain spurs can be calculated from the formula:

$$K_b = F_b (K_t + K_{sz}) + M + tf_b = K_a + A + tf_a \tag{2.5}$$

where K_b is cost of expropriation/metre length; K_f unit cost of excavation; K_{sz} unit cost of transporting earth; M cost of supplementary structures in the cut (e.g., lining sides of cutting, drainage etc.); f_b annual maintenance cost of cutting per metre; K_a eventual expropriation cost of tunnel site per metre; A construction cost of the tunnel per linear metre; f_a annual maintenance cost of tunnel per linear metre; t standard time of refund; $F_b = kh + Sh^2/2$; S the slope; k the bottom width of cutting and h the limiting depth thus obtained. Table 2.1 would give an idea of the initial construction costs of a few tunnels in different countries and over a period.

On railways (as on any traffic tunnel) the overall economy of a project in construction and operation of the tunnel is dependent on making the best use of the ground configuration. This is especially significant in hilly terrain.

2.1.2 Topographic Consideration

The geology of the area also affects route selection to a major extent apart from the topography. In hill tunnels normally, as a rule of thumb, changeover to tunnel is made when cutting depth exceeds 20-25 m for normal cutting slope of 1.5 to 1. In soil requiring flatter slopes, even lesser overburden will justify tunnelling. In rock, practical tunnelling work problems mostly govern the balance of advantages and choice of critical depth.

Topography of the area has a bearing on the choice of the ruling gradient and consequently the length of the line or road. On the other hand, the operational requirement for route governs the adoption of minimum gradient and curvature standards. The original Alpine tunnels were deep and long because such choice offered immediate operational savings. On the other hand, on the Canadian Pacific Railway (Vermont Project) across the Rocky Mountains, very sharp gradients and sharp curves were adopted because traffic volume initially would not be sufficient to effect large saving by adopting flatter gradient. (Pequignot, 1963)[2].

In addition, the alignment should be such that it has ease of construction with minimum geotechnical obstructions. Some major geotechnical considerations to be kept in mind are:

(i) Avoid passing through main boundary thrusts (MBG)
(ii) Should not be parallel to 'main boundary faults' and thrusts
(iii) Avoid passing through shear zones or parallel to the strike of rock formation

Hence at this stage itself, some preliminary geotechnical study should be done with the help of a geologist with good knowledge of the area.

It should be kept in mind that the starting and end points of tunnel where portals have to be located, avoid steep slopes and not prone to landslides etc.

On the Kicking Horse pass, the original line was with 1 in 25 ruling gradient. It was later reduced to 1 in 45 looping the line two spiral tunnels driven into the mountainside when traffic became high enough to justify the cost. This added 6.4 km to the length of the line.

Iranian railways built prior to 1938 have a number of 'figures of eight' with spiral tunnels and gradient up to 2.8%. (The ‘figure of eight’ type alignment was also adopted on the Darjeeling Himalayan Railway in India near Ghoom.) On the Tehran-Tabrin line (700 km of single-line railway, the

Table 2.1 Costs of Selected Old Highway and Railway Tunnels

Location & period of construction	Length, km	Shape	Diameter or width/height, m	Lining	Rock material	Cost/m at time of construction, US$
Railway Tunnels						
Mont-Cenis 1857-72	12.70	Horseshoe	8.00/7.30	Brick & ashlar masonry	Volcanic rock	910
Simplon I, 1985-1906	19.80	Horseshoe	4.90/5.40	Ashlar masonry	Mixed rock	800
Simplon II, 1914-1915	19.80	Horseshoe	4.90/5.40	Ashlar masonry	Mixed rock	400
Great Apennine 1923-24	18.60	Horseshoe	8.70	Ashlar masonry	Marl, limestone	1,200
Moffat, 1924-1927	9.90	Horseshoe	7.40/4.80	Concrete	Limestone	1,550
Karbude, 1995	6.51	Segmental	4.92/6.24	Concrete	Basalt	10,000
Highway Tunnels						
Pennsylvania turnpike 1939-40	10.60	Semi-circle vault	6.90/4.30	RC	Marl, slate sandstone	1.165
Holland N.Y., 1920-27	5.08	Circle rock debris	6.00/3.95	Cast iron	Silt mixed with	9.500
Mersy 1925-34	3.18	Circle	19.00/5.70	Cast iron	Fissured rock debris	11,100
Lincoln N.Y., 1934-35	4.68	Circle rock debris	6.45/4.00	Cast iron	Silt mixed with	10,000
Memorial turnpike 1954	0.54	Semi-circle vault	7.20/4.30	RC	Sandstone and slate	6,200
Baltimore, 1954-57	2 × 3.77	Double circle	6.60/4.20	Steel sheet RC lining	Silt, sand and day	6,650

radius of curvature had to go down to 250 m. This line has been proposed to be rebuilt for high speed as a double line with a radius of 5000m and gradient of 1.0% except in hill areas where it would be 1.4%.

2.1.3 Other Considerations

Tunnels should be avoided in undermined regions, i.e., in mining districts, because of the unpredictable nature of settlements and forces to which the tunnels may be subjected. Apart from this, in choice of alignment, additional considerations are hydrological factors and construction requirements. Questions of accessibility and ventilation during construction and availability of local facilities for isolated work have to be considered.

2.2 GEOMETRICAL PARAMETERS

2.2.1 Horizontal Alignment

In providing the surface-to-surface tunnels, it is preferable to have straight alignment for the maximum length possible. The reasons for this are need for shortening tunnel length, economy in operation, better visibility and simplification in construction and surveying operations, including setting out. It also affords better ventilation. It is always preferable, however, to have straight alignments on both approaches since they must connect the railway or road alignments coming over the slope of the hill, which cannot be expected to always be straight. If curves are unavoidable, the radius of curvature should be restricted. The Russian practice is to restrict the curve radius to 250 m on main line railways and to 200 m on branch lines.

The recommended practice in India is to normally restrict the minimum radius to 450 m for main line BG railways and to 225 m on MG lines even though there are 6° (R = 290m) BG tunnels and 10° (R = 175 m) MG tunnels in some sections. On roads also, in order to afford better visibility to drivers, sharper curves should be avoided. For major highways, radius less than 300 m is avoided. On navigation and waterway tunnels, the flow conditions and losses in pressure head permissible will dictate the minimum radius. Location of points of entry and exit are governed by site conditions. Subaqueous tunnels are generally built normal to flow in waterways and should be as straight as possible. In underwater tunnels the approach roads may be parallel to the waterway but curves would have to be provided over the approach cuttings. In these cases sharp curves are unavoidable.

2.2.2 Gradients

1. Rail and other Resistances

As far as traction is concerned, the ruling gradient permissible in tunnels would be appreciably flatter than in open air, owing to reduced adhesion and increased air resistance on moving vehicles in tunnels. For example, the adhesion coefficient f on a hillside exposed to the sun for steam traction varies from 0.17 to 0.18. It may drop to as low as 0.15 to 0.16 in deeper cuts or in moist atmosphere at altitudes of around 1000 m. The atmosphere in the tunnels is usually saturated with moisture due to ground water infiltration. This moisture precipitates on the rails or on the road pavements, causing a significant reduction in the adhesion coefficient between wheels and rails or road surface. The reduced f for rail was found in experiments to be 0.11 to 0.12 for steam traction.

Some measurements taken for air resistance in the deep cuttings and inside the "Simplon tunnel', (which consists of 2 single-track tunnels spaced at 12 m apart) and provided with artificial ventilation, are as indicated below:

Train velocity, V (km/h)	50	60	70
Air resistance, open cutting (N/kN)	3.2	4.1	5.0
Air resistance in tunnel in Direction of draught (N/kN)	5.0	6.4	8.0
Air resistance in tunnel against the draught (N/kN)	7.5	9.5	12.0

Adapted from Reference 2.

Figure 2.1 gives the results of experiments made on air resistance in underground tunnels in London[3] (Turner, 1959). The air resistance is more in a tunnel than in the open area. It increases with the speed of the train and with reduction in the free space between the vehicle and the tunnel section. On the other hand the resistance due to curvature in tunnels is the same as on the surface line. Hence while fixing the ruling gradient for tunnels, all the factors mentioned above should be taken into consideration.

2. Drainage Requirements

Inside mountain tunnel, at ends gradient of at least 0.8% for the first 250 m falling outward at either end should be provided to ensure rapid drainage and to minimize the adjacent height of formation. The minimum recommended gradient for drainage inside the tunnel in other portions is 0.2%. In long tunnels, arrangements should be made to collect the water in sumps at suitable intervals and pumping out.

3. Operational Requirements

The main factors, apart from the above considerations, in fixing ruling gradient are the hauling power of the locomotives used on the line, density of traffic which would determine average load on train; speed proposed and line capacity requirements.

In early days, ruling gradients as steep as 4% had been adopted but present trend is to go in for 1.33 % to 1%. This subject is dealt with in more details in Section 3.1

2.3 PARAMETERS FOR HIGHWAY TUNNELS

The desirable maximum gradient is 3% and gradients exceeding 4% should be avoided. In very difficult conditions a 5% gradient may be acceptable but a gradient exceeding this should be exceptional. Such exceptions are 6% in Glyde tunnel and 5%in Tyne tunnel. The sharpest radius provided is 128 m. Gradients should not exceed 6%. If such limits are likely to be exceeded, it would be better to provide approach loops to ease the conditions.

Suitable vertical curves taking into consideration speeds and visibility lengths for drivers on roads have to be provided.

The roof of the tunnel should not obstruct the driver's line of vision for minimum lengths required for corresponding speeds. Alignment facing directly into the sun at the point of emergence into the open should be avoided as the glare will affect driver vision. This applies equally to railway tunnels.

2.4 INVESTIGATIONS

2.4.1 Types of Investigations

Investigations for tunnels comprise the following aspects:

Route location
Topographic Survey or Alignment Survey
Geological Investigations
Hydrological Investigations
Seismic Studies.
Environmental and Social impact studies and Mitigation measures
Traffic requirement and profile studies
Assess construction requirements, facilities available
Potential modes of failures
Risk Analysis

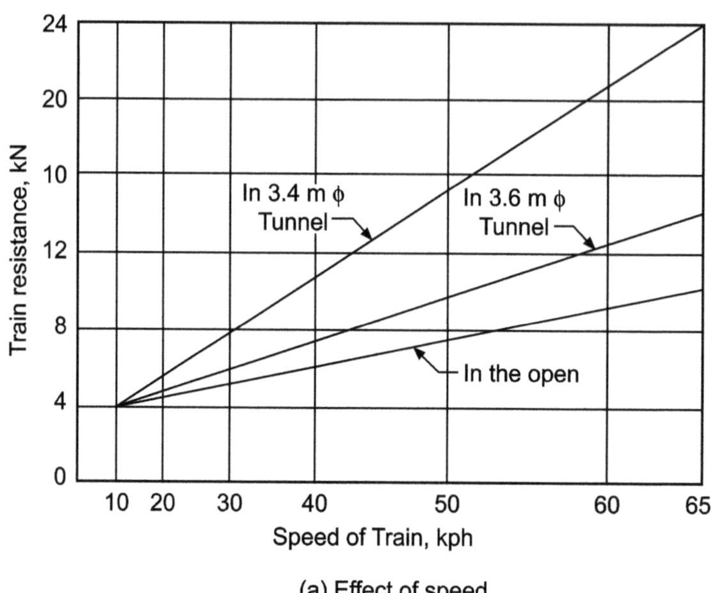

(a) Effect of speed

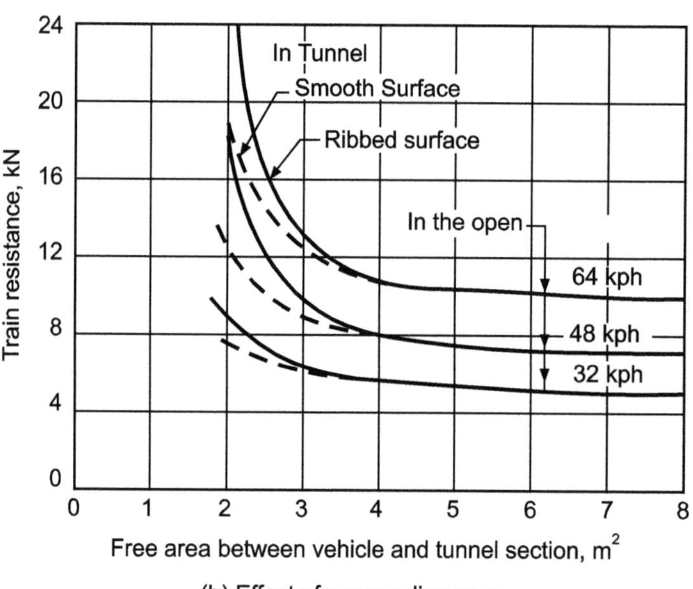

(b) Effect of surrounding area

Source: Adapted from Turner, 1959[3]

Figure 2.1 Measurements of Air Resistance in London Subway.

All these have to be done with care and accuracy from the very start. It should be noted that overall costs of these investigations, are small compared to the cost tunnelling. Even an hour's delay, for example, in correcting alignments or other interference by a survey team or engineer for correction or other difficulties during construction of heading would cost heavily due to idling of men and equipment.

e.g., Unit bid price of tunnel — Rs 100,000 per metre

Average advance per day — 12 m

Cost per hour $= \frac{12 \times 100,000}{8} = \text{Rs.}\,150,000$

Average cost of heading to be one-third cost of tunnel.

Loss per hour delay = Rs 50,000.

2.4.2 Alignment Survey

The engineer has to carry out the following surveys well before commencement of work:

(a) Preliminary location survey
(b) Primary control survey on line
(c) Fixing control/check points.

Any checking inside the tunnel during construction should be planned for the weekends when no tunnelling work is performed. The recommended accuracy requirements in the survey for a rail line are:

Triangulation	:	Closing error not to exceed 1:50,000
Vertical control	:	Establishment of benchmark to the requirements of second order class I
Primary traverse	:	Angular measurement to the nearest second of arc. Stationing length measurement nearest 0.3 mm. Benchmark levels nearest 0.3 mm.

Additional vehicle clearance should be provided in designing the tunnel profile to allow for error arising due to the following during construction:

(a) Inconsistency of primary surface survey.
(b) Errors encountered during transfer of line from surface to heading. (These two can cause error of up to 1 mm for every 50 m of heading).
(c) Inability to keep the tunnelling equipment of indicated alignment during construction. (While some correction is possible in this respect in rock tunnelling, resetting lining in soft-ground tunnelling is out of the question.)

On BART (Bay Area Rapid Transit) tunnels, an error of up to 1:40,000 in alignment fixing was allowed. Singapore Metro specified a requirement of achieving 1: 100, 000 accuracy at break through.

2.5 GEOLOGICAL INVESTIGATIONS

2.5.1 General

The geological investigation is the most important phase of preliminary work in tunnel location and design. A good geological investigation helps in anticipating the type of soil that will have to be gone through and planning construction methodology. This study will primarily determine the method of working the tunnel. It will also give an idea of the type and requirement of construction equipment and availability of materials in terms of quantity. It is necessary also for providing basic data for designing the lining wherever required. A geological survey combined with a hydrological one will indicate the possibilities of seepage or inrush of water channels and enable taking adequate precautions to avoid flooding hazards, slips etc. The rock pressure encountered is controlled by the geological structure of the rocks. Some idea of this is required for designing the lining and also the temporary supporting structures. Some rock structures have the potential for stress relief failure ('popping' rock) and may cause bending stresses on supports. It is therefore necessary to have some idea of the deformational and strength properties of each material likely to be encountered, the direction and intensity of the *in-situ* stress and its relation to rock strength properties. The presence and influence of faults, shear zones and seismic activity potential also need to be ascertained.

The rock or soil through which a tunnel is excavated forms the main construction material for the tunnel designer and builder. Difficult ground conditions per se do not cause difficult construction problems, provided they have been properly identified and taken into consideration in the design. It is the unanticipated problems that cause redesign and consequent delays causing cost overruns. General experience is that tunnels for which sub-strata conditions have been thoroughly investigated are completed on schedule and without much cost overruns. The knowledge obtained during investigations should be complete and correct. In fact, when nothing is known, one is better prepared with a safer design and methodology than when incomplete or inaccurate information is available, giving false confidence and wrong preparedness.

Exploration should be planned to proceed from ‘general’ to ‘specific’. It should be understood that the amount of detail required by the designer and construction engineer should determine the extent of exploration and the 'funds' available. Exploration coverage should not be reduced below the minimum required for arriving at conclusive and reliable data. It would be advantageous to have a well conceived and controlled boring programme. It will be preferable to have less number of deeper well planned bores than as it would yield more useful data than sketchy data collected from larger number of ill planned or shallower or badly carried out bores. Location, direction and depth of bores, drifts, and /or exploration shafts should be planned.

Geological exploration should be extended to:-

(i) Investigation of top cover,
(ii) Determination of extent and quality of sub-surface rock, and
(iii) Surface drainage conditions,
(iv) Position, type and volume of water and gases contained by the subsurface rocks;
(v) Determination of the physical properties and resistance characteristics to driving of the rocks encountered.

2.5.2 Aim of Geological Investigations

In short, the purpose of geological investigations is (Szechy, 1970)[1]:

(a) Determinations of the physical characteristics of the soil and rock through which the tunnel will pass and support;
(b) Ascertaining age and origin of rock;
(c) Determination of mechanical and strength characteristics of the rock/soil so as to obtain design parameters;
(d) Defining the stability nature of soil for providing the engineer with possible conditions that would be encountered during construction;
(e) Collection of data regarding (i) subsoil water conditions, (ii) presence of gases and (iii) rock temperatures; These would affect conditions of work as well safety measures to be taken during construction and stability thereafter and maintenance planning.
(f) Help in deciding on methods of construction and equipment required and planning the operations;
(g) Minimizing likely uncertainties for the designer and construction agencies.

2.5.3 Sequence of Geological Investigations

Geological investigations perforce must follow a logical sequence, each stage improving upon the accuracy of the former but being more and more complete by itself. Finally, the results have to be studied together and decisions taken. The various steps in geological investigations are:

	Stage of work	Details
(i)	Preliminary study and interpretation by geologists	– Study of literature and records – Aerial photographic study – Surface reconnaissance
(ii)	Detailed geo-technical (design stage)	– Borings and test exploration pits accompanied by (a) Recording bore log (b) in-situ testing (c) Laboratory testing, Driving drifts/shafts to supplement borings (if necessary), accompanied by a, b, c above Full-scale testing, wherever necessary.
(iii)	Construction stage	– Driving pilot tunnel accompanied by a) Sample collection and testing b) Strain measurements and convergence measurements for study of rock movements and relief stresses and use of pressure cells and extensometers

The investigations can be divided into three stages as listed above:

(a) At Feasibility study stage, various available reports and literature giving an idea of the morphology, petrography, stratigraphy and hydrology of the area have to be studied and local geologist consulted. This should be followed by field reconnaissance. A trained observer can draw conclusions even by identifying the vegetative plant types. Geophysical explorations by way of electric-resistivity or seismic methods can also be helpful in knowing the rock-soil boundary and delineating fault and shear zones, ore bodies, geological structures etc.

(b) The next stage of study is the geotechnical (subsurface) investigations, done simultaneously with the *planning and design*, but prior to construction. These studies update and augment the information previously gathered, in particular on the physical strength and chemical properties of the rock to be penetrated. Determination of gas occurrence, sub soil water level and rise in rock temperature in respect of both locations and extent are done at this stage.

(c) The third stage involves detailed geological investigations done during *construction*. This means running a pilot heading in advance of the working face to explore actual rock conditions, increase in moisture, interruption by springs etc.

Details of coverage in different stages are discussed below.

(i) *Preliminary Studies*

The 'Preliminary Studies' will cover study of the geological history of the region, structure and age of the crust and various component soils and rocks. The rock (and soil configuration) as originally formed; do not take a permanent set all at once; cooling of the earth's solid crust and accompanying contraction keep the deposited rock layers in continual motion. Various layers of rock are subject to compressive forces which cause deformation, creasing, ruffling and distortion, resulting in folding. Atmospheric action causes weathering of the rock, erosion, sedimentation etc. Typical structures of a rock formation are indicated in Figure 2.2. The main fold formation will generally be as indicated in Figure 2.3[2] .

Faults developed over a period or due to subsequent distortion/ disturbances may cause sudden changes in the structure of the rock in a particular section. These result in (and sometimes are affected by) tectonic forces in the building up of various geological formations. In some zones these folds are prone to disturbances and have to be avoided as far as possible for tunnel location. Such zones also aggravate conditions which aid seepage and inrush of water wherever inclined folds occur. Construction of reinforced box sections may be necessary in such locations, involving additional expenses. There is possibility of encountering greater 'hazards' during construction.

Those geological formations in which the earth's crust is broken under the action of tectonic forces into large separate blocks pose the least threat to tunnelling. Such blocks slide over each other along the bedding planes without greatly fracturing the adjacent masses. A formation of this type generally affects a short section of the tunnel only and the difficulties can be overcome with relative ease. Some of the joints may be filled with dry deposits or may be open and may convey water. Faults of this kind are not regarded as dangerous even when considerable movement occurs along the surfaces. Furthermore, such formations are rare and the main fault is usually accompanied by a number of minor sub-faults, similar to the sets of sliding surfaces. Conditions can be regarded as favourable wherever the main fault face is not extensively fractured by the sub-faults. However, the surface of

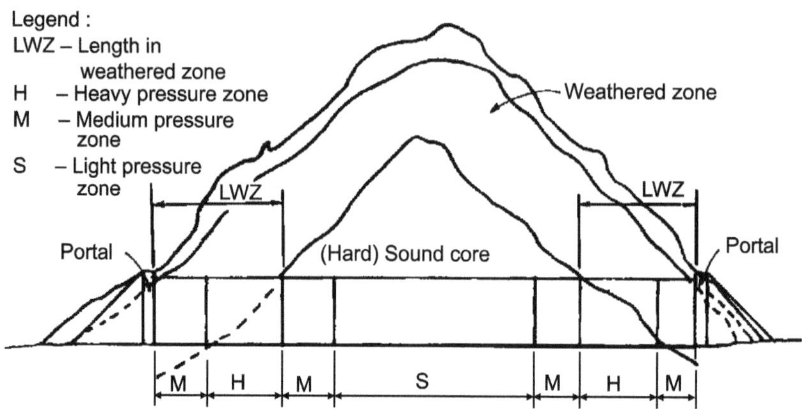

Source: Pequinot, 1963

Figure 2.2 Typical Rock Formation.

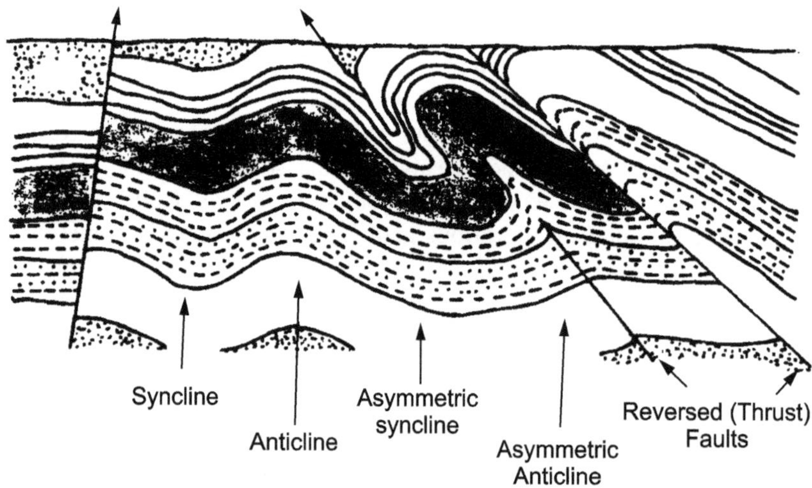

Source: Pequinot, 1963[2]

Figure 2.3 Typical Folds and Faults.

dislocation *per se* usually becomes more or less shaly due to the secondary movement of adjacent rock masses. Information on these matters can be obtained by a study of the available literature pertaining to the area and /or discussions with the local geologist. A general report can alternatively be obtained from the geological survey personnel/experts regarding the general structure of the rock and soil formation, its age etc. This should be followed by a site inspection of the tunnel region in the company of geological experts.

This inspection will cover observations of surface formations, tracing of past landslides, nature of vegetation, presence of springs, shape of blocks of rocks or presence of isolated boulders. All information that can help in reconstructing the geological history of the region and assessing the geological nature should be collected, with particular attention paid to the pattern of earth movements. Such movements are usually indicated by the surface unevenness in the pattern of ridges, hills and valleys.

If adequate data is not available, including aerial photographs (mosaics) of the area, an aerial survey may have to be conducted. Aerial photographs are useful in geomorphic analysis as they give good insight into the engineering properties of rock. For this purpose a skillful evaluation of rock response to the natural environment is called for. Various techniques used in aerial reconnaissance include vertical, oblique, colour and infrared photography as well as side view radar. Infrared and remote sensing techniques are also available. Infrared photography is one of the methods used in aerial reconnaissance using remote-sensing devices. Film sensitive to radiation in the infrared wave-length is used. This technique helps in recognising features that exhibit marked differences in heat radiation characteristics.

The National Aerial Remote Sensing Agency at Hyderabad has done remote sensing studies of various areas in India and also has facilities for doing aerial reconnaissance. It should be noted that detailed interpretation of aerial photographs requires the services of a specialist geologist. Satellite imageries for the location also can be used. Such photos/ imageries can be used otherwise by a designer also with some experience, for interpretation of topography, drainage pattern, land use, location of potential construction materials and their sources and lines of communication etc.

(ii) *Site Investigations (Geological)*

These include geophysical (seismic and dynamic) and electrical resistivity soil investigations done during Feasibility stage. In the first method seismic refraction and reflection surveys are conducted. This is based on the principle that velocity of an elastic wave passing through a material is a function of the material, structure, composition and in-situ stress condition. The velocities vary (increase) with density, compaction and water content of materials since seismic waves follow the same principles of propagation, refraction and reflection that light waves do. The results obtained are more reliable above the subsoil water level. They can be used for: (i) identification of general type of material (soil, type of rock); (ii) location of anomalous conditions, e.g. weathered zones, shear zones and buried valleys; (iii) location and depth of hard rock and (iv) locating boreholes for detailed exploration.

In the second method electrical resistance of various soil layers is measured. Changes in potential across known distances between electrodes when a current is applied between them are used for evaluation of material types. Wet clays and silts and some metal ores are good conductors. Dry sands, gravels and crystalline rock without metal ore are poor conductors. Mineralised (pure) water is a better conductor than saline water. Thus some general idea of the nature of the soil, presence of water etc. can also be obtained from electric resistivity tests. As mentioned earlier, these measurements help in drawing conclusions concerning variation in the type of rock layers as well as water layers contained between. These investigations can be carried out by specialists in the field in a relatively short time and the general geological pattern reconstructed.

Valuable information on the nature of subsoil can be obtained when resistivity surveys are done in conjunction with a seismic study. Typical characteristics of some subsoil materials in response to such surveys are given in Table 2.2.

Results obtained from such studies will help in deciding on location of tunnel. Detailed studies are called for at final location, design and construction stages, as detailed below.

(iii) ***Detailed Exploration***

These site investigations done at design stage (more particularly in mountainous areas of recent origin and rock formations subject to various disturbances such as faults) require exploratory borings, followed by driving exploratory shafts and drifts. The latter choice depends on the type of strata and importance of the structure. Boring should be of the wash-boring (percussion boring) type or rotary percussion boring or rotary-drilling type in which samples can be collected at various depths and proper soil layer identification done. However, in less important and ordinary soils, dry or wash borings may be done when limited to about 100 m. Other normal or mixed soils necessitate rotary-percussion boring.

The core drilling method enables cores of rock to be extracted at various depths/layers for study but reliable cores can be extracted only from solid rocks. In some countries special television cameras have been developed (Grundig-Fernauge) for inspecting the orientation and original condition of the rock layers inside boreholes.

Boreholes should be located on the sides of the proposed tunnel alignment, staggered alternatively on either side. Bores, if drilled deep along the alignment, can leave holes above, which may result in grout material, if any grouting is used, escaping through during tunnel boring. In case casing pipe is used in holes passing through the tunnel area, there is also a possibility of some length getting stuck in rock and be left behind. Such pipes will cause hindrance to tunneling, leading damage to machinery, hazards etc.

The areas requiring detailed exploration by way of boreholes along the alignments are: (*i*) *portals*, (*ii*) *topographic depressions above the tunnel*, (*iii*) *water-bearing zones*, (*iv*) *shear zones and* (v) *rocks with a tendency for deep weathering*.

Table 2.2 Typical Characteristics of Subsoil Materials

Subsoil Material	Electrical Resistivity	Seismic Velocity (Wave propagation)
Dry gravel	High	Low
Dense rock	High	High
Pure water	High	Medium
Saline water	Very low	Medium
Dry compact boulder and cobbles	Very high	Moderate
Saturated boulder and cobbles	Moderate	Moderate

Borehole spacing of 300-500 m is sufficient for the preparation of preliminary designs. But for detailed design and before taking up the work, boreholes should preferably be at 50 to 100 m intervals. In geologically disturbed regions a dense network of boreholes is required. This is particularly necessary for designing underground railway tunnels in areas where a great deal of variation in tunnelling conditions may be anticipated. The boreholes should preferably be located at a lateral distance of 110-150 m off the contemplated tunnel axis (staggered on either side). The holes should be backfilled or even concreted after taking samples and any other study. This precaution is necessary to prevent water seepage from the upper water-bearing layers through these borehole's into the tunnel and also to prevent ingress or escape through the boreholes of compressed air if the compressed-air method is likely to be used for driving the tunnel.

Boreholes in rocks are generally limited to 100 to 150 m depths, with a few deeper ones upto 300 m normally. In exceptional cases deeper boreholes are made, e.g., at the Great Apennine tunnel between Bologna and Florence seven boreholes of 390 m depth were made. Recently, Indian Railways did investigation for their longest tunnel across Pir Panjal Range in Kashmir by *drilling a few holes* 640 m deep. In general, the principle is to sink the borehole 20 to 50 m deeper than the contemplated tunnel bottom level. Apart from depths and general spacing, the direction of bores also is important. *Some of these may have to be located at an angle so as to obtain data of all rock types, which may not be possible if all of them are driven vertically, since rock bedding planes will mostly be at an angle.* Boreholes near portal locations will normally be vertical. See Figure 2.4. During drilling operations the engineer and the designer should inspect the site thoroughly to offer proper guidance to the field staff in this regard.

Any hard rock boring programme should be directed towards obtaining information on the following:

(i) Defining the geologic stratigraphy and structure through which the tunnel has to pass.
(ii) Determination of physical properties of rock materials.
(iii) Study of fracture patterns (horizontal, vertical, inclined, confined etc.)
(iv) Measurement of depth of subsoil water level and porosity of rocks/ soils.
(v) Evaluation of blasting/excavation requirements.
(vi) Evaluation of support and lining requirements.

A study of the following rock properties is necessary during geological investigations (Bickel and Kuesel, 1982)[4].

(i) Orientation of rock stratification (whether horizontal, sheet-like, moderately inclined, steeply sloping, reversed or overfold).
(ii) Thickness of individual layers, regularity of sequence of rock layers, or changes in mountain types.
(iii) Mineralogical composition (detrimental components).
(iv) Crystal structure of rocks (uniformly grained or porphyric).
(v) Bonds between the individual grains (strong, weak, direct and indirect).
(vi) Hardness and workability of rocks.
(vii) Structural form of rocks (massive, stratified, shaly)
(viii) Deformations suffered during the orogenic process (cleavages, crushed zones, faults) or other effects (weathering, mylonitisation, kaoinisation)
(ix) Probable bearing and tensile strength of the mountain (not rock) at various tunnel sections.
(x) Stability of the mountain, character and magnitude of probable rock pressure.
(xi) Bulk densities and dead weights of component rocks.
(xii) Durability of various rock types encountered.

In some bores means will have to be provided for inserting downhole testing and logging equipment.

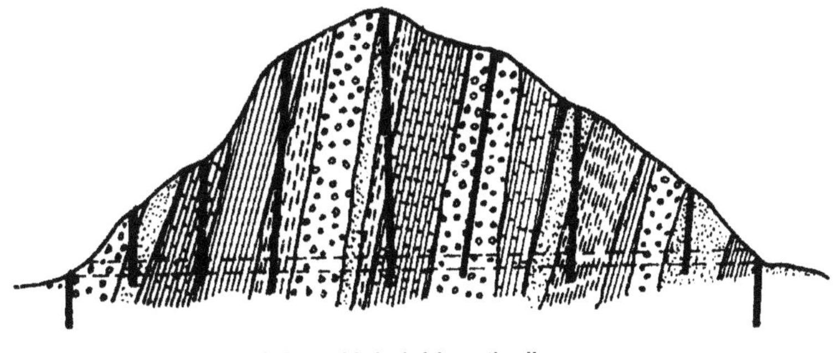

(a) Bore Hole laid vertically

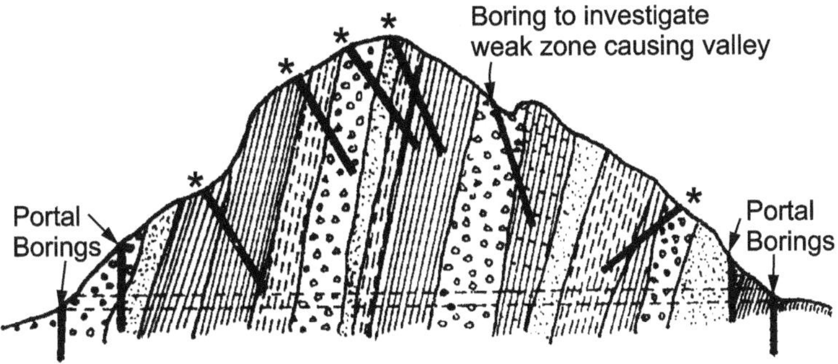

(b) Preferred layout of bore holes

Source: Bickel and Kuesel, 1982

Figure 2.4 Layout of Bore holes[4].

(iv) *Tests to be conducted during Exploration*

In ordinary soil, in-situ measurements have to be made in the boreholes at intervals for pore-water pressure, flow of subsoil water and state of compaction (vane shear) test. In rock, the in-situ measurements are made for pore-water pressure, flow rate of subsoil water, joint and fissure pattern, and in-situ stresses in the rock.

The samples collected at different levels (especially in the tunnel zone and for some depth above and below) are subjected to the following laboratory tests:

<u>Soil:</u> Particle size grading
Mineral composition
Density

Porosity
Moisture content (from undisturbed sample)
Shear strength (from undisturbed sample)
Angle of internal friction (in granular soil)
Plasticity (in clays, silts and mixed soils)

Rock: Lithology
Density
Porosity
Water content
Strength: Crushing, Tensile, Shear
Any identifiable plane of weakness
Abrasiveness
Rock quality distinction (RQD)

(RQD is the percentage of core borings recovered in lengths not less than 100 mm.)

2.5.4 Shafts and Drifts

A detailed/final tunnel design and execution plan has to be based on accurate information about the physical, structural and chemical properties of the rock layers to be penetrated and on the hydraulic, gas and temperature conditions prevailing in them. For this purpose, at construction stage, especially in long tunnels, exploration shafts may have to be sunk, which are generally done in the vertical direction and as an exception in an inclined direction, these are also sunk slightly away from the tunnel alignment but in such a way that they permit subsequent use for constructional or even later for operational purposes. The exploratory shafts are the best form of geological investigations as they permit direct inspection of the bedding and dip conditions as well as the thickness of the layers. The main drawback of this type of exploration is the cost. There is also some danger in using only the information obtained from the shafts. Since the tendency is to locate them far apart in view of the expense involved, such shafts cannot adequately indicate the changes in rock structure, quality etc. over the intermediate section. Over-looking these changes, particularly where the soil conditions, stratification and hydraulic conditions are not uniform, could lead to unpleasant surprises during construction.

2.5.5 Pre-Construction Investigations

As part of preconstruction investigation, the geological nature can also be determined by running horizontal pilot headings between the individual shafts. This, however, forms part of the future tunnel itself and hence this method should not be resorted to unless a particular section appears to be especially dangerous or a great deal of uncertainty exists. The usual practice is to drive these headings immediately prior to construction and generally to advance them always a few hundred metres ahead of the working face. Without doubt, exploratory headings are the most accurate means for determining the geological conditions and for supplying the most reliable data for correct design and proper estimation of construction requirements, equipment, time involved etc. Rock pressure measurements can be made more accurately through the drifts and whenever the lining and support structures for the full bore are to be designed, these measurements can be taken into consideration in advance.

2.5.6 Present Approach for Large projects

Geophysical Studies[5]

In tunnelling projects, the geophysical studies have to be aimed at collection of information on Rock types, shallow faulting, shear zones, saturated pockets, and fracture systems and also the physical characteristics of the rock/ soil. They all cannot be obtained by borings only since the alignments may run very deep and boring itself is time consuming and cost prohibitive. With the advancement in the technology and survey methodologies, geophysical studies are now possible for developing a continuous profile of a terrain in respect of all the above, except the actual physical characteristics of the soil to be tunnelled through. Hence, present day investigations combine both geophysical survey and a few borings. They are supplemented by probe holes drilled through the excavation face during tunnelling operation. The geophysical studies include 2D high resolution seismic methods (conventional reflection and tomography processing) and electrical resistivity imaging[5].

'Seismic reflection, refraction, and refraction tomography surveys utilise seismic waves (compressional and/or shear) to map sub-surface' structure and layering. For example, they cover fault mapping, voids, bed thickness, depth and stratigraphic continuity, bed rock mapping and continuous velocity studies. Recent developments in seismic instrumentation and survey techniques in land and marine surveys and data processing have helped in application of seismic reflection and refraction surveys for 'near surface investigations'.

Other geophysical methods like resistivity profiling are used as complementary to seismic profiling. During processing, data from both are integrated to arrive at very detailed 2D site models. From these, it is possible to generate Geological plan, Geological LS and Geologic cross sections at required locations. Such integration is necessary since there are limitations in any specific geophysical study. For example, seismic velocities of hard limestone vary from 5,000 m/sec to 6,160 m/sec, while the velocity range in hard granite is about 4.000 m/sec to 6,100 m/sec. On the other hand, resistiviry value of granite varies from 25 to 58 Ω while that of limestone is 45 Ω. Thus a combination of seismic survey and Resistivity imaging can help in better prediction of type of rock.

Seismic refraction survey methods can be grouped under Conventional and Advanced seismic refraction tomography. The former which can detect P and/or S waves can be used to find depth to bed rock, overburden thickness, and to define shallow soils and rock stratigraphy. The refraction velocity can be used to 'estimate earth Rippability'. Combination of P and S wave data is used for identifying the 'velocity shear zones.' Processing of the basic refraction data is done using any one of, or a combination of, Delay-Time, Reciprocal; and 'Generalised Reciprocal method'.

Advanced Seismic Refraction Tomography processing technique is used when such survey is conducted over complicated geological structures or whenever 'a higher degree of P and/or S wave vertical and horizontal velocity delineation is called for. Refraction tomography (P wave) can delineate stratigraphy and identify fracture zones in the rock. It can provide 'velocity profile for Rippability study'.

Wave velocity information obtained in Seismic Tomographic survey help to determine variations in elastic properties, both horizontal and vertical. Where required, 2D and 3 D seismic reflection imaging surveys can be used for obtaining high definition lineation of rock stratigraphy, location of shallow faults, isolated sand lenses and cavity formation, aquifer zones and other shallow earth structures. These are similar to seismic refraction studies in respect of use of instruments but they measure the reflection of P and/or S acoustic waves from 'subsurface boundary interfaces and geologic features'. In general the seismic reflection techniques are used for geologic depth ranging from an average 30m to 900 m below ground surface.

Project Approach: Any geophysical survey for a tunnel project would be sequenced as follows:

(a) Study of Remote sensing imageries and Topographical maps of the area- to fine tune the exploration programme.
(b) Geological mapping- of the area to provide preliminary information on rock formation, strikes and dips, fractures and joints and likelihood of their below ground continuation,

(c) Calibration survey- doing Seismic refraction survey and Resistivity imaging at a location for which boring/ drilling data is available for correlation between local geological factors and seismic velocities and resistivity values. This calibration will be used for fine tuning the design of the Survey.
(d) Actual geophysical survey,
(e) Data processing- using state–of-art Paradigm: (e.g., Software used on UBSRL project study were: GeoCT Software I and II developed by Geotomo for Tomography and 2D Earthlmager Resistivity Inversion Program.
(f) Interpretation and presentation of results for Engineering Purposes- Presentation to cover : subsurface stratigraphy; presence of faults, joints, fissures, pockets of saturated zones and air filled cavities.

Survey Design: Important parameters critical to the design of a high-resolution 2D survey are: (a) vertical resolution and horizontal spatial sampling and (b) maximum signal frequency:

Seismic Profiling

Vertical resolution is a function of interval of p- velocity. For example, a nominal average p-wave velocity of 2000 m/s frequency required for delineation of a formation with minimum thickness of 15 m is 100 Hz. Hence a broad band width of 10 to 20 Hz is required for optimum vertical resolution. The energy source has to be capable of producing a broad band seismic signal. In addition high spatial sampling is required. Seismic line geometry should have adequate far-offset source-receiver range and 'CDP fold'. For example, a Source/ receiver geometry of 96 channel seismic lines to derive an accurate earth model and earth image (48 active, 48 roll-along). (The total active seismic line length should be at least equivalent to maximum depth of investigation). Geophone (single 10 or 14 Hz vertical with 50% damping characteristics) spacing should not be more than 15m. They may be used in cluster. Shot spacing can be either 15 m or 30 m for yielding 24 and 12 fold CDP data respectively. In order to record and process the data in field, a field rugged PC with seismograph control software and suitable processing software (GeoCT Field Seismic Tomography software and QA/QC analysis software).

The seismic energy used would be light explosives in form of small charge (Gel-Pack) of 500 grains loaded in a shot hole drilled 3 to 4 m deep. Man portable drills are used for drilling shot holes. Number of shot positions depends on terrain and whether 12 or 24 fold CDP data is required. Suitable safety precautions and health procedures for use of explosives will have to be observed. Shot control will have to be exercised by a portable I/O ShotPro Blaster Unit with UHF/VHF radio transmission control. The data processing would be such that they produce vertical and horizontal velocity gradient cross sections, which can be used for determining the soil/rock structure and make it possible to identify anomalous fracture and fault zones.

The quality of these images are not good in Thrust belts where they are affected by (a) physical propagation effects of the wave field and (b) insufficient modeling capability of recorded data. In case of Complex structure and geological characterization (in rugged terrain with complex geologic formation), a more sophisticated seismic tomography processing' may be performed. The data analysis for same will have to be designed to arrive at more accurate earth model and earth image.

Electrical resistivity profiling

This is done in areas subjected to slope stability problems. This profiling can delineate the soil-rock interface structure and stratigraphy. They are more relevant near portal areas where shallow soil-rock interfaces along sloping zones would be present. Equipment generally used can be a survey line with upto 84 electrodes employed. Typical array electrode is with spacing of 5m and one 84 electrode resistivity line can cover a length of 415 m. Maximum depth of investigation of such a line is 100 m (about 25% of line length). For processing the data a 2D Earth Imager Resistivity Inversion software in a field rugged PC used at site. The output would be on a *x,y,z* file for use with 'geographic and geologic mapping software.

2.5.7 Types of Rocks

Different types of rocks met with can be grouped as under:

(i) Hard and solid rocks which may be regarded as an integral mass.

(ii) Fissured rocks subjected to considerable deformation and compaction at first loading, but which display a more or less elastic behaviour when loaded repeatedly.

(iii) Soft rocks, whose internal structure is destroyed beyond a certain limit of load, and which afterwards suffer a residual deformation of increasing magnitude under each successive loading cycle.

Typical load-deformation curves for the different types are given in Figure 2.5.

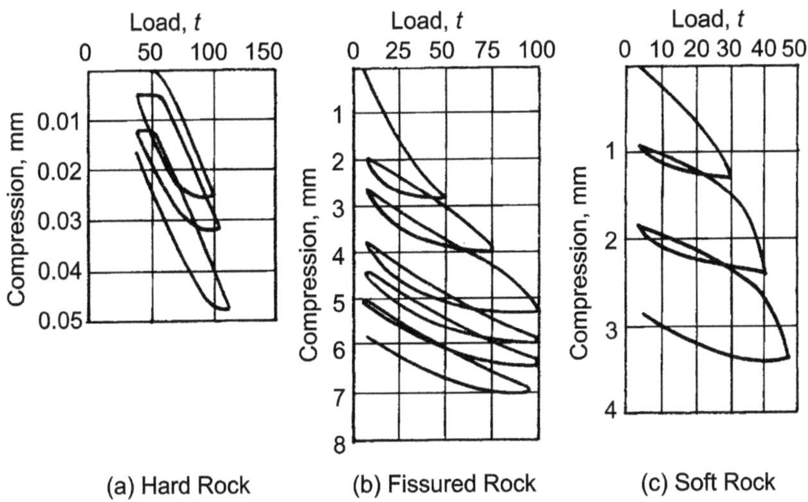

Figure 2.5 Typical Load Deformation Diagrams under Repeated Loading in Different Types of Rocks[1]

2.6 HYDROLOGICAL INVESTIGATIONS

A hydrological survey is carried out simultaneously with geotechnical exploration. The appearance of water in tunnels depends primarily on the character and distribution of water-conveying passages and subsoil water level (Szechy, 1970)[1]. A typical case met with in rock tunnelling is indicated in Figure 2.6. Groundwater and water of interrelated aquifers where the faces of the rock are saturated with the water mass extending over the entire thickness of the layer or over the major part of the layer would be the most dangerous during tunnelling. Locating tunnels in such layers should be avoided. Where unavoidable, special tunnelling methods and techniques have to be adopted for driving the tunnel through such layer.

They can be such as shield driving and dewatering by compressed air. The possibility of locating the tunnel above the groundwater table by relocating the alignment of the tunnel should also be considered. The possible tunnel levels, relative to water levels, and relative advantages and disadvantages are indicated in Figure 2.7 (Szechy, 1970)[1].

2.7 GASES AND ROCK TEMPERATURES

Another important element of the preliminary exploration is the estimation and study of gas bursts, gas exfiltrations and rock temperatures. Gas and

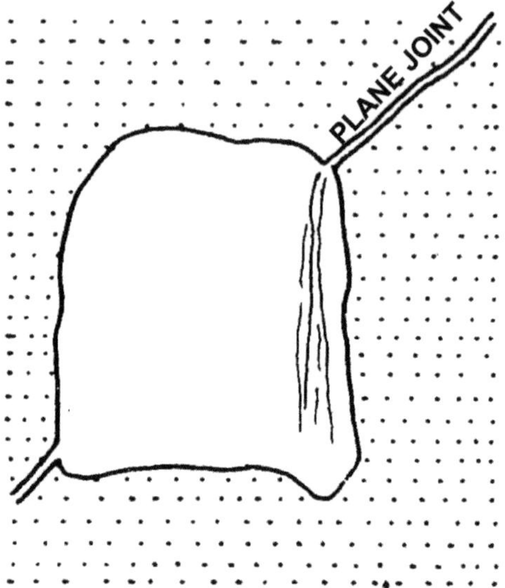

Figure 2.6 Curtain like Water Infiltration from a Joint[1]

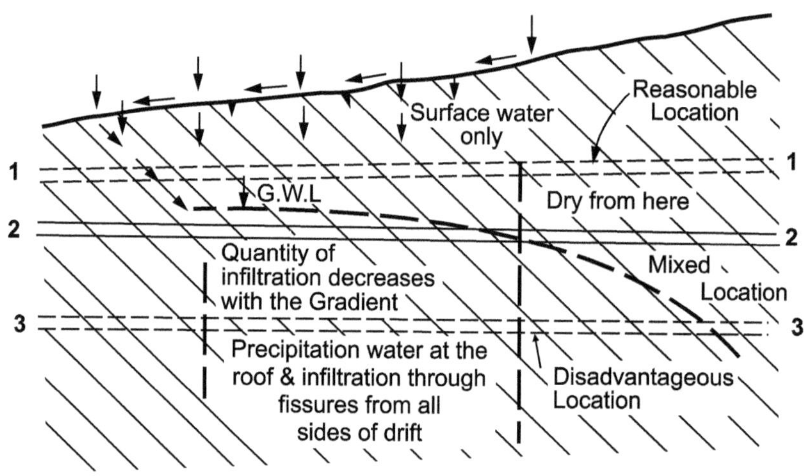

Source: Szechy, 1970

Figure 2.7 Possible Tunnel Elevations Relative to Water-levels.

temperature are very significant for the safety and health of workmen though their influence on the technical feasibility of location of the tunnel is less pronounced. Thus they call only for necessary precautions to be taken and help in selection of the tunnel working and transportation equipment.

Gases met with usually are carbon dioxide (CO_2), carbon monoxide (CO), methane (CH_4), hydrogen sulphide (H_2S), sulphur dioxide (SO_2), hydrogen (H_2) and nitrous gases, besides water vapour. *Carbon dioxide* is generally encountered in the proximity of coal layers. The detrimental effect of this arises from the lack of oxygen where CO_2 is present and suitable ventilation and supply of oxygen through this region is required. The other effect of CO_2 is its aggressive nature on the concrete of the lining and corrosive action, dangerous to steel structures.

Carbon monoxide is more toxic than carbon dioxide. It also usually occurs in the vicinity of coalfields. This gas is poisonous to workers, resulting in agitated heart pounding, headache and dizziness above 25% limit. It will cause loss of consciousness at 50% and above. 75% saturation is fatal.

Methane gas occurring in the vicinity of coal and oil fields may also be the result of decay of organic substances. This gas is dangerous since it is likely to cause explosion as it is inflammable even at as low a percentage as 2%. Safety precautions have to be taken such as:

(i) use of battery-powered electric lamps;
(ii) installation of gas indicator lamps in all drifts;
(iii) use of remote controlled electric detonators;
(iv) constant supervision of all working activities by gas experts;
(v) use of compressed-air locomotives for transportation of debris and construction material;
(vi) installation of high-pressure water main for fire extinguishers;
(vii) air-extracting ventilation at all points of accumulation (especially at the roof) and provision of ample artificial ventilation in general; and
(viii) strict prohibition of smoking and use of open flame lamps in the entire tunnel.

Hydrogen sulphide is the product of disintegration of organic substances and is generally accompanied by water inrush. It is dangerous because of its toxic effects rather than as a fire hazard. In a concentration of 0.05%, it causes sickness. At 0.1% it causes unconsciousness and at higher levels is lethal. Sulphur dioxide occurs in volcanic regions and is detrimental to concrete linings. Hydrogen is dangerous due to its inflammability. It is generally found in salt deposits and in their vicinity. Nitrous gases are the by-products of explosion fumes and are even more dangerous and detrimental to health than carbon monoxide. The lethal concentrations of various gases are indicated in Table 2.3.

Table 2.3 Lethal Concentration of Gas in Tunnel (in percentage)

	Time (min)	CO_2 or CO	H_2S	SO_2	NO or NO
Aspiration	10-20	0.05	0.05	0.01	0.01
Short exposure	20-25	0.10	0.20	0.05	0.03

2.8 ROCK TEMPERATURE

Temperatures in deeper rocks are much higher than on the surface. When tunnelling is taken through such zones, men have to withstand very high temperatures. Generally, below a depth of 20-25 m in rocks, the crust is barely affected by external influences and hence a consistent increase in rock temperature occurs with depth. The rate of increase is not uniform since the geothermal step or geothermal gradient depends on several factors topography of the terrain, stratification and dip of the rock layers etc. As observed

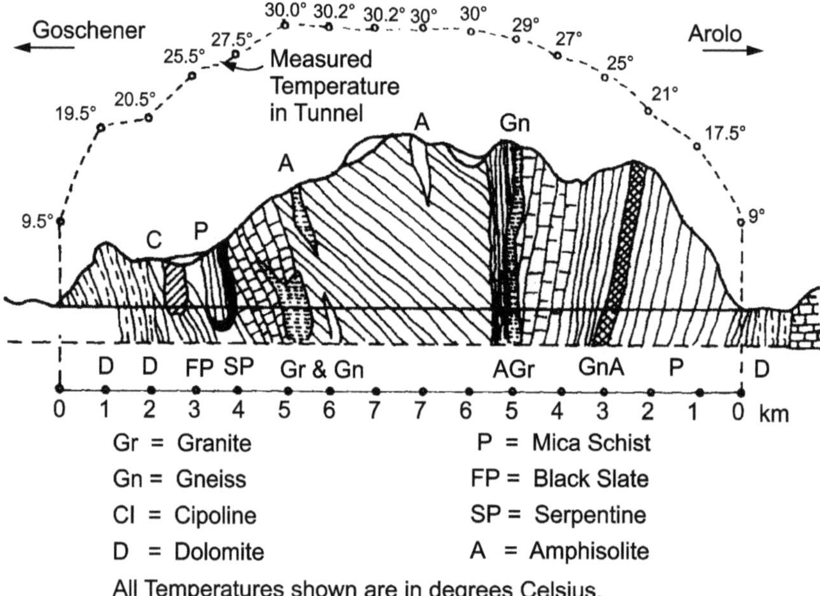

Source: Szechy, 1970.

Figure 2.8 Geological Profile and Temperature Variation in St. Gotthard Tunnel.

in the major Alpine tunnels, this varies from 27 to 144 m per °C (Andreas, 1953). The temperature likely to be encountered in the interior of the mountain is governed by the following factors:

(i) Position of the geo-isotherms under mountain ranges (geothermal step).
(ii) Soil temperature on the surface over the tunnel.
(iii) Thermal conductivity of the rock and hydrological conditions.
(iv) Elevation of the tunnel.

Temperatures as high as 63.7°C have been met with in Apennine tunnels. The geological profile on the temperature of the St. Gotthard tunnel is given in Figure 2.8 (Szechy, 1970) as an example.

2.9 REFERENCES

1. Szechy, Karlowi, (1970) The Art of Tunnelling, Akedimiai Kinda, Budapest, Hungary, 1970
2. Pequignot, C.A., 1963, 'Tunnels and Tunnelling', Hutchinson Scientific and Technical, London, 555 P.
3. Turner, J., 1959, 'Newest Trends in the Design of Underground Railways', Journal of Institution of Civil Engineers. Vos, Charles J., (1993),
4. Bickel, John O., and Kuesel, TAR., (Eds.) 1982, `Tunnel Engineering Handbook', Van Nostrand Reinhold Company, New York, 670 P.
5. Chowdhury, P.K., (2009), 'Geophysical Methods Suggested for Udhampur-Baramulla Rail link project' .ppt Technical Presentation.

CHAPTER

3

Tunnel Requirements

3.1 GENERAL

The requirements of tunnels to be provided differ with the purpose for which the tunnel is being provided. The Traffic tunnels are provided for the purpose of transportation of goods and people and their requirements can be broadly classified as operational, structural, Constructional and maintenance. The conveyance tunnels are provided for conveying materials, water or sewage. The ones for conveying materials are generally limited to industries or the smaller areas and their requirements pertain mainly to the structural, constructional and maintenance aspects. Ones used for conveyance of water, sewage etc., have in addition to be taken care of in respect of surface finish, water tightness, velocity of flow through them etc., in addition to structural, construction and maintenance concerns.

3.2 OPERATIONAL REQUIREMENTS

These requirements cover such aspects as profile, gradient, curvature, ability to withstand pressures caused by moving vehicles in a confined space, ventilation, lighting, drainage and communication and fire fighting facilities in the case of accidents.

3.2.1 Profile

The cross-sectional profile of the tunnel has to allow movement of the largest vehicle it is designed for and provide for some additional safety margin. The opposing streams of traffic have to be separated. This is done either by providing a median in a two-way highway tunnel or providing one separate tunnel for each direction.

For railways, even if separate tunnels are not available, the wheel guidance available and the additional guard rails installed between rails provide the necessary safety. The safety and lateral (sway and lurch) movement clearances over moving vehicle profiles or gauge is specified for straight alignment. On curves, the clearances are increased to provide for extra projections caused due to movement of vehicle over curves and by superelevations. An encompassing profile (internal dimensions) allows for these minimum horizontal clearances and also providing for space for other facilities, such as ventilation, ducts, lights and cable ducts/lines for highways, or for signals, traction gear and cable ducts/lines in the case of railway tunnels. These will be after allowing for lining to be provided where required.

The shape of the tunnel itself is dependent on purpose (traffic requirements in case of vehicular tunnels), location, type of soil, overburden depth and to a large extent on the method of construction to be followed.

3.2.2 Ruling Gradient

The maximum or ruling gradient to be used is decided based on the type of traffic. In a railway, the ruling gradient specified for the section of the line of which the tunnel forms part has to be decided taking into account the extra resistances on track through tunnels mentioned in Para 2.4.1. Steeper gradients enable shortening lengths of and thus less costly ghat/hill lines but put a severe constraint on train loads, resulting in reduced section capacity and speeds, which would lead to higher energy consumption. Maintenance costs also go up with steeper gradients. Based on long-term operational advantages and economies many planners now prefer flatter gradients of 1% or even 0.67% for busy freight traffic lines and lines on which medium high speed trains are likely to be run in future, e.g. Konkan Railway line in India.

On highway tunnels also the ruling gradient is dependent on type and volume of traffic to be catered for. Generally, the limiting grades at entry and exit ends are in the range of 2.5 to 3.0% while flatter gradients are adopted in the mid-section. A minimum gradient is also specified for tunnels in order to ensure proper drainage. Chapter 5 details this aspect.

3.2.3 Curvature

Curvature in railway tunnels depends on speeds to be permitted on the section and the gauge of the line. The wider the gauge, larger the radius of the limiting curvature specified. At approaches to signals, curves are limited, taking into consideration safe sighting and braking distances. Spacing of tracks inside tunnels is to be such that the induced wind pressure on the bodies of coaches of crossing trains will not cause windows to shatter nor cause discomfort to passengers travelling with windows opened.

3.2.4 Ventilation and Lighting

Ventilation needs for railway and highway tunnels differ very much. In short tunnels the force of a moving train provides necessary ventilation by pushing out foul air left by the previous train and drawing in fresh air from the rear. Special ventilation arrangements are required in metro tunnels, for providing comfort to passengers and bringing down the temperature. In highway tunnels, due to the presence inside tunnel of large amounts of CO_2 and unburnt CO as well as smoke from diesel engine vehicles, continuous forced ventilation is called for.

Highway tunnels need to be illuminated well enough for clear view of the road for the drivers. The arrangements are thus more elaborate and complicated. Main line railway tunnels need minimal lighting for use by the maintenance crew and for use in emergencies while the lighting requirement for metro tunnels is much larger. The subject is dealt with in detail in Chapter 10.

3.2.5 Clearances

The additional dynamic vehicle clearances required in railway tunnels to allow for curves and superelevations are indicated in Figure 3.1. Figure 3.1 (a) conceptually indicates how the different parts of vehicle project out due to curvature and effect of cant provided in the track. Figures (b) and (c) indicate the quantum of such shift for a Standard Gauge (1435mm gauge) vehicle.

3.2.6 Aerodynamic Problems in tunnels[2,3]

Aerodynamic problems in railway tunnels arise due to piston action of the train through a narrow tube. This pressure induces fluctuations in pressure inside the tunnel and on periphery of coaches. Due to narrow space availability between the train and tunnel surface, the reactive air that passes by the periphery of the coaches is at high velocity, causing shearing force along the sides of train and tunnel surface. One would feel its effect by way of rattling of doors and pressure felt, especially in ears if windows are open. On a double line single tube tunnels, the effect will be felt more pronounced

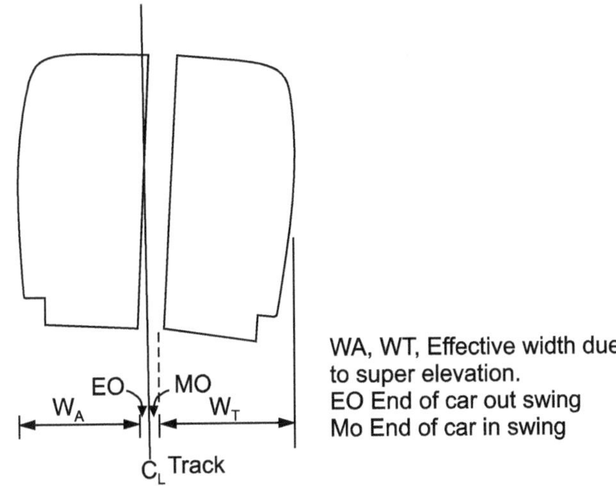

(a) Additional dynamic vehicle clearance for curves

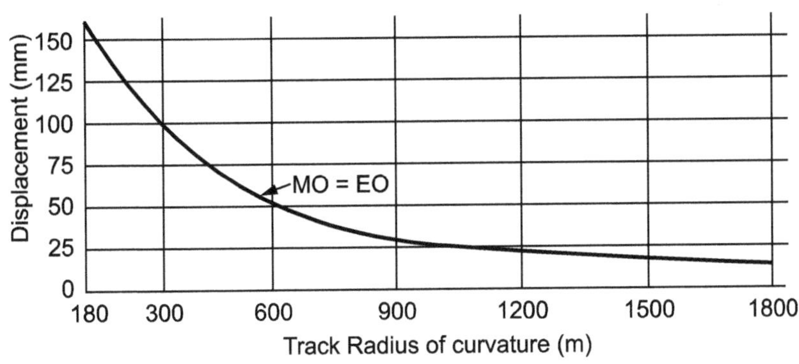

(b) Change in dynamic vehicle width due to curvature

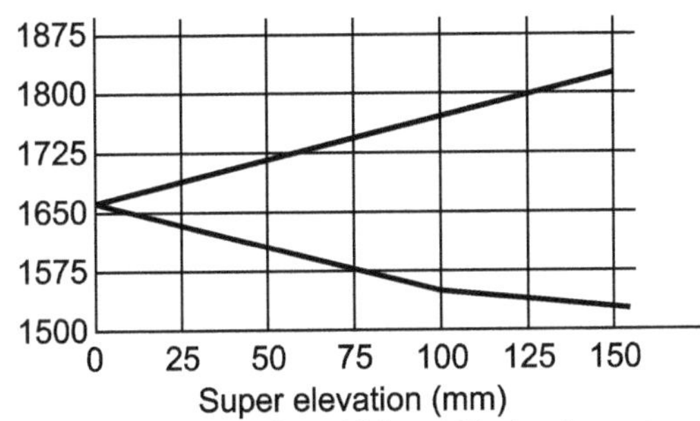

(c) Change in dynamic vehicle width due to super elevation.

Source Morton, 1982

Figure 3.1 Additional Dynamic Vehicle Clearances for Railways[1].

when another train passes in the opposite direction. At higher speeds, the aerodynamic effects will be more pronounced. The velocity of air passing by the train increases in proportion to square of velocity of the train and hence the effects of such pressure fluctuations become more pronounced in high speed rail tunnels. Japanese were the first to notice the adverse effects, in form of peeling of paint in the train, discomfort felt by passengers and damages caused to tunnel lining.

In case of high speed trains, as it enters the tunnel, there is an abrupt change on the practically stationary air in tunnel, which sets up some abrupt changes[2]. Air ahead is compressed and is pushed through the tunnel at a speed close to sound. This 'column of air' is reflected back at the far end portal as 'a pulse of rarefaction' which causes a sudden reduction in reduction in pressure as it reaches back the train. A mechanism of pulses is set up and stream of air passes through the space in the periphery of train to the rear where a suction effect is created. The difference in pressure between front and rear of the train causes a braking effect on the train and also a frictional drag on same. These effects are more pronounced in longer tunnels. The pressure pulses, if too great, cause discomfort to passengers in their ears. The effects of such pulses are interrupted at any opening on sides of tunnels and at shafts, reducing their adverse effects. Larger spaces by adopting higher clearances around the train also reduce the adverse effects. Rapid changes in the pressures within the tunnel have to be avoided. A limit of 0.5 kN/ sqm. is set for frequently repeated pulses, such as the ones occurring in a tunnel with a number of shafts and cross passages. USDOT Handbook recommends a limit of 0.41kN/ sqm and upper limit for same. Upper limit for any change in pressure is set at 3 kN/ sqm.

The pressure waves (and consequently problems) occur when[3]:

- trains enter or exit them
- trains pass sectional variations, if any
- trains pass any openings to cross passage to adjacent tunnels; pass shafts; pass other openings like intermediate adits

The pressure behaviour is as follows[2]. In a 3 km long single track tunnel without intermediate openings, when a 350 m long train runs at 200 kmph, it will take 54 secs. for any point of the train to pass through the tunnel. In first 7secs, a pressure of 4 kN/sqm occurs, dropping down slightly as tail enters. After further 7 secs, it drops down steeply to (–) 1kN/sqm, when wave of 'rarefaction' enters the tunnel at far end. After this it fluctuates through a 'diminishing range'. This drop of 5kN/sqm is not acceptable and provision of intermediate shafts or cross passage to adjacent tunnel is needed. (High speed lines will have to be double lines and hence an adjacent single track tunnel is a reality.) Alternatively, speed will have to be restricted to 150 kmph or provision of larger section, as has been done by German Railways (75 sqm as against Japanese practice of 64 sqm). Provision of flaring at portals (trumpet shapes) will help reduce pressure wave amplitude at entrance. But, if the exit portal is flared, there can be discomfort to passengers. Each of these remedial methods has advantage and disadvantage and a judicious choice has to be made, considering density of traffic and costs.

In a double track tunnel, the adverse effect at entry is reduced at entry points due to larger area availability. Problem arises when two trains pass each other. Then the aerodynamic effect causes discomfort to passengers, peeling of skin (paint) and rattling of doors. When a freight train with empty containers passed a high speed passenger train, there was a tendency for empty containers to lift off the wagons, if not held down by other means. It is desirable to go in for twin single tubes with interconnection at intervals, rather than a larger section with double line, from passenger comfort and safety points of view, specially in case of high speed lines and subways.

3.3 STRUCTURAL REQUIREMENTS

3.3.1 Supporting Structure

The temporary support structure inside a tunnel has to first support the surrounding soil from caving in and to facilitate easy installation of lining (permanent structure). Preferably, it should be capable of being incorporated as part of the permanent structure. During construction, when compressed air is used for drilling ahead, such support should be airtight and should withstand induced pressure. In service, it has to withstand the vertical load from soil above, and the side thrust from the most saturated condition of soil around.

Lining or permanent structure has to transfer all the superimposed load, the self-weight of the structure itself and the vehicle loads, including impact to the base soil below, without causing any relative settlement along the line. It should be capable of resisting the weathering action over time and chemical action due to fumes and gases emitted by passing vehicles. In subway/metro tunnels it should be leak proof so that ingress of moisture into the tunnel is eliminated or minimal. In the case of tunnels in open country also, the body should be watertight but where there is heavy seepage or where a tunnel cuts through a spring which can build up pressure around if not let free, provision should be made to let such water into the side drain of the tunnel at appropriate places and the tunnel suitably graded towards adits for quick clearance. In seismic areas tunnel structure is designed to withstand seismic forces also in combination with other loads. The different combinations of loads for which a subway tunnel structure is to be designed are given in an example at Annexure 4.1 and in Para 8.10. Similar principles apply to highway tunnels also.

3.3.2 Portals

Tunnel portals are the entry points to the tunnels at either end. Except in case of subways and those used in urban areas for roads, the surface of the ground at the entry points will be having a natural slope. The Adits/ approaches will be in a cutting. From economic and stability considerations, their depths are restricted to about 20 to 25 metre. The portal is a solid structure (face wall) in form of a retaining wall across the tracks depending on depth of cutting. The slope of ground above is protected by a headwall support over the tunnel, as shown in Figure 3.22. It should be such that it can properly retain any sliding mass of soil from above. It is necessary to investigate structure of the rock and soil conditions above the tunnel at this length. Normally conditions met with at these locations will be presence of weathered, fractured and loose layers of rock/ soil where the tunnel pierces through. The type of soils and rocks underlying such layer should be examined to see how the separating sliding planes are sloping/ and if they are towards the cutting, such sliding planes will provide no resistance for the soil above to slide down, when lubricated by percolating water (in rain) etc, The protecting Head wall should extend upto the surface of the soil at the slope it would assume in worst conditions. The adits should not be located in the sliding layer and cutting depth also restricted to 20 to 25 m. The coping of the portal should reach a few decimeters above the natural slope at which the fractured/ loose layer will assume. The coping should be adequately designed to resist the pressure exerted by the sliding soil (in wet condition). The portal face wall itself has to be designed to resist the earth pressure building behind. There has been a case or two where the tunnel failed at entry due to such pressure before the required structure could be completed.

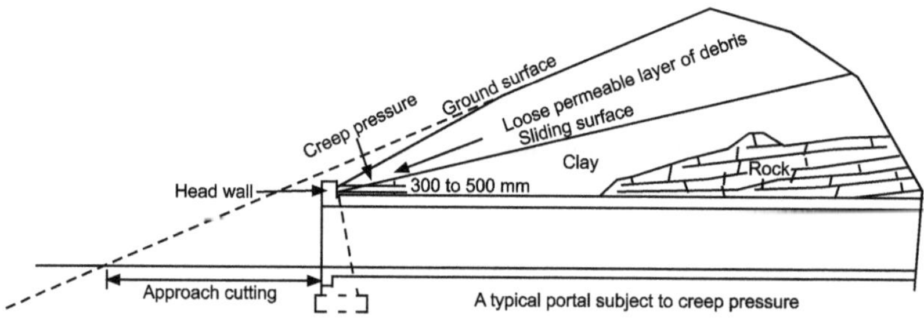

Figure 3.2 A Typical Location of Portal and Forces acting on same.

The portals should be located, as far as possible, in firm ground. They are like end abutments of a bridge and have to be designed to resist the pressure developed by the soil behind. If, for any unavoidable circumstances, it has to be located in soft ground, in order to restrict the pressure developed by the ground behind, the overburden over the tunnel at entry should be restricted and / or it should be supported in front with wing walls/return walls extending into the approach cutting (Figure 3.3). Alternatively, the tunnel body itself is extended into cutting till it reaches beyond the toe of the ground slope on the approach formation.

Figure 3.3 Portal with Wing walls in soft ground- Pir Panjal Tunnel North Portal.

3.3.3 Invert Types

Invert of a tunnel refers to the base of the tunnel, over which the utility (road, rail track, canal etc..) are laid. There are two ways of providing same viz., directly laying over the exposed base or to provide a structural slab spanning the face walls and creating space below for ventilation, drainage and other utilities. First method is used mostly for railway lines and the second method for highway tunnels.

In the first method also, there are two types. One is to just provide kerbs and drains alongside the side walls and lay the track on ballast between them directly on exposed grade. This is possible in case of tunnels through hard rock and was the traditional practice. In case, the grade is on disintegrated rock or soil, a concrete base is provided at grade first over which track is laid. This practice is now being given up in favour of provision of ballastless track, by laying a lightly reinforced slab and providing track over a plinth or embedding PSC sleepers in concrete. Except in hard rocks, where the base of the side walls and supports are keyed into rock and provided a base, struts spanning tunnel supports or a RCC base slab has to be provided below or as part of base for resisting horizontal thrust. Figure 3.4 shows the different types.

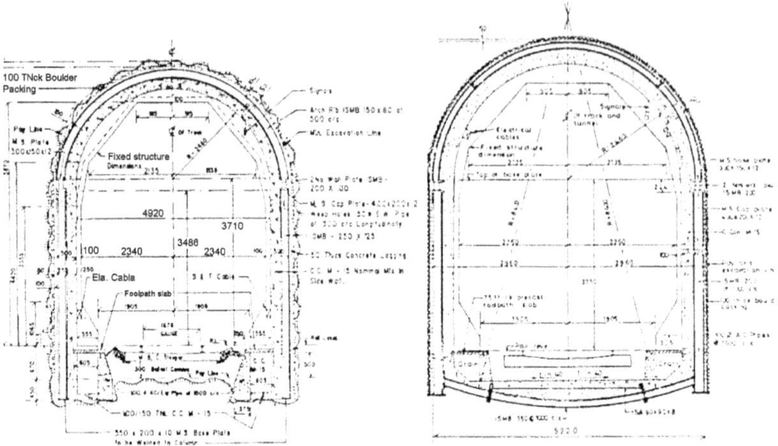

Source : ICJ, February 1994

(a) Ballasted Track on Grade-Rocky base

Source : ICJ, February 1994

(b) Ballasted Track on Concrete Base + Strut

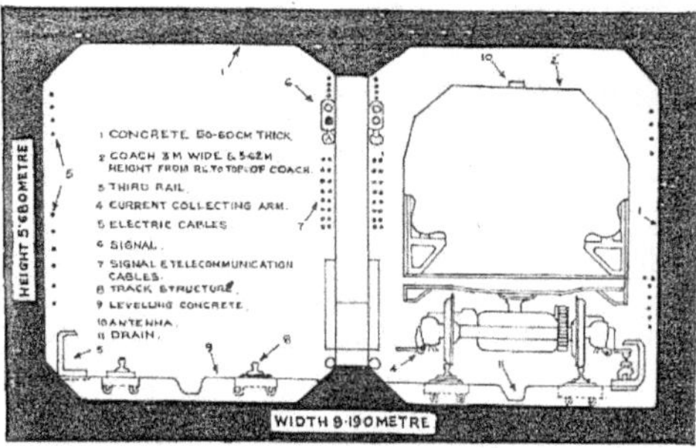

(c) Ballastless track on slab-

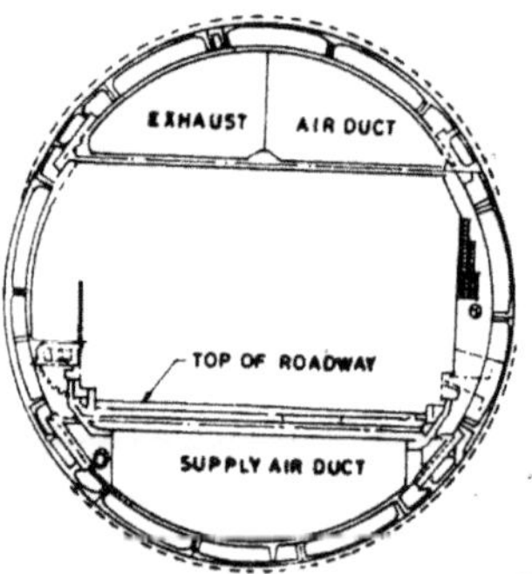

(d) Road on Structural slab with duct below

Figure 3.4 Different Types of Invert in Vehicular Tunnels

3.4 CONSTRUCTION AND MAINTENANCE

Type of soil, end use and construction methodology govern the choice of shape of tunnel. Thus construction methodology plays an important role. In open country and in hard rock, generally an elliptical shape or a segmental arch over a rectangular section is adopted. In soft rocks and medium cohesive soils also, such a shape is adopted so that the top portion is excavated and supported by arched ribs first and the lower portion extended by benching. In non-cohesive mixed soils or sandy/mixed strata, generally the shield tunnelling method is used. This calls for a circular profile. In subway tunnels for which the cut-and-cover method is used, mostly rectangular sections are adopted. If in cohesive soils, shield tunnelling or boring using moles is adopted (with or without compressed air) and a circular profile has to be adopted. The dimensions of tunnel have to be the minimum necessary for traffic requirements and at the same time large enough for use of mucking machines, trollies / trains or trucks for clearing away the tunnelled soil. In long tunnels shafts are provided at intervals to clear the muck. They are spaced at intervals of about 1 km to 2 km and at sharp bends.

The minimum maintenance requirement is good drainage. As already mentioned, a suitable grade in the line of the tunnel should be provided and in long tunnels, if possible, sumps are provided, cross-drains are drilled through and/ or suitable pumping arrangements are made. The minimum ventilation requirement will call for some shafts at intervals in long tunnels, generally over 2 km long. Inspection squads and maintenance men in railway tunnels will need some shelters in form of niches in which to take refuge when trains pass, and for this trolley refuges have to be cut into alternate sides at about 90-m intervals. For example, on the Pir Panjal tunnel built recently in Kashmir a 3 m wide paved pathway has been provided for the full 10.9 km length of tunnel for maintenance and emergency evacuation vehicles to be taken in.

Subway tunnels do not provide for such facilities as maintenance is to be done when there is no traffic. The track, these days, is made ballastless, thus serving as a paved way for emergency evacuation of passengers. Such provision minimizes maintenance requirements also. Subway tunnels have to provide for increased safety requirements, as they deal with large passenger traffic. Stations also being in same level experiencing large movement of people in opposing directions, this becomes more important. They need good ventilation and lighting arrangements. They also call for a high degree of cleanliness. Such tunnels are, in most cases, partially or fully below sub-soil water level and in granular or clayey strata. The lining or shell has to be completely waterproof, to combat consequent seepage.

In highway tunnels footpaths serve the purpose. On Konkan Railway some of these refuges in a long tunnel cut in rock had been used for reversing and passing mucking trucks coming from opposite directions, thus increasing speed of mucking operations during construction. During emergencies the staff in the case of railways or users in the case of highway tunnels must have access to some communication facilities with staff at portals or in neighbouring stations. Hence telephone plug points have to be provided at intervals. In long highway tunnels emergency equipment in the form of fire tenders and towing cranes is maintained at either end of the tunnel. Fire fighting water pipe lines can be run over the length of tunnel with fire hydrant points at intervals also in long tunnels in remote areas and in subway tunnels.

3.5 CANAL/NAVIGATION TUNNELS[2]

Navigation canals connecting natural streams or on their own formed the backbone of large traffic in seventeenth and eighteenth centuries. Where they had to cross from one valley to another in hilly terrain, taking them through tunnels became necessary. The earliest known navigation tunnel is the Maples tunnel opened in 1679. Canal tunnel construction had practically stopped after nineteenth century. Last perhaps is the one connecting to smooth water strip along coast near Marseilles and mouth of Rhone, in Golfe de Fos, an inland lagoon. Work on this 7.2 km tunnel crossing was started around 1905 and completed in 1927, work having been interrupted by the First World War. This is believed to be the largest canal tunnel, about 22m wide at springing and 15.4 m from invert to crown , The depth of water in canal itself is about 3m and it has two toe paths 2m wide each on the sides. See Figure 3.5. Large vessels including those with bow of beam of 8m and 1200 ton displacement capacity could use this tunnel safely.

They are similar to Highway tunnels, but with a level gradient. Their width has to be large enough to provide for maneuvering of the boats, though most of them have been built for unidirectional movement of boats. For example, the smallest section of such tunnel in UK is 3063 yard long Butterly tunnel 4'-10" wide and 8'-10" high and the largest 3036 yard long Netherton tunnel 15'9" wide and 27' high roof. They are mostly inverted U shaped with vaulted elliptical circular roof.

Alternatively, poling or shafting was used in manual haulage. If canal floor was lined with only clay puddle, the alternative forms of haulage by rope from ends or use of railings was resorted to. With the advent of diesel engines diesel powered tugs have been to haul a ling of boats. Steam powered tugs also had been used in some, but they resulted in the soot covering the roof and sides, which had to be periodically cleaned.

Walkways were provided for use by men or horses for toeing the boat. Paddling and rowing is not permitted in such tunnels for want of adequate width for rowing. Where space cannot accommodate such walkways, some form of railing or chain links have been fixed on short brackets for men to hold and move boats. Some form of control system between the portals has been provided to regulate the movement. In some wider canals, bi-directional traffic could be provided for smaller boats. Height has to be adequate to provide for vertical projections from the craft like chimneys and use of ores, as necessary. An additional and important requirement is that the floor and sides of the canal have to be watertight. Earlier ones were lined suitably with clay and puddle. Such tunnels cannot be provided in porous grounds. Channel section in tunnels in rocky soils are lined with concrete and well finished. In case of large width compared to height of the tunnels calls for adoption of vaulted structures. They are not safe to be provided in terrain subject to tectonic movement.

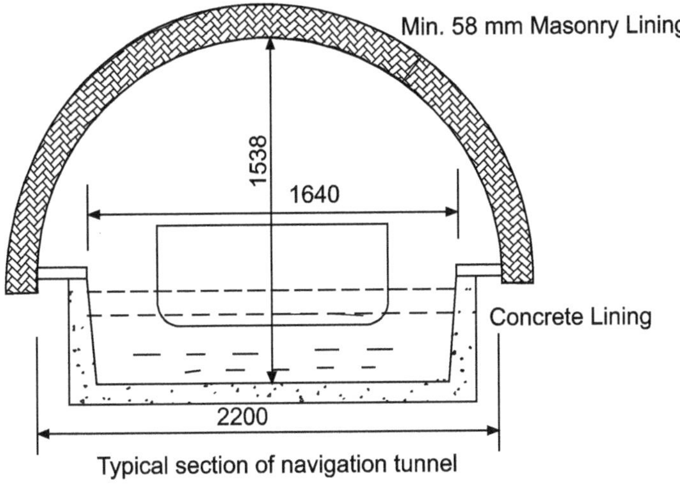

Adapted from Szechy 1970

Figure 3.5 A Typical Large Navigation Tunnel[2]

Curved alignments are avoided for better sighting and easy maneuvering of boats. Driving the tunnels is done by heading and benching or by driving through a number of drifts and widening, depending on type of soil. Long ones are done by sinking a number of shafts at intervals and opening out a number of faces, Some of these shafts are lined and retained for natural ventilation. Canals in long tunnels are generally divided into a number of sections with provision for inserting temporary shutters so as to isolate a section at a time for maintenance purpose, by pumping out the water for cleaning and any repairs. Long tunnels should be provided adequate lighting for the boatmen to toe and maneuver boats safely.

3.6 ADDITIONAL REQUIREMENTS IN WATER CONVEYANCE TUNNELS

3.6.1 Hydro-electric Power Plant Tunnels[2,3]

Water has to be conveyed to power plants located down the hill in the valley from storage reservoirs at higher levels. This can be done directly to the turbines in the power house through *Pressure tunnels* drilled through the hill or by pressure conduits or pipes laid on the sloping surface of the hill. In the latter case also, tunneling is done through any intervening ridge from reservoir outlet to the top of the conduit. In latter case, the tunnel is known as *Discharge tunnel.* The pressure tunnel will be subjected to comparatively higher outward pressure than the resistance that can be offered by the surrounding rock and overburden. Hence they have to be designed as a pressure pipe and the lining has to be suitably reinforced. A circular section is preferable for same, which will be easier to drill also. A horse shoe type also can be used, but adequately reinforced. Such pressure tunnels can be provided only through solid rock and are not suitable for rocks with fissures and cracks. They have to be highly water proof also. On the other hand, Discharge tunnels are designed and constructed like a railway tunnel and a horse shoe shape with inverted base is preferred. They also have to be made water proof. Figure 3.6 shows a typical layout of such tunnels serving a Hydro-power plant.

3.6.2 Water Supply Tunnels

Water supply tunnels are provided for conveying water from storage reservoirs or streams to the treatment plants or ground storage at distribution centres across an undulating country. They are similar to the Discharge tunnels mentioned above. In most cases, they will be at shallow depths and can be built by cut and cover method, in which case they can be designed like subway tunnels. In many cases in past, they have been built with RCC base, brick side walls and arch roofs with water proof lining inside.

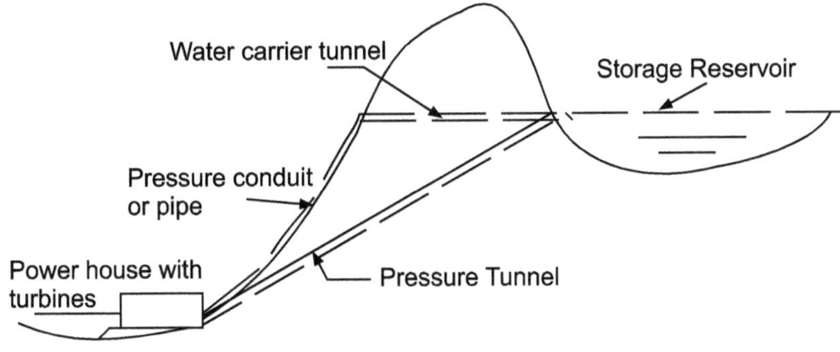

Figure 3.6 Typical Water Conveyance and Pressure tunnels for a Power plant

3.6.3 Utility Tunnels

It is desirable to carry utilities like liquid / gas pipe lines and cables underground housing them in larger conduits in from of circular or rectangular tunnels. Two typical examples are shown in Figure 3.7. Conveyor belt systems in power houses and large industries may also be housed in rectangular tunnels similar to the cable utility tunnel shown in the figure. The minimum internal dimension of any such tunnel should be 2.5 m for facility of maintenance and inspection personnel to move in them freely.

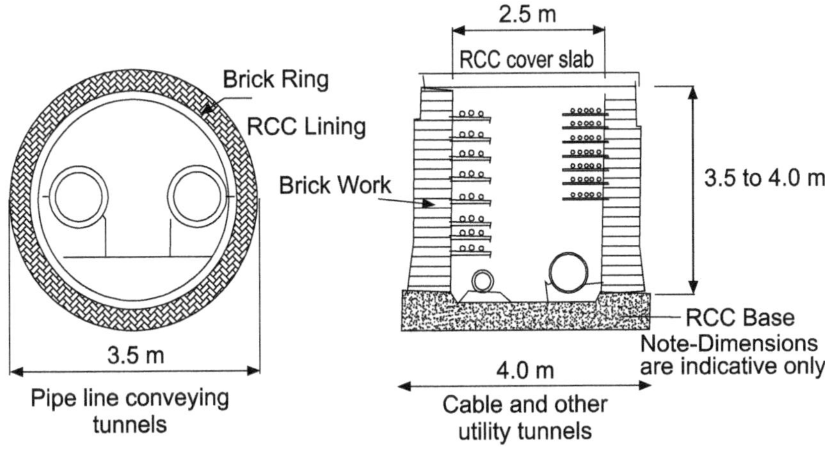

Adapted from Szechy, 1970

Figure 3.7 Typical Utility Tunnels[2]

3.7 REFERENCES

1. Morton, D.J.,(1982) "Subway Construction' in Ed. Bickel, John.O. and Kuesel, T.A.R., 'Tunnel Engineering Handbook', Van Nostrand Reinhold Company New York, (1982), pp 417-444
2. Szechy, Karlowi, (1970), The Art of Tunnelling, Akedimiai Kinda, Budapest, Hungary, 1970, pp 117-120
3. Narayanan, G., (2009), 'Aerodynamic Problems in Tunnels in High Speed Lines', Proceedings of International Seminar on High Speed Corridors and Higher speeds on Existing Network, Mumbai, Institution of Permanent way Engineers (India), New Delhi , 2009, pp3.1- 3.7.
4. Agarwal, M.M. and Miglani, K.K. 'Global Experience of Design, Construction and Maintenance with Special Reference to Indian Railways', National Technical Seminar, on Management of P.Way Works through need based Outsourcing and Design, Construction and Maintenance of Railway Tunnels, Institution of Permanent way Engineers (India) pp 315-356
5. Pequinot, C.A., (1963) "Tunnels and Tunnelling'- Hutchinson, Scientific and Technical, London

CHAPTER

4

Design of Tunnels

4.1 GENERAL

A tunnel design has to cover the following elements: (a) geometry of alignment, (b) tunnel profile or cross-section, (c) design of supports and lining and (d) design of transition (applicable to water conveyance tunnels). Geometry of alignment includes the following: (i) elevation or level, (ii) gradient and (iii) curvature.

Though many of the design factors are common to all tunnels, there are some special requirements for traffic tunnels.

4.2 ALIGNMENT

In surface tunnels which pass through hilly regions, alignment is mainly governed by the topography. Other general considerations which affect choice of alignment are the characteristics of soil through which a tunnel is to be driven, its position in relation to the groundwater level and ruling gradient, which in turn, is governed to a major extent by traction and operation considerations. Placing a tunnel much below the groundwater table should be avoided. It is advantageous to locate such tunnels in an impervious layer under an adequately thick cover, which thickness should be sufficient to prevent the inrush of water and also to prevent the escape of compressed air if it is proposed to use compressed air for tunnelling operations. If, however, traction and operating considerations require and it is otherwise advantageous, the tunnel can be taken at an elevation below the groundwater level after taking all necessary precautions. In subaqueous tunnels there is no choice as they have to be taken below flowing or still water. Tunnel boring operations will then invariably have to be done using compressed air. Such tunnels have to be placed below adequate overburden to minimise loss of air. They have to be placed well below the normal scour level, if the bed and flow conditions would induce noticeable scour. In still water or creeks not subject to scour, depth can be reduced through construction of tunnels by the precast unit /caisson sinking method.

4.3 GRADIENT

4.3.1 Railway Tunnels-Minimum/ Maximum Requirement

The gradient of any section is dependent on load and traction considerations. From the drainage consideration also, a certain minimum gradient may have to be provided in the tunnel portion for easy drainage of water entering the tunnel. This condition is particularly applicable to tunnels located below groundwater level and subject to seepage. The minimum gradient in terms of drainage is 0.2 to 0.3% so that the drainage water can flow out automatically. In terms of traction, the ruling gradient permissible has to be less than in open air owing to reduced adhesion and increased air resistance inside tunnels. The values for electric traction are compared in Table 4.1. From the energy point of view, the saving in energy over a horizontal line is 20% and 25% when using 2% and 3% starting-down gradient over 145-m and 125-m lengths respectively. A rising gradient of 2% results in a saving of 20% in braking energy. These two examples give some idea of the effect of gradient from the operational point of view.

Table 4.1 Change in Ruling Gradient in Tunnels

Ruling gradient (%)		Reduction
Open line	**Tunnel**	
1.00	0.65	0.35
1,50	1.02	0.48
2.00	1.40	0.60
2.50	1,70	0.75
3.00	2.12	0.88

However, the other governing factor is the capital cost involved in providing a flatter gradient which requires a longer length of track for negotiating the difference in level between the two points and consequent increased maintenance cost also. Therefore, at the survey/investigation stage, a very careful study of the ground contours is necessary for considering the overall effects. A number of alternatives have to be examined. This depends on lengths over which steep gradients are required and how far apart such lengths are situated. Even if some banking assistance is required over a short length, the overall saving, in terms of saving in energy and time of travel, can very well be worth the expenditure involved in cases in which such lengths are short and occur over the same stretch. In India, on main line ghat sections a gradient of 2.85% has been used with banking assistance, though a gradient of 1.67% is generally preferred. Presently, a ruling gradient of 1% is preferred for routes with heavy traffic. In metro tunnels 3% and 4° gradients can be used with increase in number of motor coaches. It is preferable to give a rising gradient towards the station at either end and to keep the station level higher. This results in economy and also helps in deceleration and acceleration of stopping trains.

4.3.2 Examples

The advantages of tunnelling in terms of route location can be understood from the following two examples (Pequignot, 1963)[1].

On the line between Skokomish and Wenatchee of the Great Northern Railway (USA) at the time of opening in 1892, there were eight reversals (switchbacks) provided for traversing the steep mountain slopes. Electrification was done in 1909 through the tunnel for increasing traction capacity. In 1929, by providing a single long tunnel across a spur, all the seven reversals were eliminated. There were still a few steep gradients to be negotiated. This tunnel was near the summit and the line thus ascended and then descended. In 1929, a new tunnel of 12.50-km length was bored at a lower level, reducing the elevation by about 150 m. This reduced the length of the line from 69 to 54.5 km, i.e., by over 20% (Figure 4.1). Replacement of electrical traction equipment (which was worn out and due for replacement in 1956) through the tunnel was not done. However, diesel traction on either side was continued through the tunnel, thereby eliminating need for change of locomotive at either end of the tunnel.

Another typical case of how tunnel provision helped in easing the gradient of a section is the Canadian Pacific Railway line linking British Columbia and Eastern Canada. The line had to pass through the Rocky Mountains, During quick construction in 1885, a section on these mountains was built in Yoho valley and over Roger's pass with a 1 in 23 gradient (6.4 km was on 4.5% grade) continuously. Four catch sidings were provided. It was difficult and costly to operate on the line. Increasing traffic demanded easier gradients. In 1908, the line was realigned and built with an easier gradient of 2% (making the line 12.8 km long). This was made possible by providing 2 spiral tunnels, one of about 980 m in Cathedral mountain falling 13.5 m on the circle and another about 880 m under Mount Ogden with a fall of 13.6 m on the line between the two running in reverse directions. It involved crossing the river twice. Both the tunnels were driven from both ends.

On the other hand, Konkan Railway in India had adopted a ruling gradient of 0.67%, in order to provide for operation of higher speed trains, being more of a passenger service oriented line. This has resulted in providing 76 tunnels aggregating a length of 79 km (about 10% of the route length)[3].

4.4 CURVATURE

The curvature resistance inside a tunnel is the same as that outside and hence the principles governing fixing of the curvature with respect to proposed speed of trains are applicable to the tunnel also. However, the need to avoid boring a tunnel with a number of sharp curves involving difficulties in alignment and other lurking errors should be borne in mind. Various practices prevalent have been discussed in Para 2.3. There are two spiral curves set at a gradient in the example on Canadian Pacific Railway mentioned in Para 4.3.2 above. On Indian Railways, 6° curves on Broad Gauge and 10° curves on Meter gauge have been common in the past. It is not uncommon to find reverse curves on the alignment.

Konkan Railway was the first ghat section railway in India to adopt 1.4 degree as limiting radius, since that line had been designed for 160 kmph speed. On the Jammu- Srinagar line, sharpest curvature adopted is 2.75° and steepest gradient 1%. Section Katra- Qazigund is over the most difficult terrain and there are 35 no tunnels in a length of 129 km, totaling 103 km of tunnels alignment. Longest tunnel in this section, and in fact in India, is across Pir Panjal range 10.96 km long. It is on a straight alignment. An ideal alignment for a tunnel is that a major portion, if not the entire length, is on a straight line.

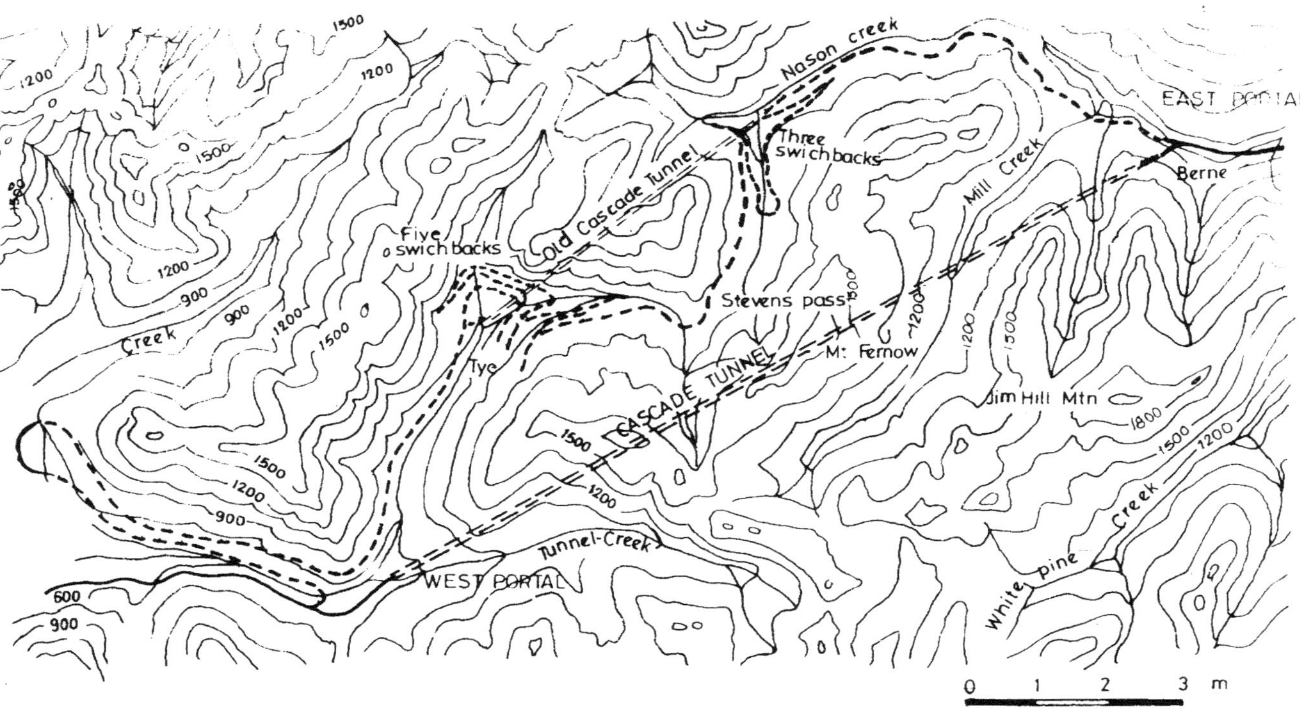

Source: Pequinot (1963)

Figure 4.1 Cascade Tunnel-Old and New Alignments.

4.5 DETERMINATION OF CROSS-SECTION

4.5.1 The choice of a cross-section for a tunnel is dictated by one or more of the undermentioned factors (Szechy, 1970)[2]:

(a) Clearance and moving dimensions specified for the vehicles moving and the goods transported in the tunnel;
(b) Type, strength, water content and pressures of soil;
(c) Method of driving;
(d) Material and strength of tunnel lining as well as internal loads acting on it; and
(e) Need for accommodating a single or a double track in the tunnel and other line side equipment.

Typical dimensions for Railway tunnels adopted in various countries are given in Table 4.2.

Table 4.2 Dimensions of Railway Tunnels in Some Countries

Country	Width m	Height of arch above rail level
Double-track tunnels		
Italy	8,00	6.00
France	8.00	5.80
Germany	8.20	6.20
Australia	8.20	6.40
Single-track tunnels		
Italy	4.60	5.00
Australia	5.50	5.00
Switzerland (Ricken tunnel)	5.20	5.80
India	4.72	6.00

The dimension given is for 1435 mm gauge in all except for India, which is for 1676 mm gauge.

4.5.2 Tunnel Clearance

The general principle adopted in providing clearance for railway tunnels according to the Central European Railway Association is that it should be at least 300 to 400 mm larger than the clearance required for structures on the open line. This is for additional safety against constructional inaccuracies, deformation of section due to rock pressure, rock displacement or water inrush. A tunnel section has to provide adequate ventilation and smaller air resistance. Air resistance, as detailed in Chapter 2, increases with velocity and if a tunnel has to be provided on a high-speed line, this aspect has to be taken into consideration even while deciding on spacing the track on a high-speed line. Air resistance will have considerable effect on two trains passing each other at high speed. Clearances permissible for railway and highway tunnels in Europe are indicated in Figure 4.2.

The most usual type, which has proven economical and easier to drive through, is the horseshoe shape for surface-to-surface railway tunnels,

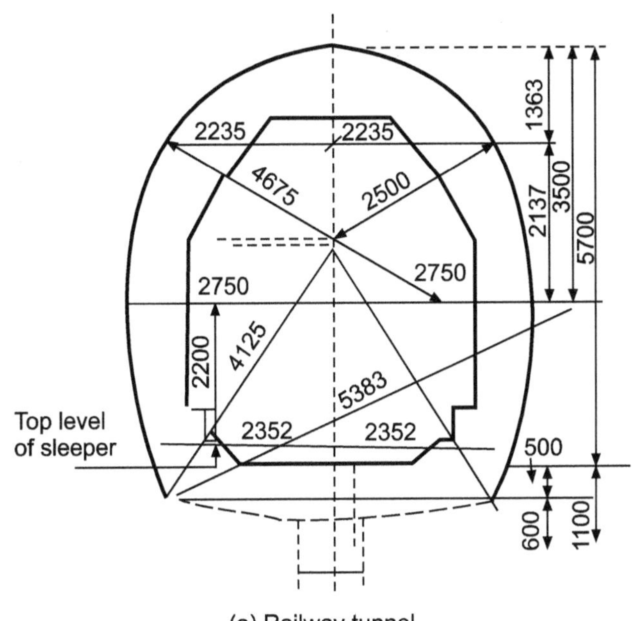

(a) Railway tunnel

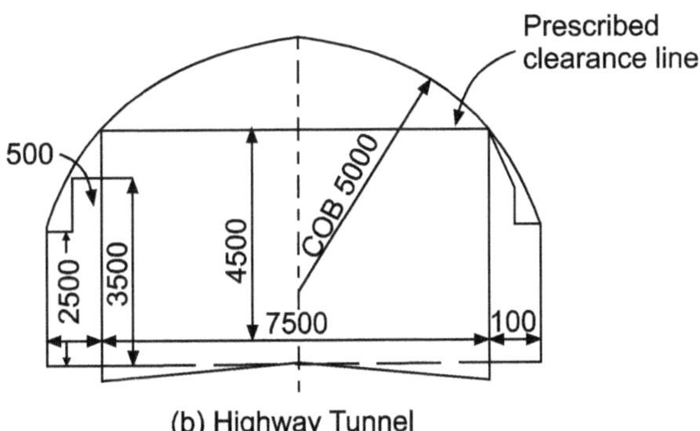

(b) Highway Tunnel

Figure 4.2 Permissible Clearances in European Tunnels.

especially if one tunnel has to be provided for each track, and occasionally for a double track. The main consideration is that the tunnel should have a section that provides the minimum clearance required for vehicles passing each other, without endangering a passenger who extends his hand out of the window, after allowing for oscillations and throw of vehicles on curves. The absolute minimum profile specified for a single-line railway tunnel in India (equally applicable to bridges) for MG and BG, which are the major gauges on the Indian Railways, are indicated in Figure 4.3 (a). The present practice is that on a new line the profile required for the BG is provided at the construction stage itself, irrespective of gauge of construction, so that later there is no need for widening when the line has to be converted to BG. It will be seen that for this profile, a horseshoe cross-section would be the most suitable.

4.6 GEOMETRIC SHAPE OF TUNNEL

4.6.1 Choices Available

There are five principal geometric shapes available for tunnels: (a) circular, (b) D-shaped, (c) horseshoe, (d) egg and (e) rectangular. In addition, there is a modified horseshoe type and a modified egg shape. The latter is sometimes known as egglipse. The modified horse-shoe type provides for a flatter base for ease of construction and for ease of change from that profile to a circular shape whenever both sections have to be adopted in contiguous lengths. The egglipse shape comprises of two arcs of a circle joined by straight lines or smooth curves. Thus it resembles an egg shape: the bottom and top profiles (i.e., base and roof) are arcs of a circle. Figure 4.3 (b) shows some of the shapes used in Europe for traffic Tunnels and clearances in them.

4.6.2 Factors Affecting Choice

The choice of shape depends on the following factors: (a); geological, (b) hydraulic, (c) structural and (d) functional and (e) method of construction. A comparison of different shapes is given below[3].

Circular Shape: Structurally the most efficient as all pressures are converted into hoop tension/compression, and thus it is best suited for resisting heavy inward or outward radial pressure. Hence it is ideal for conveying water under pressure with inadequate cover of overburden and in soft soils in case of deep traffic tunnels. But it needs Shield or Tunnel Boring Machine to execute the same. It is best suited for soft and medium soils but with use of special cutters and bore through rocks also. It needs careful

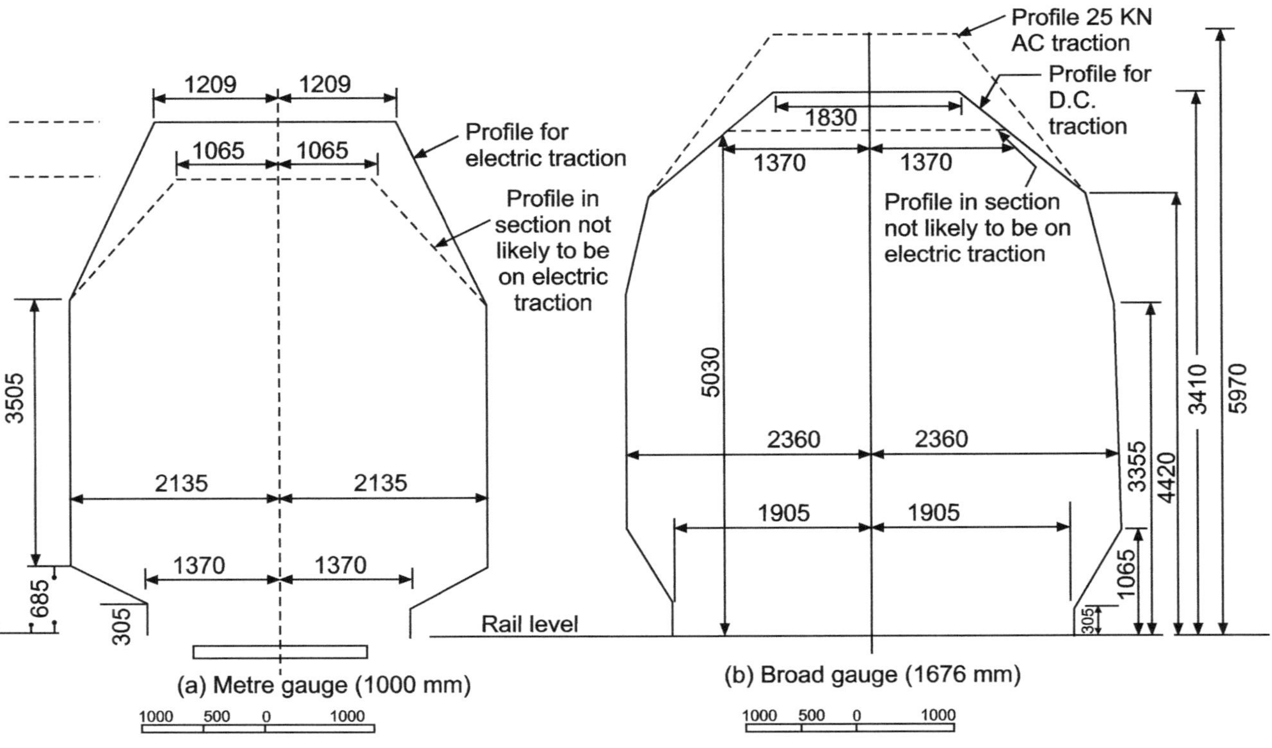

Figure 4.3(a) Minimum profile for Railway Tunnels in India.

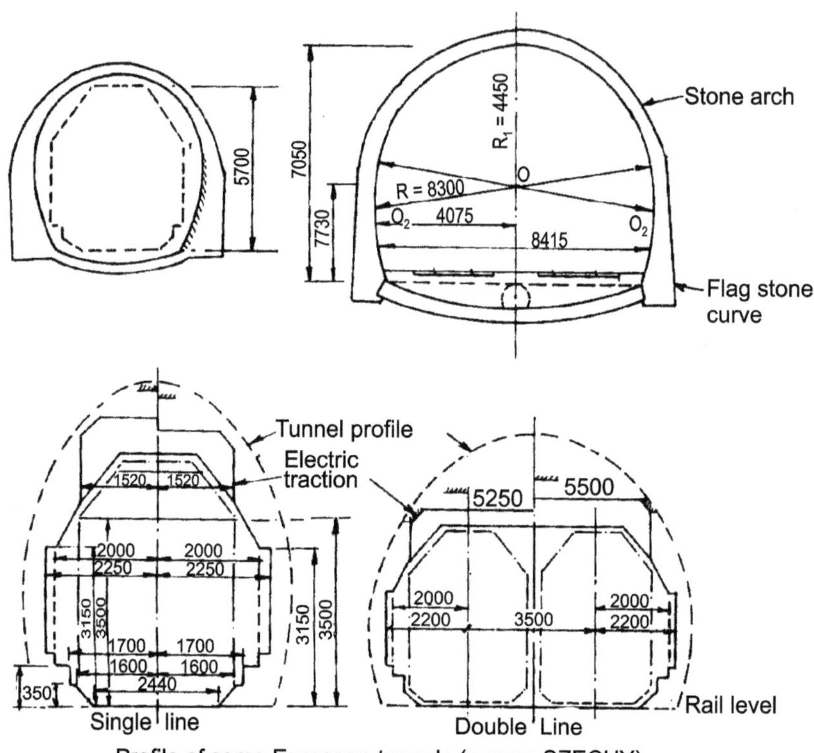

Profile of some European tunnels (source-SZECHY)

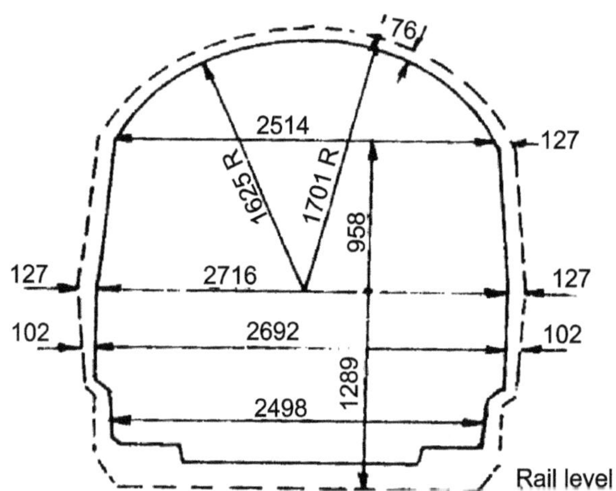

Structure Gauge for London Metro Extension

Figure 4.3(b) Different Traffic Tunnel shapes and Clearances

control of alignment during boring, since any misalignment or dipping is very difficult to correct.

D-Section: Easily constructed and gives more working floor space. Structurally also it is quite efficient due to its arch action. But the lining sections are heavy, especially on the sides, which usually makes it uneconomical. However, it is good and economical under massive igneous, hard, compact, metamorphic and good-quality igneous rocks where lining is not required. It is preferred in particular for water conveyance as it gives a large A/P (Area: Perimeter) factor for flow. The D-section is best suited for cut-and-cover type of construction. It is suitable for navigation canals and wide roads.

Segmental: Suitable for rocky formation (soft and hard). It has an arch shape for roof with vertical sides. The arch shape of roof provides direct resistance to the external forces, but the sides being just vertical have to resist side thrust by bending action. Bottom also needs heavy support to resist uplift forces. Thus it is structurally not so efficient. It is used for traffic tunnels, both for railway and roadway. It is easy for construction. It is done by drilling and blasting methodology full face or by heading and benching.

Horseshoe and Modified Horseshoe Shapes: A compromise between circular and D-shapes. The former shapes are stronger than the latter in providing resistance to external pressure. The modified horseshoe shape is easier to construct and also more economical in use of section, particularly in water conveyance. In rock it reduces quantity of blasting, especially at difficult corners. For traffic tunnels, it is superior to the circular shape since the space of excavation below the road-bed is reduced. It is generally preferred for transportation tunnels in hard strata and for water conveyance when there is adequate rock cover or when quality of rock is adequate to resist internal water pressure.

Polycentric: It is an extended form of Horse Shoe shaped tunnels. As name suggests, it is polycentric i.e. is made up of curves with different radii and centres. Its action is similar to that of horse shoe type and is fairly efficient. It is suitable for medium soils and can be done using NATM. Suitable for Road (multi- lane) and Rail lines (Double line).

Egg and *Ellipse* Shapes: Structurally it is most efficient as all forces, from above, pressure from sides and uplift from below are resisted by the arch action of the ellipse in all four directions. It is suitable in stratified soft and closely laminated rocks and when external pressures and tensile forces on the crown are likely to be high. These are suitable for Single track

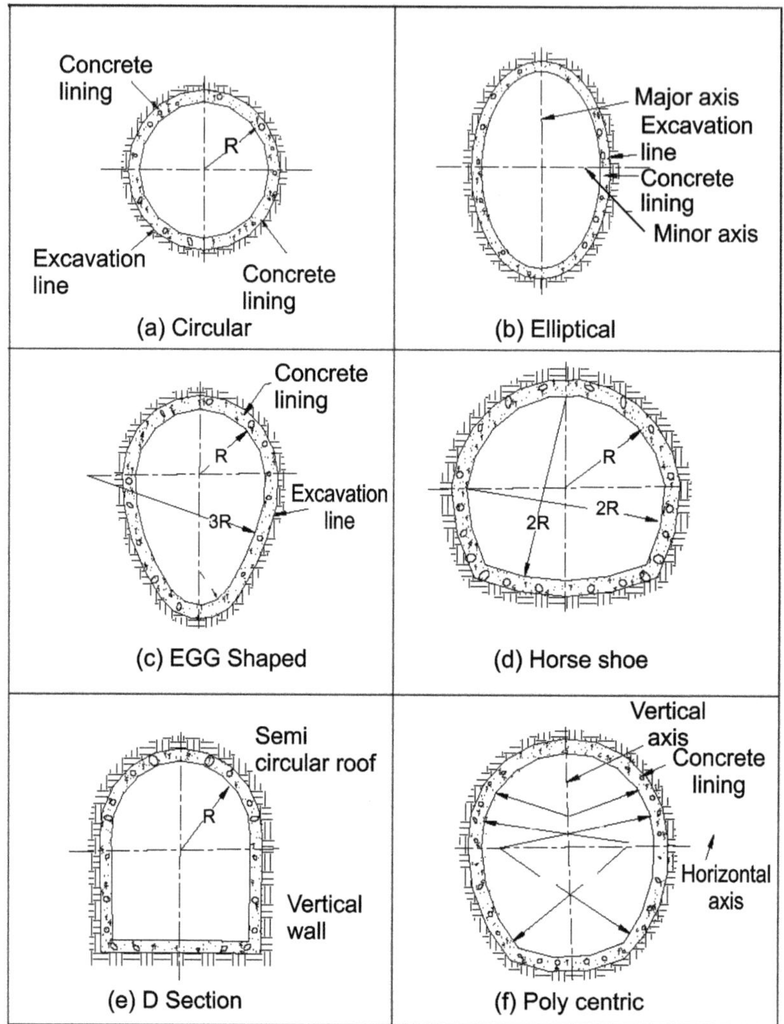

Figure 4.4 Different Forms of Tunnels[3].

Railway tunnels but not too often as the shape is not economical for providing the minimum side clearances required. The shape is advantageous for sewage tunnels which have to carry flow under gravity since the velocity reduction at low levels of flow is not much. The ellipse is hydraulically more efficient. Figure 4.4 shows different shapes for comparison.

Rectangular Shape: Structurally the least efficient but certainly the most convenient shape to build when the cut-and-cover method of construction is used. All forces are resisted by members by bending action. It is generally used for underground metros where cut and cover methodology is used and for stations, even if circular section is used for main lengths. It is simple for analysis and for construction also. However, this shape is not economical in the use of material and overall economy can be derived from ease and speed of construction.

Typical cross-sections of railway tunnels in Europe are given in Figure 4.3(b). While a horseshoe type is quite adaptable for the common method of driving used for a surface-to-surface tunnel, a rectangular section, especially for double lines, is more advantageous when the cut-and-cover method is adopted. The circular profile is most common in subaqueous railway tunnels and subways and is very rarely used for surface-to-surface tunnels in hilly tracts. It is the best in soft and medium soils for which tunnel-boring machines (moles) are used for driving the tunnels.

4.6.3 Selection of Shape

It is more economical in solid rock to go for a horseshoe section even for a double-track tunnel. In soft soil it is more economical to adopt two single track tunnels of circular section. Furthermore, in soft soil it is simpler to use the driving shield method or a 'mole', the most advanced method of tunnelling adaptable for use of machinery. The axes of such single tunnels in soft ground should be spaced not closer than 25 to 30 m if both are being driven simultaneously. If one is driven later, additional spacing should be provided to preclude the effect of the load that may already be coming on the earlier driven tunnel on the driving of the second tunnel.

4.6.4 Influence of Tunnel Lining on a Cross-section

The material of the tunnel lining must also be taken into consideration when choosing the cross-section. Materials capable of resisting compressive stresses only (stone and brick masonry and concrete) are limited to structures composed of rocks and of robust side walls supported on good soil and arch roofs. They should be used to carry purely compressive stresses or tensile stresses not exceeding the tensile strength of the mortar used. Examples are horseshoe and elliptical sections (the major axis of the last mentioned may be vertical or horizontal) and to a limited extent circular.

Materials capable of resisting tensile and bending stresses alike (reinforced concrete, steel and temporarily timber) can be used for lining sections of any desired shape, designed in the most economical manner and permitting the fullest possible utilization of space. Examples are rectangular sections with a flat or arched roof and thin shell-like linings.

Some basic principles that govern the design of supports and lining are given here without going into details. In general, under Indian conditions the type and thickness of lining of tunnels in rock have been mostly determined based on past experience. Earlier, masonry lining was popular but now plain or reinforced concrete is more common. Lining for soft strata and supports in disintegrated, disturbed and soft rocks are designed as appropriate.

4.7 ROCK PRESSURE[2]

4.7.1 Types of Rock Pressure

Tunnelling through any ground causes a cavity and removes the support against load from above and side thrust on the soil around the cavity. Load from above, which the removed soil had been transferring down, has to be transferred by the sides of the opening acting as abutments. Due to this there can be some heave from below also. This is conceptually shown in Figure 4.5.

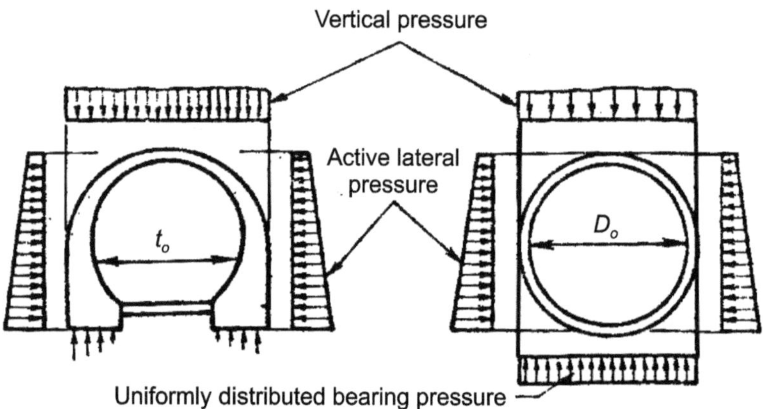

Source : Szechy, 1970

Figure 4.5 Distribution of Forces at a Tunnel Opening

Tunnelling through rock induces three types of rock pressure: (a) loosening pressure, (b) genuine mountain pressure and (c) swelling pressure. Rocks can be classified into three types in respect of pressure conditions: (a) solid rocks, (b) pseudosolid, soft and weathered rocks and (c) loose rock.

Smaller the cohesion of the soil mass, more the loosening pressure resembles silo pressure. The pressure caused by mass above the opening due to boring and removal of soil is similar to that of a soil mass piled up in a silo provided with a slot at its bottom. When the slot is opened, the pressure drops suddenly to a minimum value but gradually increases again as the slot is opened wider. The latter value, however, does not attain the intensity of the original geostatic pressure acting on the cover. If the slot is left open, i.e., if the material above is left unsupported, a wedge-shaped mass will gradually drop out from the overlying rock mass into the cavity until equilibrium corresponding to the changes in stress conditions is re-established.

Solid rock transfers the load acting on it by beam action to the side supports. In loose rocks and soils, load transfer to the undisturbed lateral section has to rely on friction developing during mass displacement. Deflections and bending stresses decrease with increasing height above the cavity, as the spans become smaller and smaller due to the cantilever support provided by the lower layers. In counteracting loosening pressure the most effective method of construction is to excavate the cavity as far as possible and with as little settlement as possible, followed by support at the earliest possible time with a rigid and permanent structure.

In the case of countering genuine mountain pressure, the type of rock must first be ascertained. The methods required for supporting the excavated cavity against genuine mountain pressure are entirely different in solid rocks from those required in pseudo solid rocks. The occurrence of genuine mountain pressure in solid rock is quite exceptional. It reveals itself in propping which, however unpleasant during construction, does not affect the final stability of the tunnel. A lining of moderate thickness provided tightly against the exposed surface of solid rock with rapidity ensures the required solidity. A rigid lining is required all around, more so at the roof. On the other hand, in pseudo solid rocks, such as clay, clayey shale, phillite shale, and crushed and modified gneiss, early provision of lining has been found to lead to failure whenever a plastic state of stress developed subsequently. Temporary supports must not be designed to resist genuine mountain pressures. There are no practical means available to resist these tremendous pressures and therefore the development of a protective zone must be awaited. It is a question of time and space. This purpose can be achieved by providing easily replaceable temporary supports. Prudence dictates that the pressure of the material is allowed to build up along the unlagged sides to be dissipated gradually by continuous yielding. This also can be done by providing a flexible lining in form of a shotcrete layer supplemented by rock anchors first. Solid lining can be provided later after section takes a set. Methods have been developed to measure rock movements and pressures, in the meantime, which help in design of the solid (permanent) lining to be provided. Readers may refer to Pequinot (1963) and Szechy (1973) for more details on such tests.

4.7.2 Theories Regarding Rock Pressure

Several theories for determination of rock pressure have been proposed. The following based on various assumptions regarding displacement and equilibrium conditions, are of particular interest (Szechy, 1973).

(a) *Bierbaumer's theory*: The theory of Bierbaumer was developed during construction of the great Alpine tunnels. According to this theory, a tunnel is acted upon by the load of a rock mass bounded by a parabola of height h equal to αH, α being a reduction coefficient and H being depth of over-burden (see Figure 4.6(a) and 4.6 (b)). Two methods, yielding almost identical results, were developed for determination of the value of this reduction coefficient. One approach assumes that upon excavation of the tunnel, rock material tends to slide down along rupture planes inclined at $(45° + \phi/2)$, where ϕ is the angle of internal friction of the soil. The second approach arrives at a reduction factor based on friction acting on the sliding earth mass.

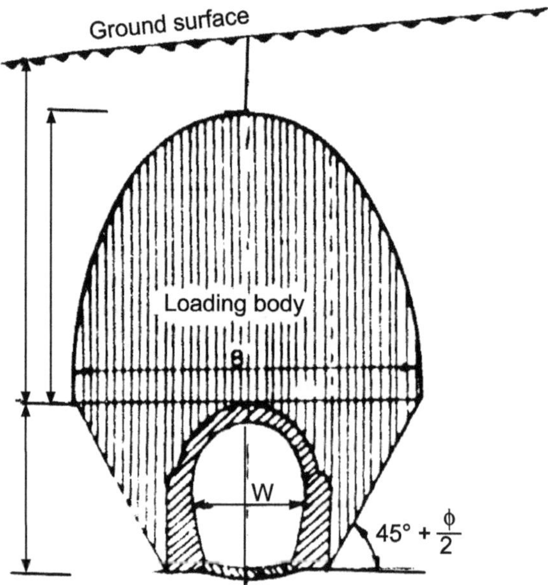

Source Szechy - 1973

Figure 4.6(a) Rock Pressure Bulb after Bierbaumer.

(b) *Terzaghi rock pressure theory*: This theory was originally developed for cohesionless, dry, granular soils but can be extended to cohesive soils as well. In correspondence with actual conditions, Terzaghi assumed that the moisture content in sandy soil is sufficient to secure the cohesion value necessary for maintaining the vertical position of the face in minor headings. The sand masses around the cavity are already disturbed by excavation, and movement continues even when the temporary supports are installed after the tunnel has been excavated to full section. These displacements suffice to lead to the development of a set of sliding planes characterising the state of imminent rupture in the sand. It is therefore justifiable while determining the width of the earth mass suffering displacement, using the inclination (45° + ϕ/2) of the plane of rupture associated with active earth pressure. Such width is obtained as:

$$B = 2[b/2 + m \tan (45° - \phi/2)],$$

where b = width of tunnel, m = height of tunnel.

Displacement of the earth mass is counteracted by the friction developing on the vertical shear planes. The vertical boundary planes of displacement may therefore be represented by the verticals drawn at the ends of the element of width B (see Figure 4.7(a)). At large depths the arching action of part of overburden reduces effect of loads transmitted as indicated in Figure 4.7(b).

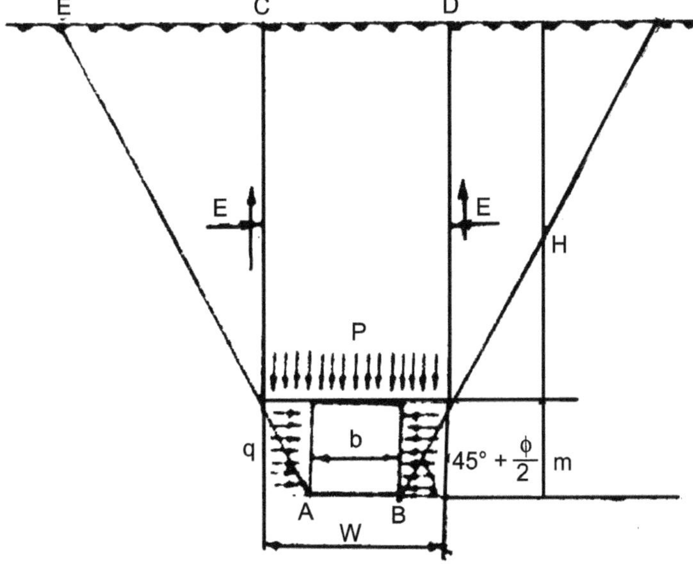

Figure 4.6(b) Assumption Model of Bierbaumer Theory.

(c) A number of theories neglect the effect of depth of overburden, the oldest being Kommerell's theory. According to this theory, displacement or deflection of the supporting structure is representative of the displacement suffered by the disturbed soil mass. As a consequence of this displacement, the mountain material is 'relaxed to a height equal to the height of the soil column capable of filling this space' by loosening. Where the mountain material is loose or soft enough for lateral pressures to also be anticipated, the curve enveloping the loading body should be

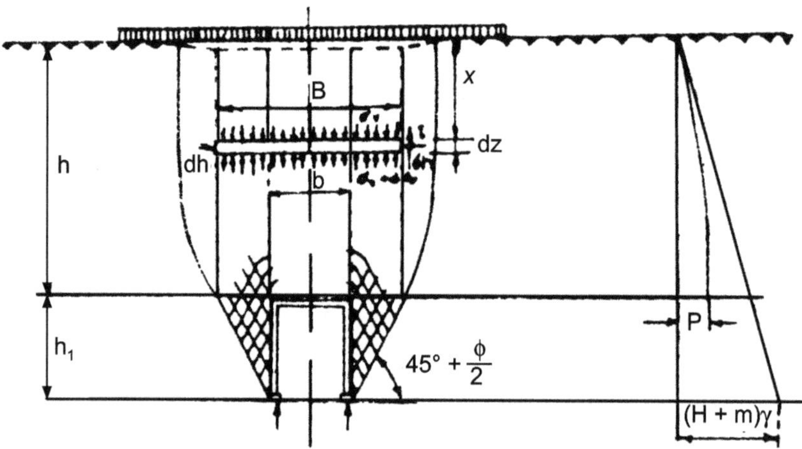

Figure 4.7(a) Basic Assumptions of Terzaghi's Rock Pressure Theory.

started from the points of intersections of the extended roof line and straight lines drawn to a slope of (45° + ϕ/2) from the lower corner points of the cross-section. Otherwise the curve should be the equivalent of a stress free body contained below a zone of arching of soil. For covers of over 30 to 50 m, tunnel lining should be designed for this condition.

If there is no lateral pressure, a parabola or an ellipse of a calculated height h can be fitted over the tunnel to represent the burden zone (Figure 4.8). If there are lateral pressures to be considered, the sliding surfaces at angle (45′ + ϕ/2) will extend from the bottom line of the walls up to the top of the section. Above this level, the zone of burden will once again be contained within a parabola. The height h of this zone can be calculated by a number of methods. In case of entrance zone the depth will be low or soil loose above tunnel.

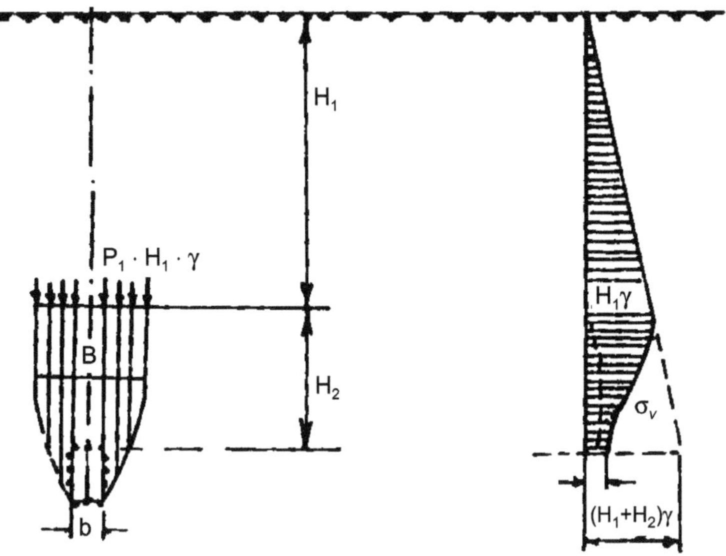

Figure 4.7(b) Rock Pressures at Greater Depths (after Terzaghi),

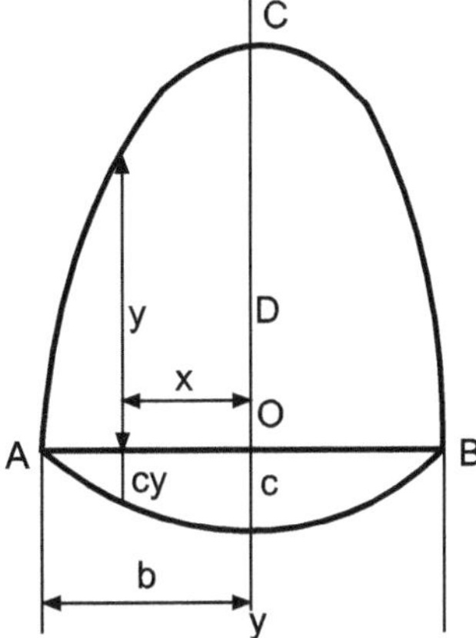

Figure 4.8 Shape of Kommerell's Pressure Diagram[2].

Then, depth of overburden '*h*' to be considered will be for full depth. In larger depths the parabola principle will apply. The vertical loads are calculated by dividing this area into narrow strips (Figure 4.9). In most cases the two dividing lines are vertical and construed as representing either side of the tunnel. The tunnel is then to be designed to withstand the vertical pressures due to the central strip and the lateral pressures as derived from

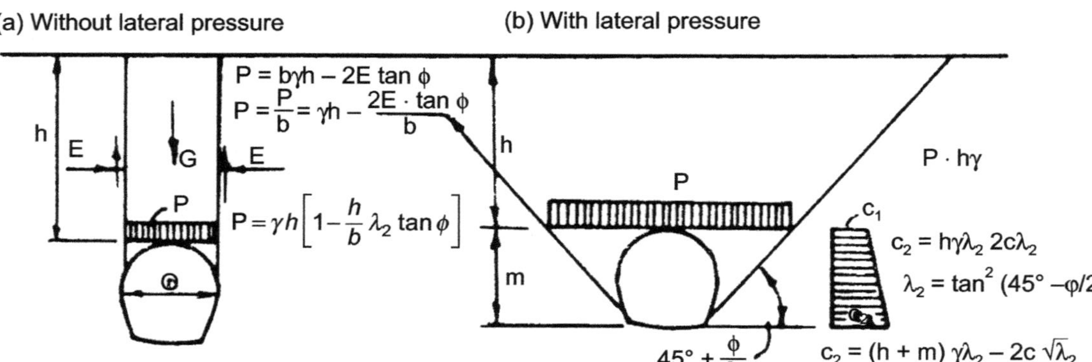

Figure 4.9 Assumed Pressures at the Entrance Zone.

the pressure diagram due to the two areas on the sides. Figure 4.10 shows pressure diagram for rocky strata.

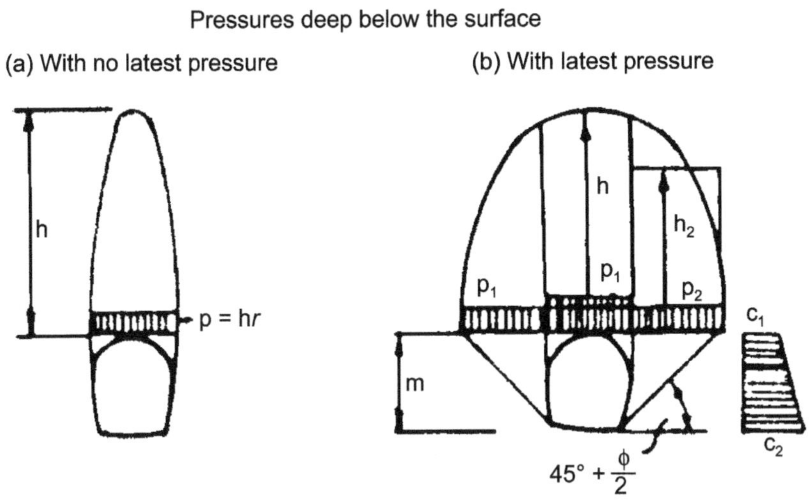

Pressures deep below the surface

(a) With no latest pressure

(b) With latest pressure

Figure 4.10 Rock Pressures with Large Depth of Overburden.

Except for the adit and exit areas, it makes no difference whether the surface is sloping or not. In those cases, the diagram of vertical pressure will be Trapezoidal and the lateral active pressures on either side of tunnel will not be equal. In solid ground with low pressures, passive pressures must not be relied on unless there is a distance of at least 9 to 12 m between the exterior face of the walls and the surface.

Fault Zones: Special consideration must be given to tunnel sections passing along dividing lines between two different rock layers and/or faults. These lines will constitute the critical sections, in which tunnel pressure will not be symmetrical and lateral pressures will also be present in solid rock. Pressures in solid rock depend largely on the condition of the material. For example, there will be no pressure at all in sound rock free from fissuration. On the other hand, adjectives such as 'laminated', 'shale structure', 'block formations', ‘interwoven’, 'cracked', 'weathered' etc., are all indications of a tendency to detrition and warrant increases in the assumed lateral pressures. Moisture content has a similar effect as far as design loads are concerned because it 'lubricates' the cracks, which would cause an increase due to its own weight, thereby influencing pressure coefficients.

4.8 DESIGN OF TUNNEL LINING

The design practice for the linings differs with the type of soil and the shape of the tunnel section. In principle, the lining has to withstand *Load* of overburden (and water if water table is above); *Horizontal pressure*—due to soil and also water if water table is above Self-weight of lining and track/road and live load transmitted to base, and; *Uplift pressure* from base due to the loads mentioned above being trans-mitted to soil below.

These loads cause direct thrust and bending moments to different components, i.e., roof, sides and invert. It is simpler to resolve these effects for rectangular or horseshoe type (arch and rib) structure (with a little approximation). But designing circular sections becomes very complicated. The total effect has to be worked out by summation for worst combination. Many designers have found that such precise computations are not called for, and have evolved some approximate methods for circular tunnels built using the shield method. One such method is that of Hewett-Johannesson, which is described below[2].

Hewett-Johannesson method takes into account both lateral support and interaction of soil. It is assumed that their values and distribution are such that the resultant thrust line coincides with the centre line of the lining and that no moments are caused. Since it is an approximate method based on an arbitrary assumption, it is limited to sections composed of segments. The joints can be treated as hinges preventing development of moments but at the same time joints are tight enough to transfer the thrusts and prevent deformation. The loads are analysed both for *short-term* case (during construction) and *long-term* case (as when the tunnel transfers its own weight and live load also to the base). They comprise of overburden load, earth pressure, water pressure and uplift from base due to overburden and superimposed load (self-weight + live load). The lateral resultant earth pressure will lie some-where between active and passive earth pressures. The passive earth pressure comes into play to counteract the forces induced by the vertical loads which tend to make the lining shorter in the vertical diameter and longer along the horizontal diameter. This elongation is counteracted by the passive resistance of the earth (Szechy, 1970). Hence the effective earth pressure will have a value somewhere between the active and passive earth pressures, i.e., the coefficient of horizontal earth pressure k will be

$$\lambda_a < k < 1/\lambda_a$$

where $\lambda_a = \tan^2 (45° - \phi/2)$

This value k has to be chosen such that there is no resultant moment in the section. In soft and saturated soil $k = 1$.

The critical stress in the section σ_{kr} is given by

$$\frac{\sigma_f}{(1+\sigma_f/\sigma_k)}$$

Where σ_f = yield stress of lining material or joints and
σ_k = buckling stress of lining elements bearing against compact soil, which is given by equation

$$\sigma_k = \frac{2}{F}\left\{\frac{\text{CEJ}}{\text{C} - \mu^2}\right\}^{1/2}$$

F = unit cross sectional area of the lining;
μ = Poisson's ratio.

In case of deep-seated circular tunnels in cohesive soils, alternative practice has been to design the circular section to withstand the external loads for hoop tension and compression, treating it as a monolithic structure. Due to the external loads, the ring will undergo deformation, becoming shorter in vertical direction and expansion along horizontal diameter. The elongation will be countered by the horizontal earth pressure which will be somewhat between active and passive pressures. There will be some bending moment induced in the ring, but the earth pressure coefficient is chosen so that there is no moment in the section. In practice this has been found adequate.

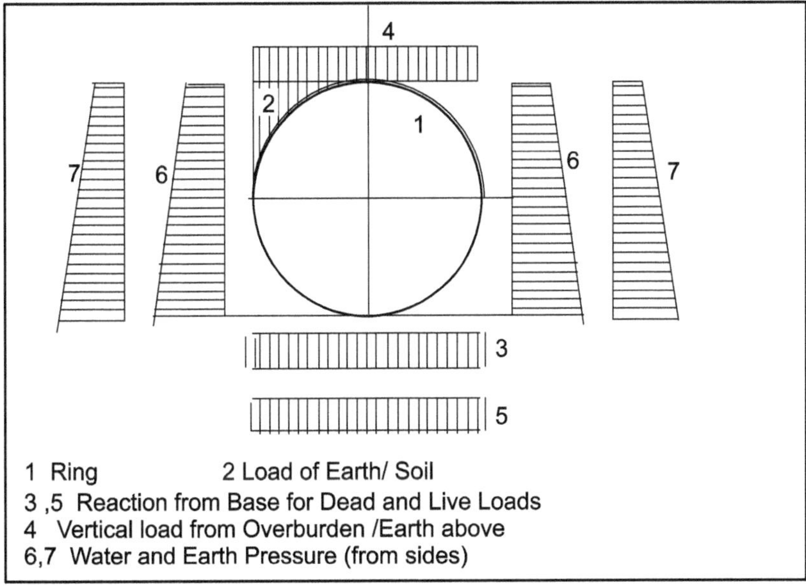

Adapted from : Szechy, 1970 (*pp* 357)

Figure 4.11 Forces acting on a Circular Tunnel (after Hewitt- Johannesson).

Rectangular box sections are generally adopted for metro subways and roads in shallow depths. These are invariably constructed using cut-and-cover method. The vertical superimposed loads above will be the full depth of earth filling over the roof, surcharge water up to subsoil water level and also the effect of live toad on surface(after allowing for due dispersion through soil), if it is used for road or railway. The horizontal forces will be active earth and water pressure on sides. Base has to be designed for uplift pressure of surcharge load, self-weight and live load within tunnel transferred to soil.

A Design example each following conventional methods are given in Annexures 4.1 to 4.3 for a Box section, a Horse shoe section and a Circular section to illustrate basic principles and to indicate pattern of resultant moments and forces different parts of the structure. These methodologies are applicable globally and need modification to suit the dimensions and local external forces in specific cases.

4.9 DESIGN PRACTICE OF RAILWAY TUNNELS

4.9.1 Loads and Forces to be provided for

Roof load: Before designing the type of supports to be provided in a lined tunnel, it is necessary to make an estimate of the roof load, i.e., the depth of overburden. As would be noted from the foregoing discussions, the standards adopted depend on local conditions. Broad indications depending on the strata of rock met with in India and generally accepted in Indian Railways have been:

(a) For hard stratified rocks, the roof load is taken for depth equal to half the width of the tunnel.

(b) For massive rocks moderately jointed, depth of overburden is 0.25 to 0.35 times the width plus height of the tunnel.

(c) For moderately blocky rocks, 0.25 to 0.35 times the width plus height of the tunnel is assumed for depth of overburden.

(d) For highly blocky rocks it is taken as 1.1 times the width plus height of the tunnel. For rock tunnels, this is almost the maximum roof load.

Side pressure in all the above cases can be deemed negligible and provision made only where the rock occurs in stratification dipping towards the alignment.

(e) In the case of crushed rock or squeezing or swelling type of rock, side pressure of considerable magnitude is met with and must be provided for.

All the loads mentioned here can be reduced by half if the tunnel is permanently located above the water table.

Szechy suggests a maximum value of 2.5 B, B being width of tunnel. IS:5880 (Part V)-1972 specifies that strata load to be considered should be 1.10 to 2.10 times (B + H_t), where Ht is height of tunnel.

In practice it may not be possible to predict to any degree of accuracy the type of rock that would be met within the range given in the previous paragraph. Most tunnels exhibit weathered broken rock at the exit and entrance and better conditions in the interior of the hills. The dip or inclination of the rock, while being favourable in some lengths of individual tunnels, becomes adverse in curved sections of the tunnel. The blocky nature of the rock produces some unexpected domes and vaults. The presence of mica, biotite, schist and graphite produce slippery rock surfaces.

4.9.2 Choice and Design of Structure

It is generally necessary to have a ready design of permanent steel supports to cover all such conditions, in standard lengths of steel verticals, wall plates and ribs, except as dictated by requirement of additional clearances on curves. It should provide a high level of flexibility and the requisite strength of support provided for only by altering the spacing of these verticals and the ribs. Broad-flanged beams would definitely be superior to joists or channel sections, where easily procurable. Consistent with the assessed roof load, in practice the arch ribs have been spaced as close as 250 mm centres and as far apart as 750 mm centres and the verticals spaced at 1.5 m centres and where required at 1.0 m centres with a runner in between. Where heavy side pressures are anticipated, one vertical under each rib or at 500 mm centres need be provided. It would be no exaggeration to state that this high level of flexibility of steel supports contributes to rapid and efficient progress in the supporting of tunnels soon after excavation and yet with economic use of steel in many cases.

IS:5880 (Part V)-1972 stipulates that "while designing the final lining the fact that the primary lining and the steel support will also participate in resisting the forces shall be taken into consideration." To ensure this, the clause 7.5 stipulates that the gap between the strata and lining shall be fully covered by grouting and that the rock around the tunnel for at least one diameter be strengthened by grouting under pressure. The steel supports and primary concrete will take care of the external load and this should be designed by methods similar to those used in design of culverts. The IS requires that the gap between the strata and support lining is fully backfilled and grouted at a pressure not exceeding 0.2 N/mm^2 soon after support and lining is placed. The roof load is transmitted to the arch rib at points where the rib is blocked against the roof. The practice on Indian Railways has been all along to provide a 50 mm thick hard wood lagging over the ribs and to provide 'tight' packing, i.e., continuous packing over the length of the arch rib. Calculation of maximum thrust and bending moments, if any (depending on the arch shape), determination of the force of passive resistance mobilised by the rock against the active tendency of the rib (in some portions) to advance toward the rock are matters of statistical determination and any convenient acceptable method can be used. Where 'tight' packing is provided, the thrust in the rib may be deemed equal to the vertical reaction. The vertical is to be designed to take the reactions from the arch and the side pressure as deemed necessary depending on type of rock and in all cases of moorumy soil strata. It is designed as a column fixed at the base and free at the top.

In case of subway tunnels, RCC is the natural choice. RCC box structures are designed as frame units on spring supports. Bored tunnels are generally designed as flexible closed rings using precast RCC segments. Earlier ones had used CI of steel prefabricated matching segments. A brief summary of the design done for a few combinations of boxes used on Kolkata Metro is presented in Annexure 4.1.

A typical calculation of arch rib and vertical forms Annexure 4.2. Where heavy side pressures are expected as in the case of squeezing rock condition (as in shale or highly weathered rock), invert struts will also have to be provided to withstand the thrust and later embedded in invert concrete.

4.10 MODELING APPROACH TO DESIGN

ITA (International Tunnelling Association, Working Group on General Approaches to Design of tunnels has presented a report on international design procedures for tunnels[5]. It takes into account the factor of the ground around actively participating in providing stability to the tunnel opening. The report covers in detail the general approach to design including site investigations, geotechnical probing, in-situ monitoring (during construction), and a flow chart for design procedure in tunneling. It gives alternative approach to structural modeling which should be based on the following criteria-

- Deformation and strains
- Stresses and utilization of plasticity
- Cross-sectional lining failure
- Failure of ground or rock strength, and
- Limit -analysis failure modes

According to this document the forces acting on the tunnel from the ground above and around for different conditions and approaches will be as presented in Figure 4.12(a).

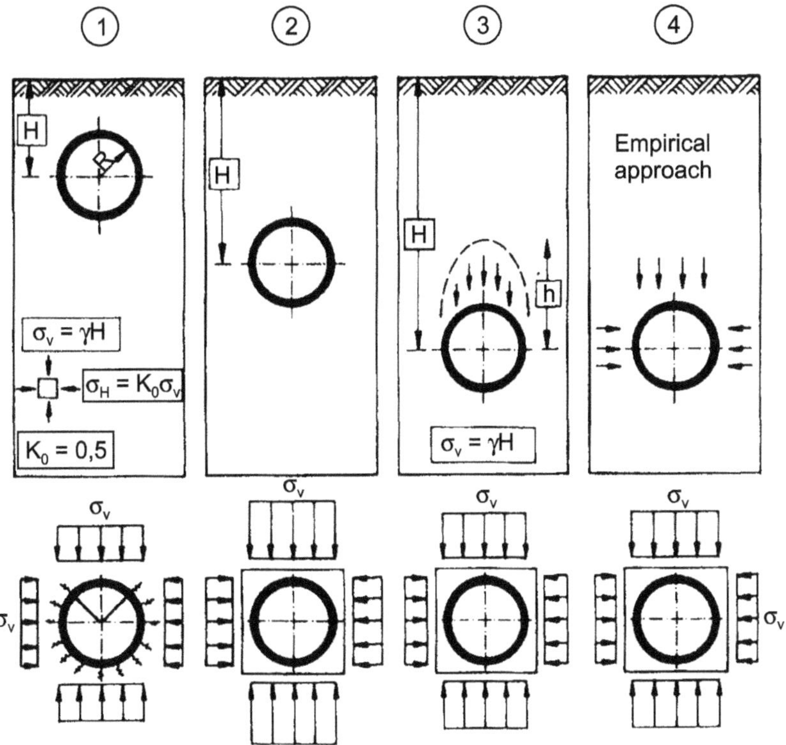

Source: Reference 5

Figure 4.12(a) Plane Strain design models for Different depths and Ground stiffness.

It presents four structural models for (plane) two dimensional structural analyses. They are:

(i) In soft ground, immediate support is provided by a 'relatively stiff' lining. In shallow depths, as in case of urban subways, a two dimensional analysis will do, neglecting three dimensional reaction as near tunnel face. Ground pressures acting on the tunnels are assumed as equal to primary stresses that would exist in undisturbed ground in cases 1 and 2. In case 1, full overburden can be taken as load. Ground reaction is simulated as radial and tangential springs, and bedded beam model approach is followed. In case 2, 'soil stiffness is employed assuming a two-dimensional continuum model' assuming existence of complete bond between ground and structure. Any inward displacement would result in reduction of pressure on lining.

(ii) In case of case 3, (in large depths) deformations in the ground before lining become active, some stress release is assumed to occur. Ground may be strong enough to allow some unsupported length near tunnel face in medium rock and highly cohesive soils. In case of high overburden, there will be some reduction in load on crown i.e., h < H, as shown.

(iii) Case 4 represents an emphirical approach (generally followed earlier), which is based on previous experience in similar grounds and same method of tunnelling, modified based on in-situ observations and monitoring of portion of tunnel already done. This is a design model subject to continuous modification based on observations.

Except in case of complicated geometrics of the underground structures, two dimensional structural models are considered adequate and in complicated cases, three dimensional models are used. Normally bedded beam models are used. Generally Finite Element analysis is used for structural design. In case of shield driven tunnels using segmental linings and fully closed circle, an approximate method can be used considering only the ring forces and consider the bending moments less important for providing equilibrium. Figure 4.12(b) shows the loading pattern and forces and moments on a circular section of a tunnel. Reference 5 may be referred to, for more details.

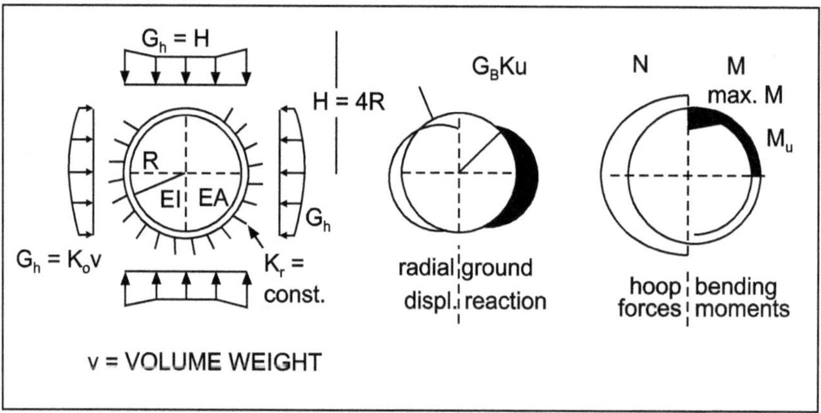

Source: Reference 5

Figure 4.12(b) Bedded beam Model for Shallow tunnels.

Latest development is use of computer models for entire design. This has been done for the tunnels under construction in a large scale by Northern Railway on the USBRL Project and by North east Frontier Railway in Assam and Manipur. Alternatives used are a 2-D Finite element program developed by Rockscience. Inc of Canada and one available on STAAD-PRO. Other commercially available programmes are SOFTiDik and C-Tunnel using Finite Element analysis. They take into consideration both thrust and bending moment on D shaped as well as Ellipse and Horse shoe forms. One typical such design, done for a Horse shoe type BG Railway tunnel is presented in detail in Reference 6.

Szechy's method for circular tunnels has been used by KRCL for the circular lining of Honavar tunnel. A brief summary of same is given in Annexure 4.3.

4.8 REFERENCES

1. Pequinot, C.A., (1963) "Tunnels and Tunnelling'- Hutchinson, Scientific and Technical, London, pp 216-240
2. Szechy, Karlowi, (1970) 'The Art of Tunnelling,' Akedimiai Kinda, Budapest, Hungary, 1970, pp 264-365
3. Lal Das, J.N. (2014), 'Design, Construction and Maintenance of Railway Tunnels', Proceedings of National Technical Seminar, on Management of P.Way Works through need based Outsourcing and Design, Construction and Maintenance of Railway Tunnels, Jaipur 2014
4. IS: 5880, Part (V), 1972,
5. Heinz Duddeck (1988), 'Guidelines for the Design of Tunnels'- ITA Working Group on General Approaches to the Design of Tunnels'- Tunnelling and Underground Space Technology, Vol. 3 No 3, pp 237- 249
6. Brahma, N.,(2014), 'Understanding Rock Mass Behaviour and Design of Tunnels using Soft Computing Techniques: A Case Study, Proceedings of National Technical Seminar, on Management of P.Way Works through need based Outsourcing and Design, Construction and Maintenance of Railway Tunnels, Jaipur 2014 pp315-356.
7. Sen, P.K., 1985, 'Computerised Analysis of Subway Boxes for Calcutta' Metro'. Proceedings of International Seminar on Metro Railway-Problems are Prospects, Calcutta-Paper 11.
8. Padmanabhan, V.C.A., 1965, 'Notes on D.B.K. Railway Project', Unpublished Indian Railways Report, 143 P.
9. Limaye, S.D. and Narayanan, S.S., (1994), 'Design of a circular lining for shield- driven tunnel', Indian Concrete Journal, February, 1994. pp 103- 106.

ANNEXURE 4.1

TYPICAL EXAMPLE OF DESIGN OF A BOX SECTION TUNNEL

A.1 General

Reinforced concrete tunnels of single-vent or two-vent box sections are commonly used for metro tunnels, especially when the tunnel is constructed by the cut-and-cover method. This type has been used for the metro railway in Calcutta. Typical method of analysis used for a box section for the underground railway tunnel is given in this Annexure*.

The box sections used for underground railway operation are of the following types:

(a) Single-storey box of one or two cells for track between stations.
(b) RC boxes of one or two storeys and with one or more cells for stations.
(c) Cells divided by RC columns spaced at about 3-m intervals in the non-station portion and by RC columns at 6-m intervals in the station portion.

Three of the different types of boxes used, with concept of forces acting on same are shown in Figure A.4.1. The concrete used was generally controlled concrete grade M-20. (Now a days richer mixes are used) The slabs and walls vary in thickness from 500 to 800 mm.

---

*Adapted from a design by P.K. Sen for Metro Rail, Calcutta. (P.K. Sen, 1985)'

A.2 Notations Used

C	—	Value of cohesion of soil
G	—	Dry density of soil
g_s	—	Bulk density of soil
g_w	—	Density of water
ϕ	—	Angle of internal friction
K_s	—	Modulus of subgrade reaction
u_s	—	Poisson ratio for soil
E_s	—	Young's modulus of soil
E_c	—	Young's modulus for concrete
B	—	Width of footing
J	—	Moment of inertia of bottom slab of subway box
S	—	Spacing of Winkler springs
E	—	Equivalent Young's modulus for Winkler spring
P	—	Load on Winkler spring
y	—	Deformation of Winkler spring
H	—	Height
P	—	Earth pressure.

A.3 Loading

Loads due to the following have to be considered for analysis:

- — Earth pressure
- — Lateral pressure induced due to building surcharge
- — Superimposed dead loads
- — Superimposed live loads
- — Train live load with impact
- — Live load due to passenger traffic in stations (mezzanine and platform slabs)
- — Machinery load with vibration effect in stations
- — Self-weight.

Earth pressure has been worked out using formulae:

$$\text{Intensity } p = \text{HK} - 2 \cdot \text{C}$$

$$\text{where } \text{K} = \frac{1 - \sin \phi}{1 + \sin \phi}$$

$$g_s = g + g_w$$

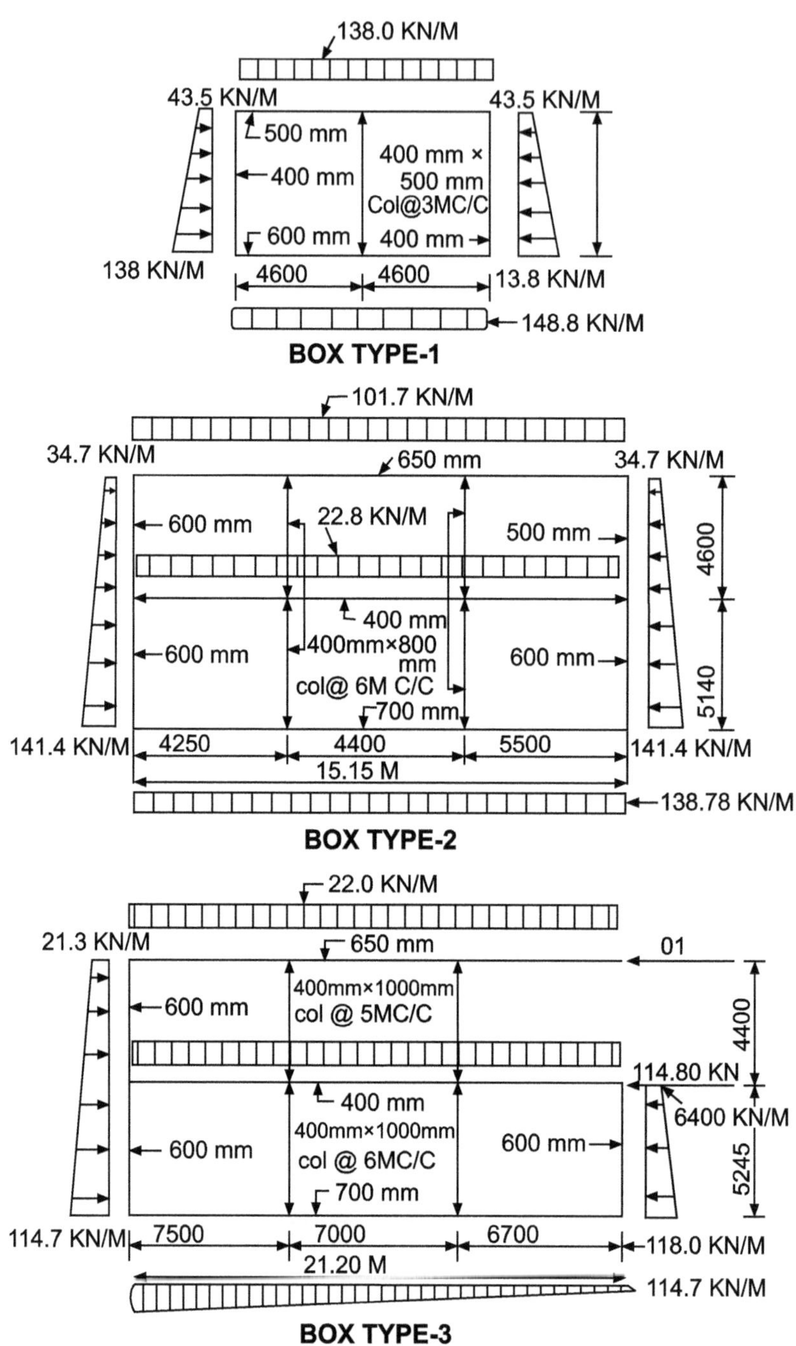

Figure A.4.1 Box Types, Loads and Forces acting on Different Segments of same.

For working out total pressure, the water table may be assumed at ground level.

Weighted average value of cohesion, bulk density and angle of internal friction may be increased or decreased by 2 standard deviations and maximum and minimum earth pressures may be considered for analysis to allow for variation in strata.

Lateral pressure due to building surcharge: Lateral pressure due to building surcharge has to be considered at locations where buildings are in the vicinity of the cutting. For rough assessment, the surcharge is taken as 55 kN/ sqm at an appropriate distance but not less than 2 m from the outer edge of the box and not less than 1.5 m below GL.- assumed to act continuously along the side of the vertical wall of the box.

Superimposed loads: Live load: Worst of the following: Surcharge due to tracked vehicles (700 kN) marching in parade with 30-m gap between tail and nose of following tanks and 2-m lateral gap of adjacent tanks or 20 kN/m^2.

Angle of dispersion in soil: 30° to vertical.

Dead load. Due to overburden of soil and water above box Plus future overburden equivalent to 0.50 m in built-up area and 1.50 m in underdeveloped area.

Mezzanine floor, platform and other slabs accessible to pedestrian traffic and/or supporting station equivalent and not continuously supported by earth design for UDL of 1 kN/m^2.

Train live load: Superimposed loads due to axle loads and longitudinal forces considered only for design of track structures and not the box as their effect here is negligible.

Live load due to passenger traffic: Platform slabs. Mezzanine slabs and all slabs accessible to passengers = 5 kN/m^2 UDL.

Loads due to machinery: As per actuals for air-conditioning and ventilation equipment with necessary increment for vibration.

Self-weight: Unit weight of concrete: 24 kN/m^3

A.4 Combination of Loads

Combinations and conditions considered:

Case I: Top slab — value of g and g_w increased by 10%
Side wall — value of g increased by 20% and 10%

— value of ϕ increased by 5°

Case II: Top slab — value of g and g_w decreased by 10%

— Side wall value of g increased by 20% and g_w by 10%

— value of ϕ decreased by 5°

In any combination, any member is designed for the worst effect. Earthquake forces are not considered.

If the box is only partly integrated with the diaphragm wall by leaving a small gap between the side wall and the former but extending top and bottom slabs to the diaphragm wall, the side wall is designed for lateral pressure due to full height of water instead of earth pressure. If the box is integrated with the diaphragm wall by treating the latter as a permanent wall of the box, the diaphragm wall is checked not only for temporary conditions but also for Cases I and II above.

Design Example

The design example given here is for a station section Box type 2 (see Figure A.4.1) Data: C = 30 kN/m³

ϕ = 0°

g = kN/m³

g_s = 18 kN/m³

g_w = 10 kN/m³

Height of overburden = 2.36 m

Add for future = 0.50 m

2.86 m

Top Slab

Weight of earth = 1.10 × 8 × 2.86 = 25.20 kN/m²

Weight of water = 1.10 × 10 × 2.86 = 31.60 kN/m²

Self-weight of slab (650 mm) = 0.65 × 24 = 15.60 kN/m²

Total dead load Live load = 72.40 kN/m²

= 20.00 kN/m²

92.40 kN/m²

say 92 kNim²

Mezzanine Slab

Self-weight 400 mm slab floor = 0.4 × 24 = 9.6 kN/m^2

300 mm slab floor – 0.3 × 24 = 7.2 kN/m^2

16.8 kN/m^2

Live load 6.0 kN/m^2

22.8kN/m^2

Side Wall

Earth pressure at centre line of top slab:

$$K = \frac{1 - \sin\phi}{1 + \sin\phi} = 1$$

Depth of C.L. of top slab from G.L = 2.36 + 0.325 = 2.685 m.

Depth of C.L. of future slab = 3.185 m.

Value of buoyant earth pressure (after reducing 20%)

= (0.80 × 10 × 0.50 + 0.80 × 8 × 3.185) – 2 × 30

= 0 since resultant negative value is not possible.

Value of water pressure (after reducing 10%)

0.90 × 10 × 2.625 = 21.30 kN/m^2.

Therefore total pressure = 21.30 kN/m^2

Net vertical load (load from top slab load from mezzanine + self-weight of side walls and columns) divided by width of box gives net vertical unit load/ pressure.

Design

Box is a closed frame totally buried in soil.

Package programme for plane frame analysis not possible.

Moment distribution method is tedious as has to be done first without sway condition and then sway correction applied.

Hence Kani's method using a set of simultaneous equations and iteration, as illustrated by Thadani was adopted (Kani, 1947; Thadani, 1964).

A set of equations, one for each member for:

End moments in terms of fixed end moment, sway, angular distributions at near and far end and relative stiffeners.

Sway and angular distortions expressed in terms of moments after multiplying by suitable factors.

Though analysis can be done manually in a semi-graphical manner, to make it quicker, a computer program written in FORTRAN-IV is used. The resultant moment diagram is given in Figure A 4.2. for some typical sections.

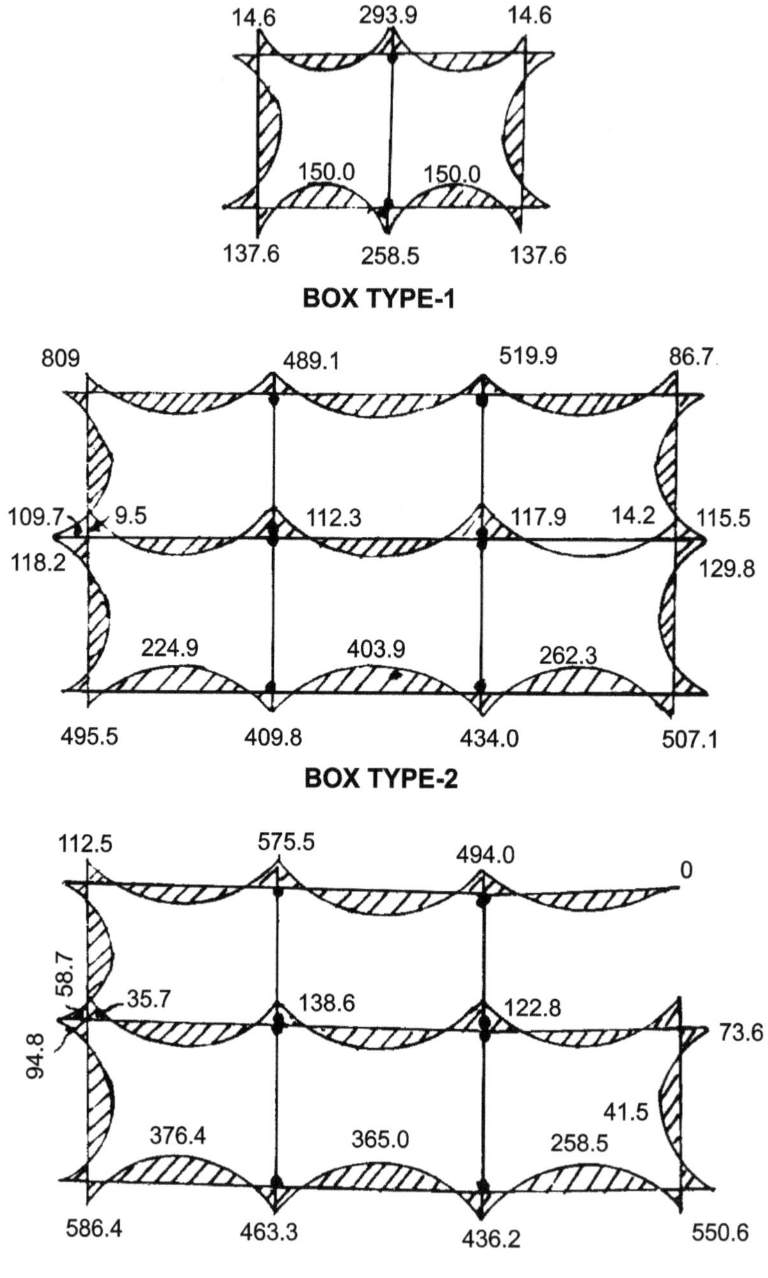

Figure A.4.2 Moment Diagrams Based on Kani's Method.

The above detailed method has some basic shortcomings:

(a) Here pressure distribution is considered uniform, which condition is not strictly applicable to Kolkata soil. Also, the section assumed thinner and more flexible than the ones designed elsewhere (Moscow or Osaka). For Kolkata soil, a parabolic distribution is more aptly applicable.
(b) The computer program is written for only 3 types; the other 5 had to be done manually, which is more laborious.

A.5 Analysis by Stiffness Matrix Method

Hence, as an alternative, Winkler spring method was tried for same design. This method, on the other hand, considers the box as supported on a series of Winkler springs (Szechy, 1970; Hooper, 1970). The relative merits and demerits of method adopted are discussed below:

Merits

(a) Analysis can be done very quickly and easily for any type of underground box using any package programme meant for analysis of plane and space frame.
(b) Realistic non-linear base pressure variation is taken into account to the greatest possible extent.
(c) Extents of data preparation and computer usage time are much less and, as such, obtaining solutions is quite economical.

Demerits

(a) As per Hooper (1970), the above-mentioned model of Winkler springs does not represent the real behaviour, in particular for a rigid structure founded on compressible strata overlying a stiff base. But a typical box like the one used in Kolkata Metro is a flexible structure. There is also no stiff base under Kolkata soil and, as such, analysis by finite element based on half space continuum or layered continuum models is very difficult and expensive.
(b) It is difficult to ascertain the modulus of subgrade reaction values experimentally for different field conditions.

But as per Bowles (1985; pi. 268), the modulus of elasticity and Poisson ratio of soil can be obtained by the formula:

$$K_s = \frac{E_s}{B(1-\mu_s)^2} \quad (2)$$

K_s and μ_s can be determined at any depth from tri-axial test data on undisturbed samples collected from that depth. K_s can also be calculated roughly from the Table given by Bowles (1985; p. 262).

Calculation for Flexibility of the Box

As per Szechy (1970; p. 473):

$$L = \sqrt[4]{\frac{4EcJ}{BKs}} \qquad (3)$$

If $l/L > 1$, the bottom slab of the box belongs to a category of short beams, it can be considered as flexible and can be analysed as supported on a series of Winkler springs.

For a typical double-storey box of the Kolkata metro (Type 2), the L value can be calculated as follows:

$E_c = 2.5452 \times 10^7$ kN/m^2

$J_c = 0.02858$ m^4

$K_s = 15600$ kN/m^3

B = 1 metre

Hence

$$L = \left\{\frac{4 \times 2.542 \times 10^7\ 0.02858}{1 \times 15800}\right\}^{1/4}$$

$$= \sqrt[4]{186.1} = 3.62 \text{ m}$$

l end span = 6.25 m

l / L = 6.25 / 3.62 = 1.693 > 1

Hence the structure is flexible and can be considered as supported on Winkler springs.

Calculations of equivalent modulus of elasticity

Length etc. for Winkler springs: Consider 1 m length of the subway box. The bottom is divided by a series of Winkler springs having an equal spacing of S. Thus the equivalent area for an intermediate spring member is S. Considering its length to be unity, the deformation as per Hook's law

$$\Delta Y = (\text{Load on Spring} \times \text{Length})/(\text{Area of Spring} \times \text{Equivalent Young's Modulus})$$

$$= (P \times l) / (S \times E)$$

As per Winkler's Hypothesis,

Pressure at structure – Soil Interface = $K_s \Delta Y$

Now load P on a spring = Area × Pressure

$$P = S \times K_s \Delta Y$$

$$= S \times K_s P/(S \times E)$$

$$E = K_s$$

The stiffness of the spring members is assumed to be zero. However, a very small moment of inertia value is fed into the computer for running the program.

Running on computer

The problem was then formulated for running on a package program called 'STRESS'. The program was written in FORTRAN-IV language for solving of plane and space frame structures by the Stiffness matrix method. Plane frame and space structures up to 900 members and 900 joints can be solved using this program.

The equivalent structural models for different types of boxes are shown in Figure A.4.3 and resultant moment diagrams are shown in Figure A.4.4.

On comparison of results from Kani's method and the matrix method assuming an elastic base, it can be seen that there is some redistribution of moments from bottom corners to top corners. Results from the stiffness method for box (1) are derived by assuming 400 × 800 mm (along the length of box) column at the centre hinged at top and bottom spaced at 3 m centre to centre. The resultant moment diagram compares well with those derived from Kani's method. The comparative S.F diagrams from Kani's method and stiffness matrix method for box type (2) are shown in Figure A.4.5. The base pressure distribution for different boxes is shown in Figure A.4.6.

Each particular problem for a single loading case took 50 secs. on an average in the CPU of the computer.

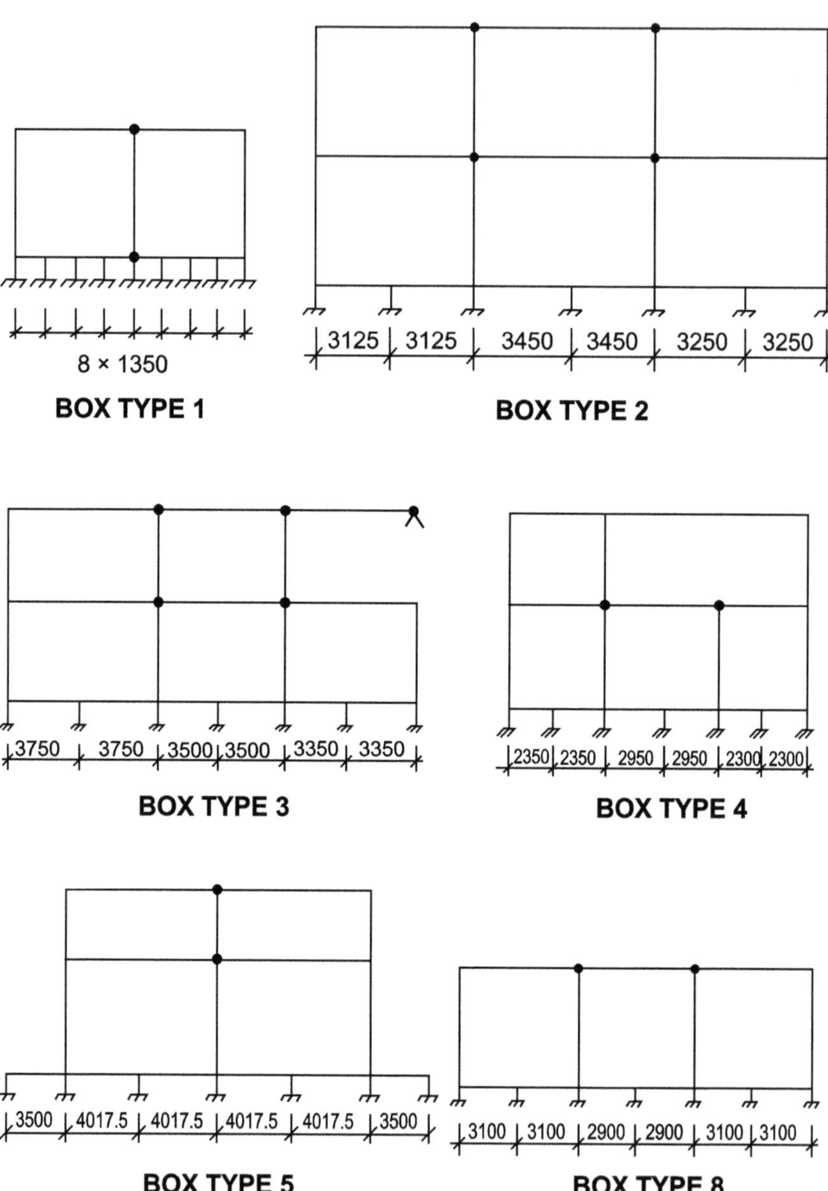

Figure A.4.3 Equivalent Structural Models for Boxes Supported on Winkler Springs,

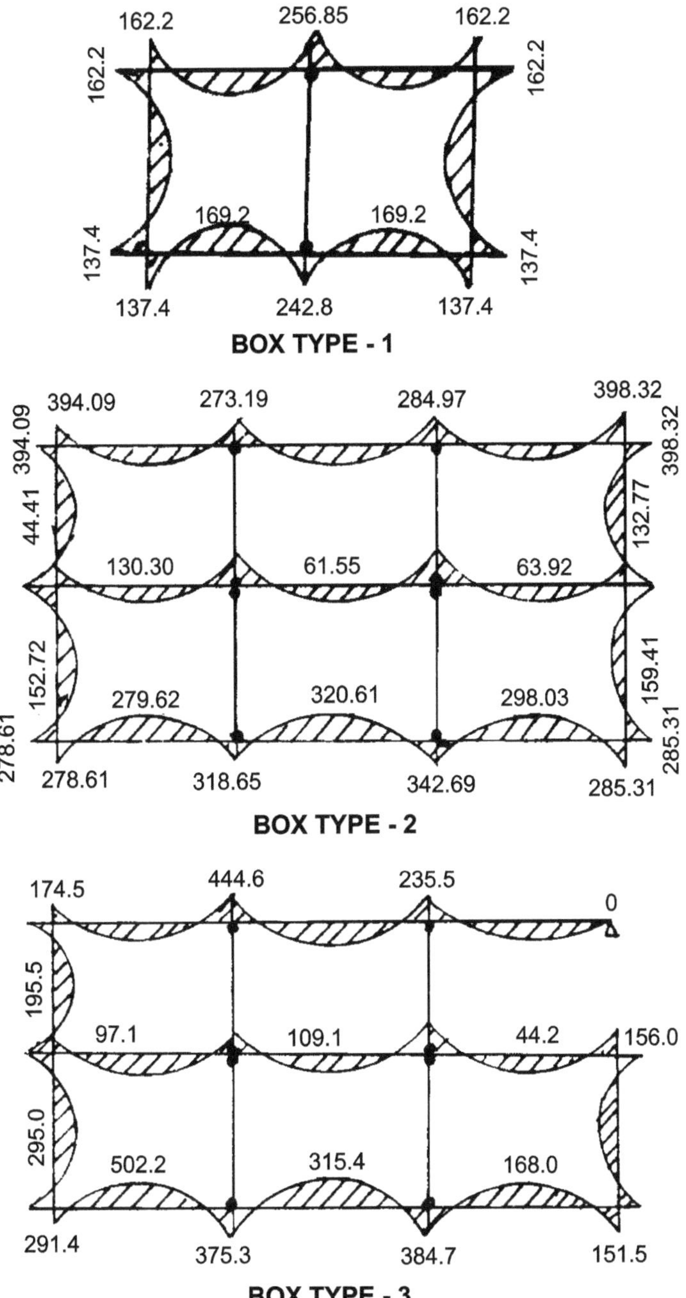

Moments in kN-m

Figure A.4.4 Moment Diagrams Based on Flexibility Method.

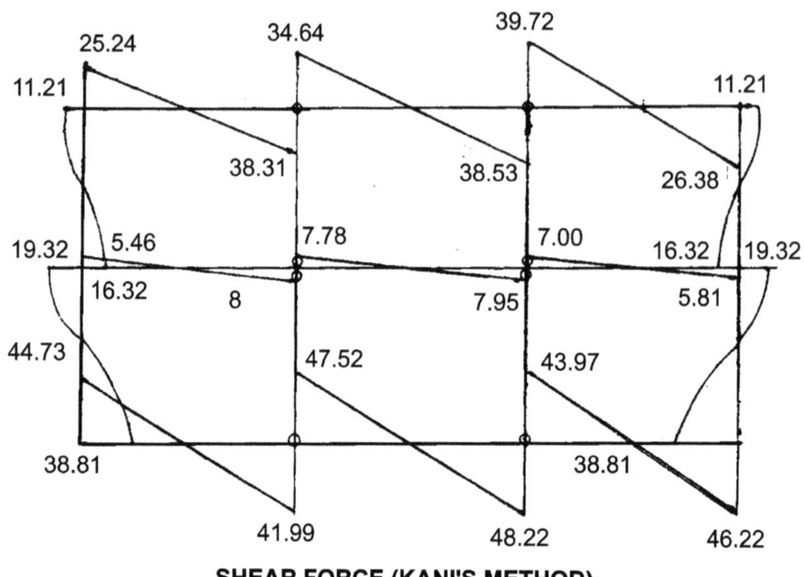

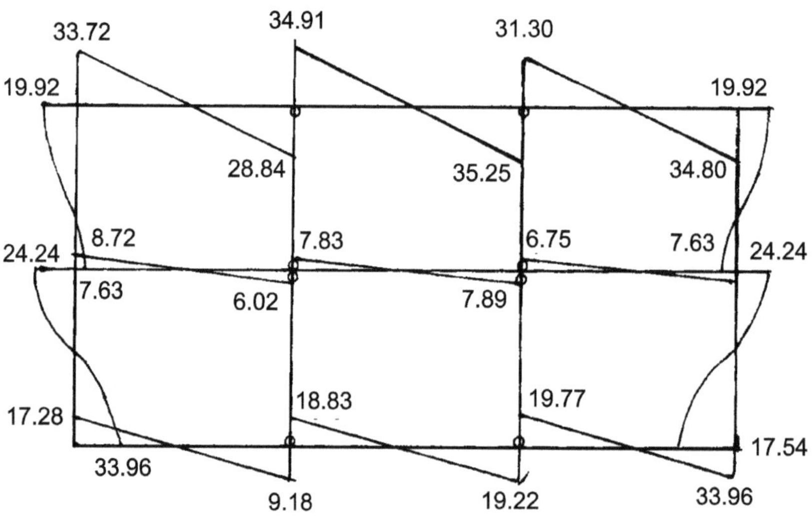

Figure A.4.5 Comparative Shear Force Diagrams for Box Type-2 (by Kani's and Stiffness Martix Methods.).

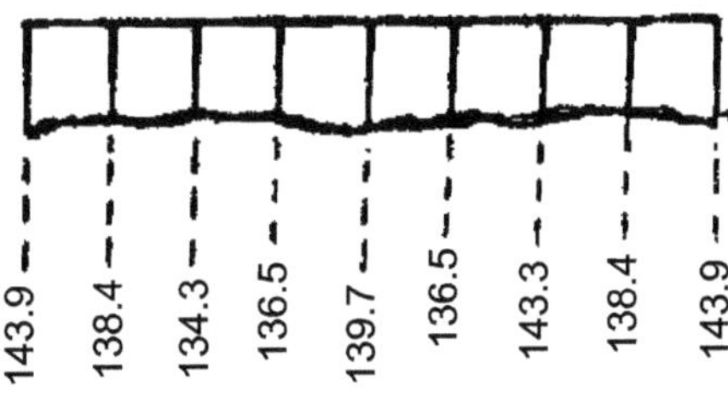

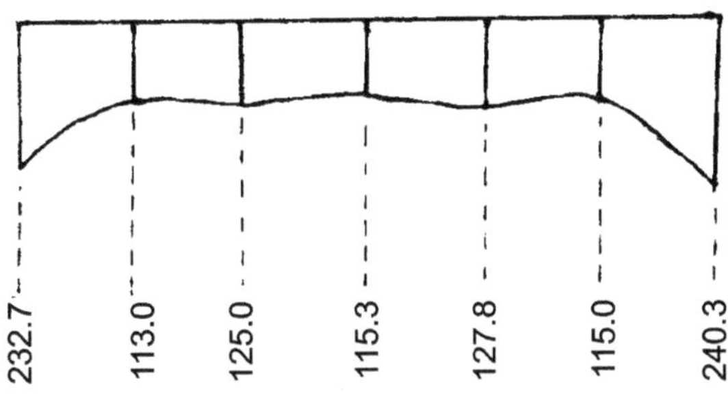

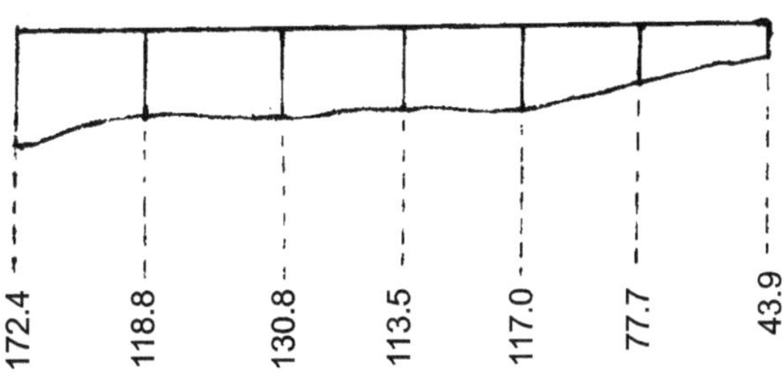

Figure A.4.6 Base Pressure Distribution.

ANNEXURE 4.2

DESIGN OF STEEL SUPPORT FOR A HORSE SHOE SHAPE TUNNEL*

Data Available

(i) The rock is completely broken and crushed.
(ii) Water table is below the bottom level of the tunnel.

As the rock is completely crushed, it will have considerable side pressure in addition to the rock load.

Calculation of Roof Load

For completely crushed rock, the roof load 'Hp' over the roof of tunnel is given by the formula:

$$H_P = 1.10\ (B+Ht)$$

where B = width of tunnel

H_t = depth of tunnel

$B = 5.86$ m

$H_t = 7.87$ m

$$\text{Hence } H_P = 1.10\ (5.86 + 7.87) = 15.103 \text{ m.}$$

As the roof of the tunnel is located permanently above the water table 'H_p' can be reduced by 50%.

---

* Adapted from a design done by M.A. Umar for DBK Rail Project in India-Refer Padmanabhan, 1965

Hence $H_P = 0.5 \times 15.103$

$= 7.55$ m or 7.60 m.

Weight of crushed rock ranges from 1960 kg to 2450 kg per m^3.

Assume an average weight of soil of 228 t/m^3

Load per square metre $= 7.6 \times 2.28$

$= 17.33$ tonnes.

The block point spacing is kept very small in case of crushed rock as small pieces of rocks are likely to fall out being loose. Actually in this case, lagging has been provided throughout the rib.

As the spacing is reduced, the thrust 'T' in the rib caused by rock load over the rib tends to become equal to the vertical reaction Rv.

Hence T = Rv.

Design of Rib Section

R_v = the vertical load on half the tunnel.

$T = R_v = (5.86/2) \times$ spacing between ribs $\times$ 17.33 tonnes.

Spacing of rib is kept between 0.60 to 0.90 m for heavy loading. Keep spacing of 0.60 m using 150 mm x 75 mm joist having an area of 22.77 sq cm. Deduct the hole area for one tie-rod of 16 mm.

Area effective = 22.17 − 1.93 = 20.84 sq cm.

Stress in the joist = T / A.

$T = (5.86/2) \times 0.60 \times 17.33$

$= 30.47$ tonnes.

$$\text{Stress} = \frac{30.47}{20.84} = 1462 \text{ kg/sq cm} < 1590 \text{ kg/sq cm}$$

Hence safe.

Provide ribs of 150 mm × 75 mm I-joist at 600 mm c/c.

Design of Vertical Post

The spacing of the rib comes to 600 mm c/c. It would have been better if continuous ribs were used in which case, the spacing of the vertical posts would have remained the same as that of the ribs. This is, however, not possible on account of the huge side pressure caused by the crushed rock on the vertical post.

Assume 250 mm × 125 mm R.S.J for vertical posts. The vertical post is to be designed for a vertical load of 30.47 tonnes and for the side pressure. This is designed as a slender column and whereas the allowable fibre stress in the arch may be taken as 1590 kg/sq cm, the fibre stress permitted for the leg is smaller on account of buckling.

I for 250 x 125 RSJ @ 216 kg/metre section along x- x axis is 6086 cm^4

A = 57.09 cm^2.

r along x-x axis = $\sqrt{I/A}$ = 10.34 cm.

The permissible stress in the vertical post is given by the formula (in FPS units)

$$= \frac{24000}{18000}\left\{17000 - 0.485\left(\frac{l}{r}\right)^2\right\} \text{ in FPS units.}$$

where

l is length of vertical post,

r radius of Gyration

l = 11' – 4½" + 2'4" = 13' – 8½" = 164.5" or 4.18 metres.

Hence Permissible stress is

$$= \frac{24000}{18000}\left\{17000 - 0.485\left(\frac{164.5}{4.07}\right)^2\right\}$$

= 21,600 lbs/in^2 or 1522 kg/sq cm.

Let M_h = Bending moment caused by lateral pressure in kg-cm.

M_c = Bending moment from all causes combined at point of maximum deflection in kg cm.

The maximum deflection is slightly below the mid-point of leg.

Maximum deflection of leg caused by moment M_h in cm

E = Modulus of elasticity, 29,400,000 lbs/ sq in. or 2072000 kg/sq cm

I = Moment of inertia = 6086 cm^4

Rv = T Thrust = 30470 kg

t_c = Stress in leg at point of maximum deflection in kg /sq.cm

H = Height of rock causing vertical load = 7.6 m

w = Weight / unit volume of rock = 2.28 tonnes/cu m

Length of straight leg = 4.18 m

s = rib spacing = 60 cm

W = Total horizontal pressure of rock on straight leg in kg

I for 250 × 125 RSJ @ 45 kg/metre run

A = 57.09 sq cm.

I = 6086 cm^4

r_{xx} = 10.34 cm

l = 3.46 + 0.71 = 4.18 m.

Permissible stress = 1522 kg /sq cm

W = (sH_p wl) / (3 × 12)

$$= \frac{0.60 \times 7.60 \times 2280 \times 4.18}{3 \times 12} = 14486 \text{ kg or } 14.5 \text{ tonnes}$$

M_h = wl /14.2 = 14.5 × 4.18 / 14.2 = 4.268 tonne metre or 426,800 kg cm.

$$d_l = wl^3 / 185 \text{ EI} = \frac{14500 \times 418\text{^}3}{185 \times 2072000 \times 6086} = 0.455 \text{ cm}$$

M_C = M_h^2 / (Mh × 1000 – 0.455 × R_v)

= 426800^2 / (426800 – 455 × 30296) kg

= 426800^2 / 413017 kg cm = 441043 kg cm

Maximum stress fc in the vertical post is the sum of direct stress and bending stress.

f_c = (R_v / A) + {M_c / (I / y)}

= (30470 / 57.09) + {441043 / (6086/12.5)}

= 533 + 906 = 1437 kg/sq.cm.

< 1522 kg / sq. cm permissible.

As this is less than the permissible stress the section adopted is safe. Hence provide 250 mm × 125 mm × 45 kg/metre R.S.J. @ 600 mm centre/ centre as vertical posts.

Design of Wooden Laggings

Span of lagging is 600 mm and is continuous over two or three rib supports depending upon the length of laggings available.

Consider 300 mm width of lagging.

Load/ sq metre = 17.33 tonnes

Load/lagging = 17.33 × 0.30 = 5.199 tonnes/m.

Neglecting weight of wooden lagging being very small,

B.M = $Wl^2/10$

= (5199 × $(0.60)^2$ × 100}/10 kg cm.

= 18688 kg cm.

z = M / f

Using Sal wood laggings with f = 148 kg/cm^2

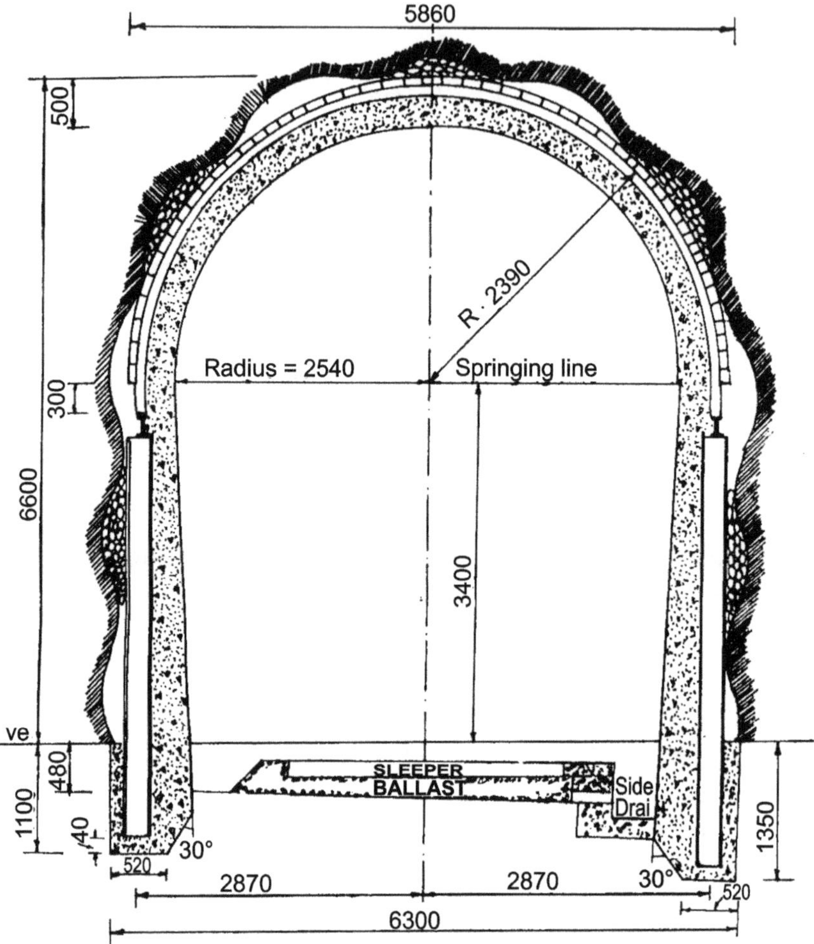

Figure A.4.2.1 Typical Tunnel Section on a Railway Track with Permanenl Steel Supports.

$$Z \text{ required} = 18688\ /148 = 126 \text{ cm}^3$$

$$Z = bd^2/6 = 30 \times d^2/6 = 5d^2$$

$$5d^2 = 126$$

$$d = \sqrt{(126/5)} = 5.02 \text{ cm.}$$

5 cm (2 in) thick Sal wood laggings will do.

ANNEXURE 4.3

DESIGN EXAMPLE OF A CIRCULAR TUNNEL

Some of the designers have found the approximate method of design by Szechy[2] gives very good results for the forces acting on the tunnel. The segmental lining used on the Honavar circular section has been designed using the same. They used the bedded beam model for the design. They used hand calculations and checked by analyzing the lining with the same forces treating it as a closed frame using a computer programme and results from both were found to be in close agreement. The design made for the Honavar shield tunnel and presented in form of a technical paper by S.D. Limaye and S.S. Narayanan[9] in the Indian Concrete Journal of February, 1994 is presented here as an example.

Forces:

The section chosen is a circle with internal dia. of 7.2 m.

The soil strata varied from coarse grained laterite to stiff clay. Percentage of clay present was 20% with c value varying from 0.15 to 0.40 kg/sqcm. and ϕ varying from 0 to 15 degrees. The effective depth of overburden depends on tunnel opening, soil strata, its properties and sub-soil water position. In this case, it worked out to equivalent of a 9.4 m height of column above crown. Lateral earth pressure comprises of two components viz., effect of surcharge in form of resistance/ passive pressure provided by the tunnel walls and the active pressure exerted by the soil over the depth of the tunnel. They are represented in Figure A. 4.3.1.

In this case $\quad Pv = 1.8 \times 9.4 = 16.92$ t/m

P_h = Passive pressure assuming k =1

P_e = computed for mean values of soil parameters given above

The lining will be subjected to bending moments and thrust caused by (i) external forces/ loads, as shown in Figure A 4.3.1 and (ii) self weight of structure.

For (ii), unit weight of RCC is taken as 2.5 t/cum and thickness of shell as 300mm, thus giving unit weight of 0.75 tonne per cum.

Equations Applied

Following equations apply to the structure treated as bedded beam. Bending moments and thrust are computed at different positions. It varies with the angle α subtended with the vertical axis, as shown in the figure.

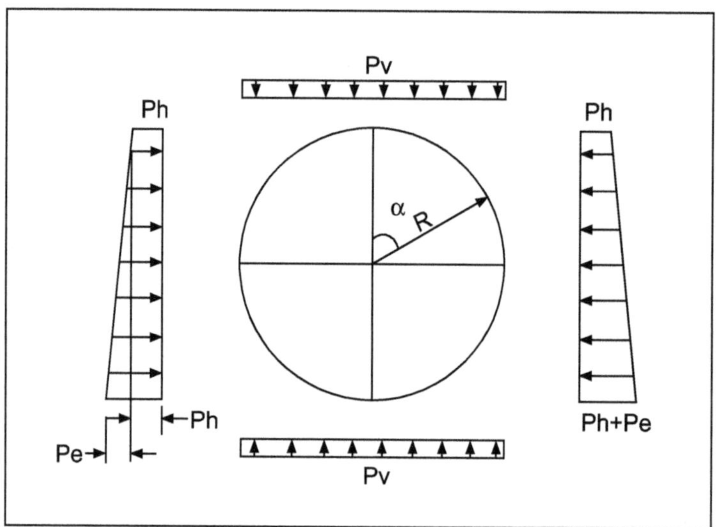

Figure A 4.3.1 Forces acting on circular lining and Deformed state of lining.

For external forces:

Moment

$$M = \frac{PvR^2}{4}\cos\alpha + \frac{PhR^2}{4}\cos(180° + 2\alpha)$$

$$+ \frac{PvR^2}{4}\{-4(1-\cos\alpha)^3 + 10 - 15\cos\alpha\}$$

Thrust

$$N = -\frac{PvR}{4}\sin^2\alpha - PhR\sin^2(90° + \alpha) + \frac{PeR}{16}\cos\alpha\,\{4(1-\cos\alpha)^2 - 5\}$$

Due to self weight, the forces and moments will be different for top half and for bottom half. Following equations are applied for same, where g = unit weight of the lining.

For top half

$$\text{Moment } M = gR^2\left\{\frac{3}{8}\pi - \alpha\sin\alpha - \frac{5}{6}\cos\alpha\right\} \text{ and}$$

$$\text{Thrust } N = gR\left\{\propto\sin\propto - \frac{\cos\alpha}{6}\right\}$$

For bottom half

$$M = gR^2\left\{(\pi - \alpha)\sin\alpha - \frac{\pi}{2}\sin^2\alpha - \frac{5}{6}\cos\alpha - \frac{\pi}{8}\right\}^2$$

$$\text{Thrust } N = gR\left\{\pi\sin^2\alpha - \left(\frac{\pi-\alpha}{6}\right)\sin\alpha\cos\alpha\right\}$$

Assuming symmetry the moment and force distribution will be symmetric about the vertical axis.

Analysis

The lining was analysed treating it as a closed frame using computer programme and it was found that the results were in close agreement. Figure A 4.3.2 shows the distribution of resultant moments and thrust in the section at different locations.

Application to a Subway Tunnel- Worked example

As an example, the Szechy method is tried for design of a Subway tunnel.

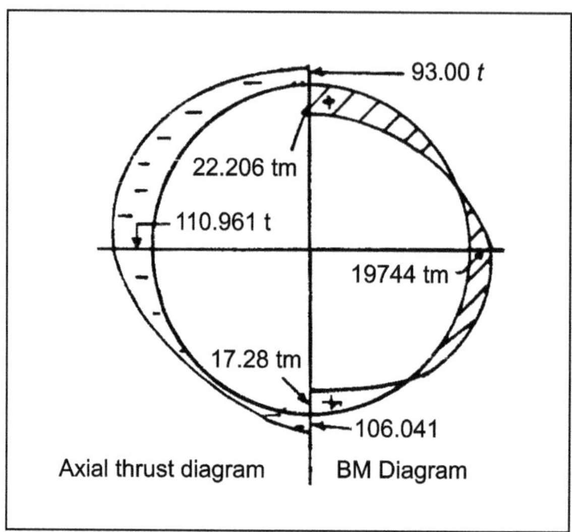

Source: Reference 9

Figure A 4.3.2 Diagram showing Resultant Moment and Thrust in the section.

The given equations have been applied to a Metro tunnel of 5.8 m ID. and loads as given below. Considering it as a shallow fill and applying the approximation of Pb = 0.5 Pv the moments and thrusts have been worked out and presented below;

Internal diameter	= 5800 mm
Thickness of shell	= 300 mm
R = (5800 + 300) / 2	= 3050 mm or 3.05 m
Unit weight of concrete	= 25 kN/ cum
Unit weight of soil	= 18 kN / cum
Pv equiv 9.4 m	= 160 kN/sqm.

Ph assumed as 0.5 Pv treating it as a shallow tunnel and applying ITA Working Group note.

Pe is not considered in addition.

Computations were made using a spread sheet, for BM and thrust at 30° spacings on the tunnel profile. Resultant BM and Thrust are shown in Figure A 4.3.3 which shows the distribution of moments and thrust on the section at different locations.

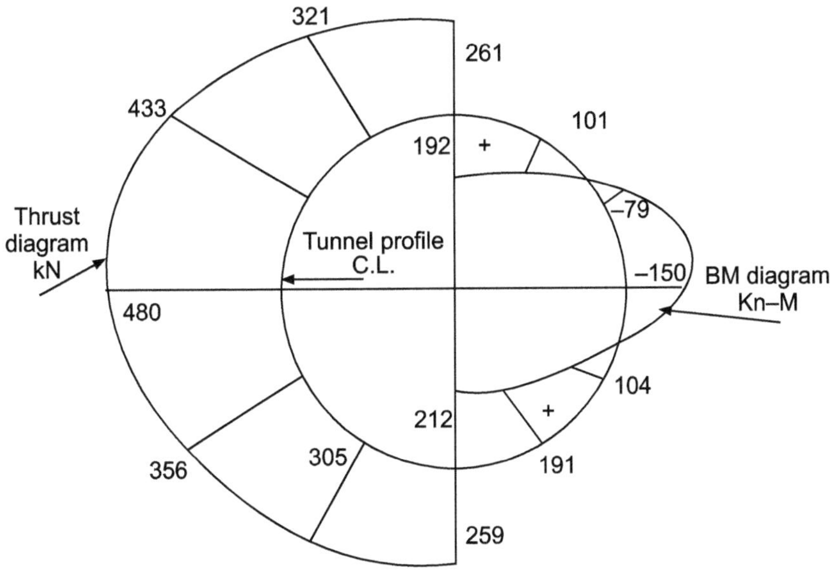

Figure A 4.3.3 Bending Moment and Thrust Distribution in a Subway Tunnel 5.8 m ID

CHAPTER

5

Survey and Setting Out

5.1 GENERAL

Survey and setting out of the tunnel alignment at the time of construction constitutes a critical phase in the construction of a tunnel. IS: 5878 (Part I) lays down the Code of practice in respect of 'Precision Survey and Setting Out'. Setting out has to be done precisely, as later correction of any error caused at this stage will either be impossible or prove prohibitively costly.

If construction is done from one end, any error in setting out the line or traverse, even just one angle wrong, can result in shifting of the exit point several metres away from where it should actually be. One can imagine the problems of connecting up such a point in hilly country. If a tunnel is bored from both ends, as is the usual case, the error in setting out may result in their not meeting in the middle due to vertical shift, or their axes being found shifted laterally or both. Extra work would have to be done to widen the ends or to introduce the sharp reverse curves that may be needed. These would have serious repercussions on speed of the vehicles plying in the tunnel. Some examples of construction with remarkably low error results are given in Table 5.1.

Normally, the alignment of the proposed line or road is marked on topographical sheets of the area, which for India are available with the Survey of India in scale 1:50,000 (and now for some areas in scale 1:25,000 also). In plain country, the alignment is then set out and fixed on the ground. Where a tunnel is involved, the position of the end portals are similarly

Table 5.1 Closing Error in Selected Tunnels

Name of tunnel	Length, km	Closing error in mm	
		Horizontal	Vertical
St. Gotthard	14.9	175	100
Letchberg	14.5	254	102
Simplon	19.7	202	87
Mont Blanc	11.6	13	200
Havenstein	8.1	50	10
Wasserfruth	3.6	50	10
Tavern	5.6	550	56

Source: Pequinot (1963) *and Szechy* (1970)

marked with reference to the nearest triangulation stations or to fixed known points.

5.2 SETTING OUT ON SURFACE

5.2.1 General

Since a tunnel would pierce a ridge or high ground the two ends will not be visible from each other. Hence the line and actual bearing have to be fixed by sighting some intervening mutually visible point or target fixed on surface.

5.2.2 Progressive Ranging

If the ground is not too undulating and not covered by too many obstructions, the simplest method is to fix these utilising one or two intervening points:

(i) The alignment of the tunnel is marked on the topographical sheet and the same is transferred to the ground with the aid of a theodolite, starting at one end, which point is identified with reference some identifiable benchmark and direction (bearing).

(ii) It is then set out from station to station fixed progressively in required direction over the intervening ground until the extended line reaches the estimated position of the tunnel station at the other end. Extension of traverse or setting out of each Intermediate station should be done by double transiting the theodolite on both circle left and circle right and the mean position obtained for the next station for further propagation.

(iii) In the first trial, some deviation/displacement of the terminal station will be seen. Let this be 'x' mm.

(iv) Let C be an intermediate point on the alignment AB and Let A be one of the stations used for set out. The point C will then be laterally shifted by y, where $y = x(AC/AB \cdot v)$

(v) The extension process will again be repeated until almost a straight line through A and B is obtained, i.e., when the line passes through A and B. Then the intermediate points are marked and pegs fixed for further reference and for later setting out of any intermediate point where it is proposed to sink a shaft (for ventilation or for the purpose of locating more working points). (Pequignot, 1963).

5.2.3 Reciprocal Ranging[1]

Alignment for a short straight tunnel can be fixed by successive approximation by a number of trial ranging whenever visibility of some intermediate points from either end is possible. Figs. 5.1 and 5.2 show two typical cases. In Figure 5.1, PQ is the tunnel length on alignment line XY. X1 is the first sight line set out from X. A point on the same visible from Y is established on the line X1 so that it is sighted from X and Y. Let point 2 be such a point of intersection established. A rod is fixed next at 3 on the sight line Y2 from instrument at Y. The sight line from X is now shifted to 3 and 4 is established on the sight line so that it is visible from Y also. The process is repeated till such time the ranging rods fixed at summit points are on same line from both X and Y. Figure 5.2 shows a modification of the method using 3 ranging rods.

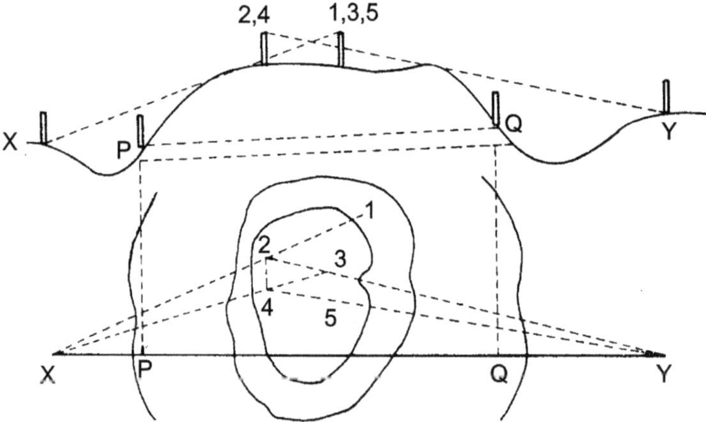

Figure 5.1 Layout for short tunnels (Successive approximation).

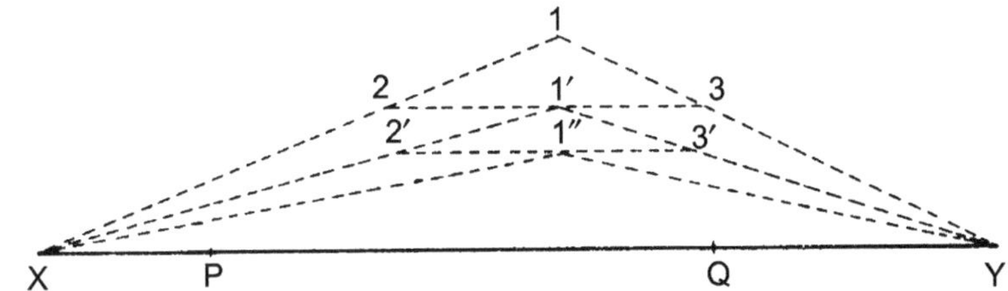

Figure 5.2 Layout for short tunnels (Successive approximation with auxiliary range pole).

Figure 5.3 shows a method by sighting an intermediate point R' visible from both X and Y. By measurement of angles and distances a and b, distance S and ∈' can be computed and point R can be established on the alignment.

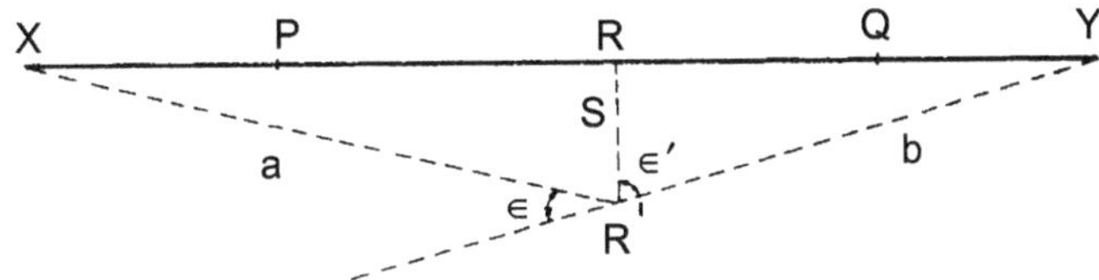

Figure 5.3 Layout for Short Tunnels (by Angular measurements).

5.2.4 Traverse Method

In heavy undulating or crowded country or in an area with intervening buildings situated on the alignment, setting out a straight line by the above mentioned methods may not be feasible. In such a case the traverse or triangulation method is adopted. The route chosen in both cases should connect a number of stations on or close to any intermediate point over the tunnel alignment, wherever sinking of shafts might be proposed. This will facilitate precise measurements to such points falling on the tunnel axis.

The accuracy of any theodolite traverse for normal tunnelling work should be such that the order of error is within 1 in 10,000 with angular error of closure not exceeding 15 $\sqrt{N}$ seconds, where N is the number of angles of a traverse. Though linear measurements can be taken with normal steel tapes, for better accuracy with use of seconds theodolite, an 'invar subtense bar' or 'invar tape' should be used.

5.2.5 Triangulation

In undulating or sub-mountainous country a direct setting out by reciprocal ranging will not be possible. Even traverse survey may not be practicable in

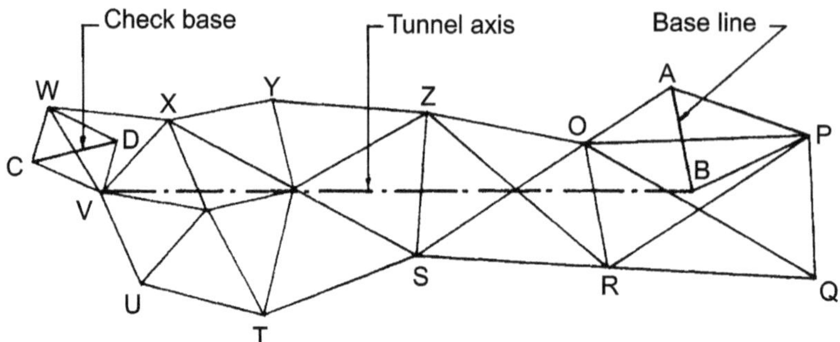

Source : Pequinot, 1963

Figure 5.4 Setting out by Triangulation.

some cases. In such a situation precision triangulation becomes a 'must'. In the two cases discussed above also, it is preferable to do triangulation survey to serve as a check on the alignment of tunnel axis.

Figure 5.4 shows a semi-hypothetical scheme of triangulation for setting out a tunnel with axis VB (Pequignot, 1963)[1]. Triangulation is based on the principle that by measuring one side and angles of a triangle accurately, the remaining sides can be calculated by principles of trigonometry. It is essential first to select and fix a suitable base tine (AB in Figure) and to measure it very accurately. This is done by subtense bar or base line measurement equipment such as invar tapes or wires. Each angle of the triangles in the layout should also be accurately measured, correct up to one second, or even finer if suitable instruments are available. The process is continued until the station at the other end is connected to the system of triangulation. At the other end again two points are connected to the system so that the distance between them can be directly and equally correctly measured, as done for the first base line. The distance between the two will also be calculated directly using the successive angle measurements made in the triangulation. If the difference between the calculated distance and measured distance is insignificant, the field layout measurements and calculations can be considered accurate. If not, corrections have to be applied to the calculated co-ordinates of intermediate stations by adopting a refined mode of adjustment, e.g., method of least squares. In the example cited, AB is the initial base line and CD the check base line as shown in Figure 5.4.

Before commencing any triangulation, the theodolites used should be checked for their permanent adjustments. The invar tape should be stretched and held applying the correct tension, i.e., same as at the time of calibration (or suitable corrections applied for any difference).

5.2.6 Procedure for Angle Measurement

The following procedure for measurement of angles is recommended:

(i) Set the instrument over station O and centre it for taking readings to P.Q.R.S (say). Set the circle and micrometer to read zero approximately while sighting station P. Move the telescope clockwise, take reading sighting of each station, operating the tangent screw to clockwise also. On closing the single round to station P, take reading again. The departure, if within permissible limits, is distributed among the various angles. If not, the process should be repeated.

(ii) Reverse the theodolite onto the opposite face and repeat process (i) starting from P but now moving in a counter-clockwise direction.

(iii) Operations (i) and (ii) should be repeated five times, with initial readings set as close as possible to 60°, 120°, 180°, 240° and 300°. Theodolite should be relevelled, preferably, after each round.

(iv) The included angles are averaged based on 6 x 2 = 12 observations to obtain the required accuracy.

5.2.7 Required Degree of Accuracy in Triangulation[1]

For short tunnels of about 500 m length, a minimum accuracy of I in 10,000 is essential. Average errors of closure for this accuracy are of the order of 5 seconds and never more than 10 seconds. For longer tunnels, much higher accuracy is needed as in the case of secondary triangulation.

The following precautions are specified in IS: 5878 (Part I) to avoid probable sources of errors and to help in accuracy:

(a) The site for measurement of the base line should be approximately level, evenly sloping, or gently undulating and as free as possible from obstructions. The base line should be as long as possible and its length should be preferably 1/12 to 1/15 of the total length of the tunnel to be driven.

(b) To set up a new reference point, the most desirable way is by a single triangle with no angle less than 45°. As a check, observations for another triangle should also be made. If a single triangle with any of its angles not less than 45° cannot be obtained, the angle may be reduced but the angle shall not be less than 30°. This shall invariably be checked with another triangle. But such triangle should not be used for further expansion of the triangulation system. If either of these two conditions cannot be obtained in any location, a braced quadrilateral should be adopted.

(c) It is advisable to have two independent sets of observations done by two independent observers using different instruments and the results calculated independently. The particular set of calculations may be taken as correct only if the final results of both are found to agree within acceptable limits of difference. The same criteria should be followed while setting out the alignment along the floor of the tunnels and checking it.

(d) In all these calculations, seven-figure log tables shall be used and the calculations for angles shall be based on tables given for values up to one second of angle.

(e) For the calculation of angles, computation forms shall be used.

(f) For base line measurements for tunnels of small lengths of about 500 m, ordinary but calibrated steel tapes may be used. For longer tunnels and complicated layouts, invar tapes or wires shall be used. At other places where the country is hilly, repeated measurements in small stretches by invar subtense bar may be made.

(g) While arriving at the true base line length, the following corrections shall be applied.

   (i) correction for absolute length;
   (ii) correction for temperature;
   (iii) correction for tension or pull;
   (iv) correction for sag;
   (v) correction for slope or vertical alignment;
   (vi) correction for horizontal alignment; and
   (vii) reduction to sea level.

Check of triangles/ polygons:

The properties of the triangles and polygons forming part the triangulation network should satisfy the following conditions:

(i) Sum of the angles of each triangle should be 180°
(ii) Sum of central angle of or hub angles in a polygon should be 360°
(iii) Sum of log-sines of the left hand angles should equal the log-sines of right hand angles looking towards the centre of the polygon.

5.3 USE OF ELECTRONIC DISTANCE MEASUREMENTS

One of the difficulties in fixing project control lines, is the wide spacings of existing National triangulation stations. Development of electronic distance measuring instruments such as the Geodometer and the Tellurometer (and now Distomat/ Total Station) has made the work of connection of tunnel alignments and project control line with available (far off) survey networks feasible and more economical, both from the point of view of time and total labour involved. These instruments can be used for checking the base line and also to run the primary survey traverse along the tunnel alignment. Their range of measurement is generally from 0.8 to 80 km. These instruments have an inherent instrumental error of ±12 mm or 2 to 3 parts per million of distance measured. These instruments work on the basic principle of calculations of the distance from measurement of time required for a light beam or radio wave to travel between the two stations and the necessary calculations are automatically made within the instrument.

5.4 LEVELLING

5.4.1 Instrument Accuracy and Precautions in Levelling[3]

The level used in tunnelling work should be capable of giving an accuracy of plus or minus 2.5 mm/km of levelling in normal country side. A modern level with 40 mm objective; magnification × 28; and a bubble with a sensitivity of about 30 s per 2.5 mm, and read by a prism coincidence system is capable of giving an accuracy of 2.3 mm/km in normal country. It is the nearest best that could be recommended (Padmanabhan, 1965)[4].

The following precautions should be taken to ensure the desired accuracy:

(a) Careful focusing of diaphragm lines and staff and elimination of parallax.
(b) Use of staves which are machine divided, preferably neither of the box type, nor more than 3 m in length.
(c) Use of micrometer device for reading fractional parts of the 0.01 division is helpful but not essential.
(d) Limiting the length of sights to a maximum of 45 m (preferably less) and equalization in length of the backsights and foresights.
(e) Use of reliable footplates as turning points, where necessary.
(j) Care in ensuring that the bubble is in mid-position at the time of reading the staff.
(k) Careful holding of the staff in a vertical position and avoidance of windy weather which makes this impossible.
(l) Levelling during the period 10:00 a.m. to 4:00 p.m. in which refraction is the steadiest. At the same time, this is the period when heat simmer is likely to be troublesome. Cool but bright weather with gentle wind is the best.

(m) Protection of the instruments from direct sunrays by an umbrella.
(n) Choice of a route which permits equalization of lengths of backsights and foresights.
(o) The level survey must be closed, preferably by repeating the levelling in the opposite direction by the same route and within a short time.

As in the case of 'angle measurements', the error (if within permissible limits) shall be distributed amongst several stations in proportion to the distance of the station from the starting point. Station benchmarks should be established at each station.

If the tunnel is long and it is proposed to have intermediate shafts for ventilation and/or to have more faces to work from, the positions of these shafts should be fixed with reference to the nearest National triangulation stations.

5.4.2 Benchmarks

Permanent benchmarks have to be established to carry the levels into the tunnels and for checking during construction. Normally, benchmarks are placed 60 m away from the centre line of the tunnel to avoid errors that may be caused due to settlement during excavation. In urban areas, where more settlement is anticipated, additional benchmarks are fixed along the tunnel alignment at spacing of 180 m.

5.5 SETTING OUT INSIDE OF TUNNELS

5.5.1 Straight Alignment

Setting out actual tunnel alignment is done from various portals and faces from which work is commenced and progressed towards each other. In the case of straight alignments this is simple. Sighting is done by back sighting on pillars aligned and constructed as far away as possible on the extended centre line on the tunnel axis plane on the approach and then transiting. If this is not possible, setting out can be done with reference to a distant pillar or triangulation station fixed on a line subtending a known angle to axis of the tunnel. In the latter case, however, it would be better to cross-check with reference to another pillar fixed on the other side (Figure 5.5). It should be noted that an initial error of 2" in angle can produce an error of alignment of 8 mm/km length of tunnel.

Within the tunnel, reference points are constructed at every 300 m to minimise errors, even though while carrying out work, to minimise errors sighting will be done for each cycle of blasting/mucking. The reference

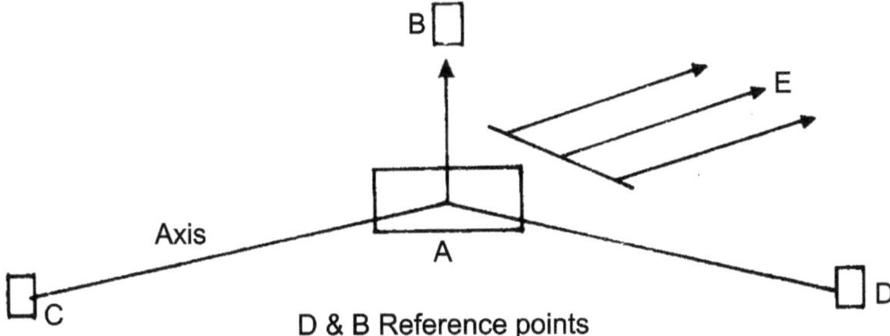

Figure 5.5 Setting out centreline of tunnel.

points mentioned above may be fixed on the roof of the tunnel (Figure 5.6) or slightly below the invert of the tunnel. This is done by fixing non-rusting plates on concrete pillars (flush with the surface) and chisel marking the line or fixing nails on the roof of the tunnel. Before marking these, repeated circle-left and circle-right observations shall be made and levels also accurately determined. In addition, reference chainage lines are marked on the sides of the tunnel at about 15-m intervals with reference to main reference plates.

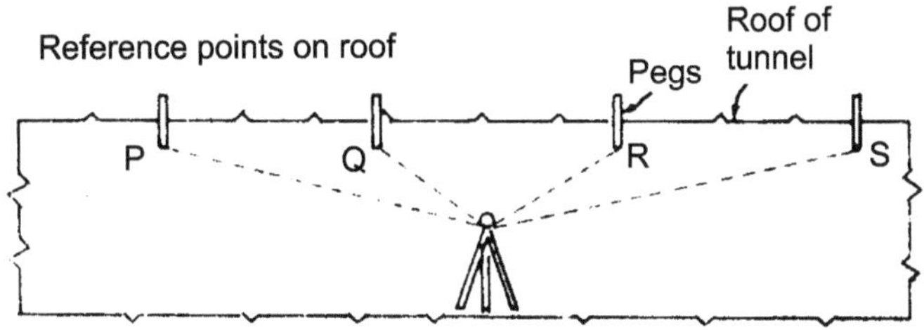

Figure 5.6 Typical Reference Marks.

5.6 TRANSFER OF ALIGNMENT BELOW G.L. AT INTERMEDIATE STATION

5.6.1 Tunnels Driven from Portals

In the case of tunnels driven from portals, to start with one work point at the portal and a backsight point on the working line are adequate for extending the alignment into the tunnel. Setting out is done by sighting back to the backsight point after fixing the theodolite over the point fixed at the portal and transiting the theodolite and extending the straight line into the tunnel. If any angle is to be set out, same type of work is done from this point by turning the theodolite suitably after transiting. The tunnel axis is aligned on the surface directly or with reference to the intermediate triangulation stations. The positions of the shafts are thus correctly fixed and the shafts sunk at the required location.

Where the tunnel is long and alignment complicated, the work is done with the help of work shafts at the intermediate points. There are two methods for transferring the lines and levels from the surface to the shafts, namely: a) by transit sights; b) by means of steel wires supporting heavy weights hung from each.

The first method is similar to the portal method. But in this case two work points are located on the working line on opposite edges of the shaft. The theodolite is set up over one and the backsight is taken to the target on the other point. Then the theodolite is transited on the opposite side as in the case described earlier. The work points which are fixed on top are extended down and across the bottom of the shaft, as indicated in Figure 5.7. The transfer of alignment through the shaft to the tunnel floor is done by suspending two or more plumb lines down the shaft and determining the bearing of the plumb lines hung by connecting to surface survey reference points. This bearing of plum lines under the surface should be the same as the bearing on the surface so that this becomes the starting direction for the underground survey work.

In the second method two work points are fixed on the opposite side of the shaft on the surface. A theodolite is set over one work point B and is used to sight the point U on the opposite side. B itself is fixed with reference to a Triangulation station A and AB set with angular measurements to S and T on the Triangulation grid. In all this, a number of repeated observations will be made and averaged to ensure accuracy. Two steel wires, each supporting heavy weights, are hung down the shaft. These are brought in line with the theodolite sight line by trial and error. The heavy weights should dip into pails of oil kept on the floor of the shaft, to give stability (dampening effect) to the wires (W1 and W2). An instrument is so set up on the shaft floor that it is in line with both wires. This instrument is positioned on the working line and is to be used to establish work points on opposite sides of the shaft. Here again there are two ways of doing the work, as detailed below.

The plumb lines are of fine piano wire and each carries a symmetrical lead weight of about 300 N (the wire being strained to half its breaking strength). In deep shafts particularly the bobs used may have projecting vanes contained in a canister with a hood to reduce rotation and oscillation of the wires. It is also customary to fill the canister with water or oil. In very deep shafts bobs weighing up to 1.3 kN are used to reduce oscillations but thicker

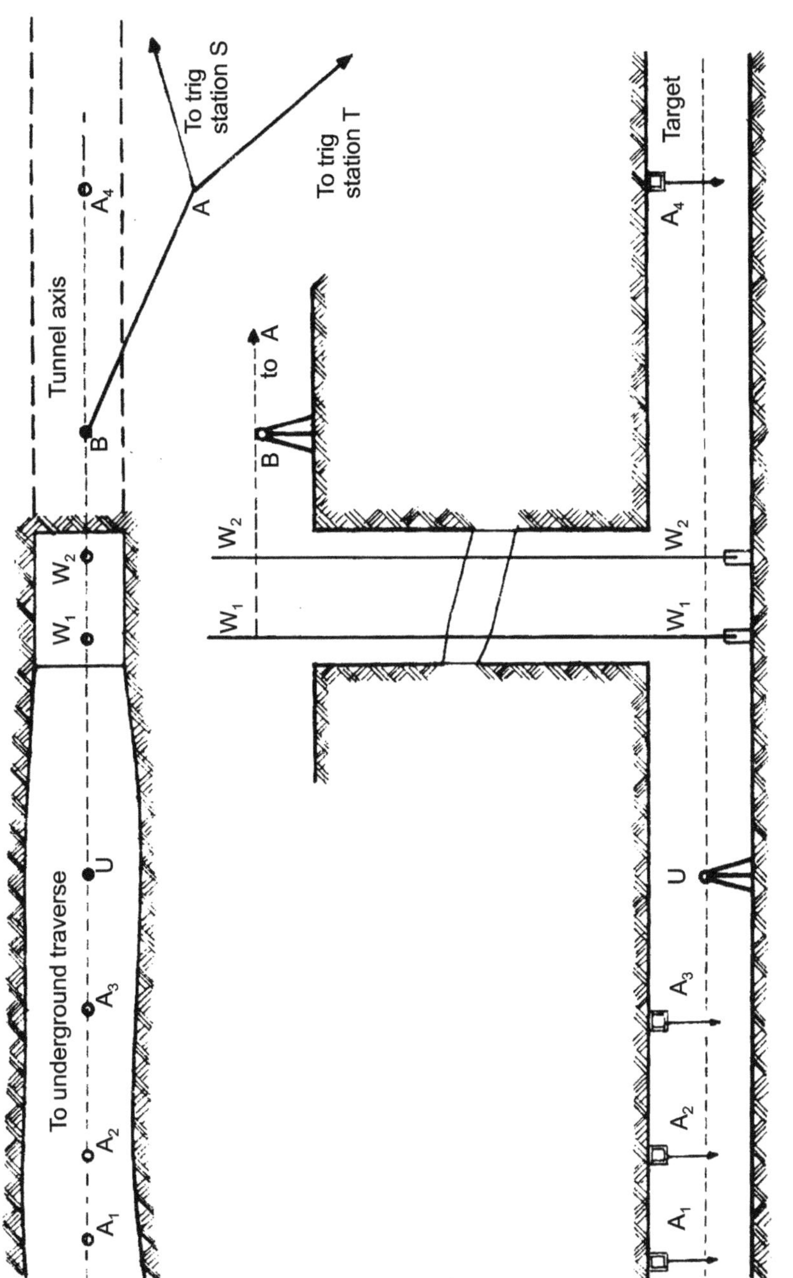

Figure 5.7 Coplaning Method.

wires are required and also geared winches to control lowering/lifting. The two wires are suspended as far apart as possible. It is very difficult to do setting out the alignment below precisely using this method, due to difficulty in preventing oscillation in the plumb/ reference wires.

Two other methods used for transferring the orientation of survey underground are coplaning method and Weisbach triangle method (Pequignot, 1963).

5.6.2 Coplaning Method

In this method two wires are aligned along the bearing or alignment of the tunnel axis at the surface and transferred down underground into the tunnel. The line joining them is then extended inside to set out and extend the tunnel axis. If this line has to be set out with reference to a triangulation station which is not in line with the bearing of the axis of the tunnel, great care has to be taken in fixing the theodolite station along the alignment on the surface. This method is diagrammatically presented in Figure 5.7. (B the point fixed with reference to triangulation station A, and bearings of AS and AT).

The process has to be reversed while fixing the theodolite station underground. The theodolite has to be fixed so that its position is in the same plane as that of the wires suspended. This is done by trial and error and sighting the wires both in face-left and face-right positions and moving the position of the theodolite as needed. In doing this, generally the theodolite is set up at a distance of 9 to 15 m from the nearer wire. The nearer the wires are to the instrument, the more the apparent misalignment when viewed through the telescope will be. If a theodolite with minimum focusing distance is used, it can be placed so close to the wires that greater sensitivity can be obtained. Final adjustment of a shifting head calls for considerable delicacy. If the distance is greater, the sensitivity reducing with distance, it becomes easier to effect an apparent though erroneous alignment. The latter can be reduced by a greater number of repeat observations in a given time to arrive at a near value, which can be as accurate as work performed close to the wire.

Once instrument position is correctly aligned, the targets can be fixed on the roof of the extended tunnel in the axis plane joining the wire (e.g., A_4) by sighting forward to give a longer base on the transferred axis. In the other directions, e.g., A_1, A_2, A_3 are fixed progressively by transiting the instruments. If tunnel direction is to be changed, the theodolite should be set up at the point of intersection and after self-alignment and correct positioning, the required angle should be set and the targets fixed on the roof of the tunnel section bored in the changed direction. Obtaining precise alignment is very difficult with coplaning method, since it is difficult to avoid vibration/ movement of the plumb lines.

5.6.3 The Weisbach Method

In this method the theodolite is not aligned exactly, but placed as close as possible to the nearer wire and nearly in alignment with W_1 W_2 as at P_1 in Figure 5.8, thus forming the Weisbach triangle $W_2W_1P_1$. The angle $W_1P_1W_2$ is measured with great care. This is the Weisbach angle from which, together with the length of the sides of the triangle, the angle $W_2W_1P_1$ is calculated by the sine formula (sin $W_1W_2P_1 = P_1W_2/W_1W_2$). And since W_1 and P_1 are small angles measured in seconds, $W_1 = P_1 = W_2P_1/W_1W_2$. Having measured the AP_1 W_1, the bearing of the wire plane can be determined. Underground, the process is reversed; the angle W_2 is computed in exactly the same manner so that, knowing the bearing of the wire plane, the bearing of W_2P_2 and thence that of P_2B may be deduced. It can be shown that:

(i) The probable error in the determination of angle W1 is proportional to that of the linear measurement and to the cotangent of the Weisbach angle. So the Weisbach angle must be small to counter possible large fractional error in linear measurement.

(ii) If the measured sides are assumed to have no error, then

$$W_1 \propto \tan W_1/\tan P_1 \propto a/c.$$

So long as this ratio is small, there is not much need for a highly attenuated angle to counteract errors in linear measurement. In other words, the instrument should be as close to the wire as possible and distance between the wires should be as long as possible. At the same time, the theodolite should be as nearly coplaned as possible, the value of the Weisbach angle being less than, say, 30 s. While the Weisbach method is easier to perform, the coplaning method is more accurate.

5.7 SETTING OUT CURVES

5.7.1 Circular Curve

Though it is ideal and desirable to avoid curved alignments in surface-to-surface tunnels, they are unavoidable, especially in long tunnels. Until the tunnel has proceeded for some safe distance from tangent points, alignment for each blast is given by using the offset method, using short-sight lengths until the tangent point is fixed within the tunnel, after which the deflection angle method is used for accuracy on curved alignments. (Figure 5.9). The sight lengths or chords (where the chord method is used) should be made as

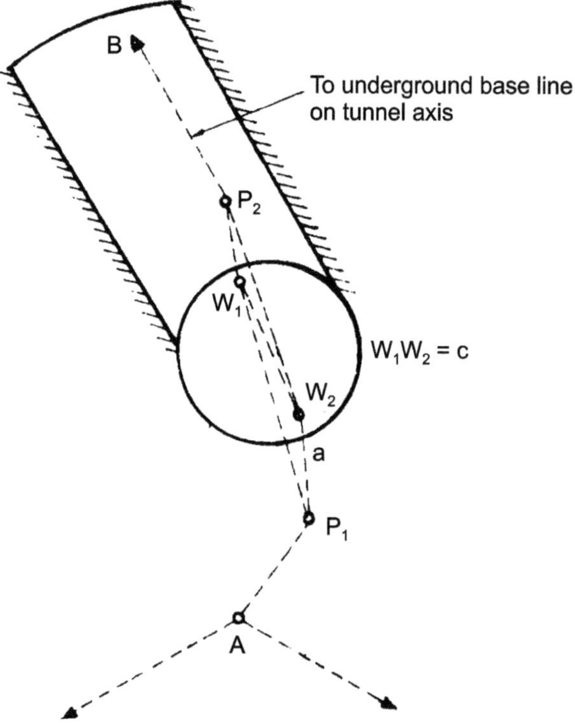

Figure 5.8 The Weisbach Triangle[1].

long as possible. The usual rule is that the sub-chord length equals Radius/20 or as large as the width of the tunnel and radius of curve can permit, e.g., for a tunnel base width of 3.6 m and R = 180 m, the angle between straights of 135°, allowing for minimum clearance of 300 mm between chord and side of tunnel, the possible chord length would work out to 46.5 m (which is larger than R/20). Adopting a 45-m chord, the deflection angle will be 7° 10' 50".

5.7.2 Transition Curves

Invariably, with increasing speeds of operation, curves on railways are now provided with transition lengths at either end so that the rate of change of radial acceleration does not exceed 300 mm/s^2/s. Introduction of transition curves requires shifting the main curve and also actual tangent points. Transition curves can be set out with sufficient accuracy by perpendicular offsets; this is feasible, however, only if the width of the tunnel is adequate for the purpose.

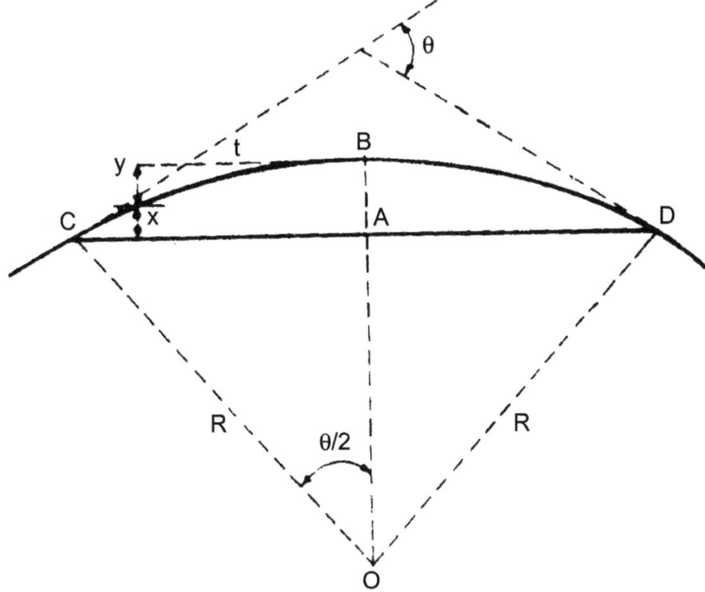

Figure 5.9 Deflection-Angle Method of Setting out Circular Curves

5.8 SETTING OUT LINE THROUGH COMPRESSED AIRLOCK

In case the line has to be extended through a pressurized airlock, care has to be taken to establish the work points inside the lock, avoiding the distortion which will take place during changes of pressure in the lock or limiting it to the minimum. For this purpose the work points are kept as close as possible to the bulkhead. The lock selected should be a stable one. In fact, a muck lock might be better than a man-lock which is supported by steel framing, since the former is supported on a concrete foundation.

Work is started from the end or from the shaft adjacent to the airlock. The method of fixing the first two work points in the shaft from the surface reference points is the same as before. For carrying out further work there are two methods. In the first a backsight is taken to the work point at the far end of the shaft (WP.4) by setting it over WP.5, the next work point in the shaft. Then, the theodolite sight is transited and three points are established over the bottom of the airlock, so that they are in line with the working line in the shaft. During this process the airlock is open to free air. After this, it is pressurized and the compressed air side door is opened. A theodolite is set up that side and moved in such a way that it comes in line

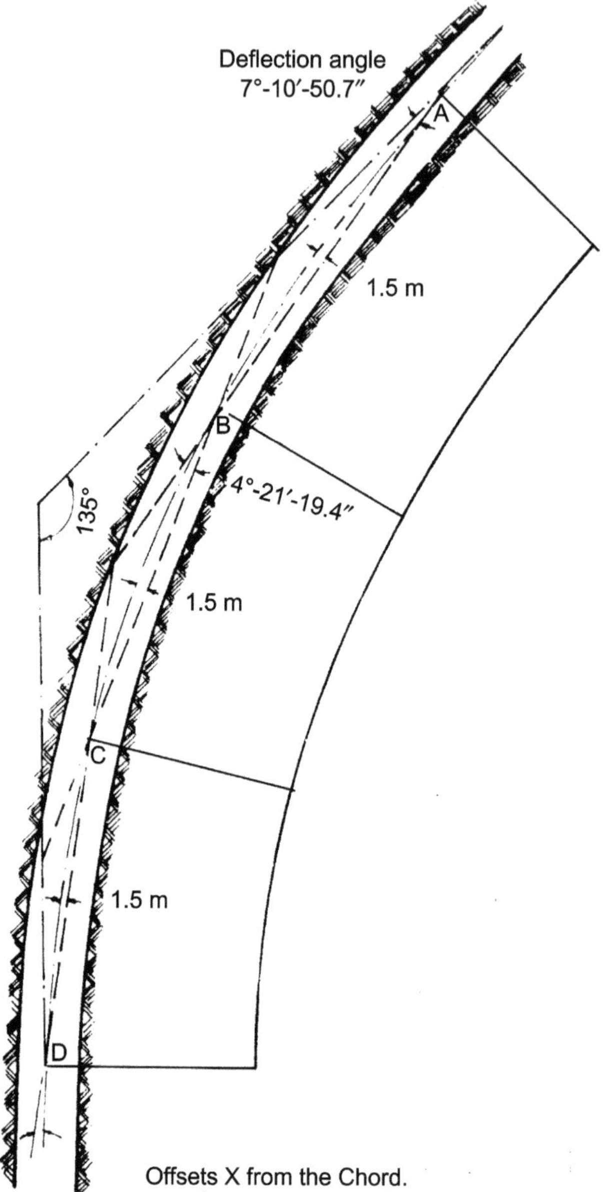

Adapted from Pequinot, 1963

Figure 5.10 Setting out a Curved Alignment with Angles, Chord and Off-set

with the three points in the airlock. It is transited and lock pressurized, after which the door on the compressed air side is closed. The remaining process is that of just extending the line as described earlier.

In the second method, the theodolite is set up over one of the working points established when the airlock is on the free air side. The backsight is taken to the points previously established towards the shaft. Then the telescope is transited. After this, the lock is pressurized. The door on the compressed air side is now opened and a foresight is established in the tunnel on the other side along the working line and it can be extended further onwards in. the normal way.

5.9 REFERENCES

1. Pequinot, C.A., (1963) "Tunnels and Tunnelling'- Hutchinson, Scientific and Technical, London.
2. Szechy, Karlowi, (1970) The Art of Tunnelling, Akedimiai Kinda, Budapest, Hungary, 1970
3. IS;5878 (Part I) 'Construction of tunnels- Part I. 'Precision Survey and Setting out' , Bureau of Indian Standards, New Delhi
4. Padmanabhan, V.C.A., 1965, 'Notes on D.B.K. Railway Project', Unpublished Indian Railways Report, 143 P.

CHAPTER

6

Tunnelling Operations

6.1 GENERAL

Tunnelling operations, apart from fixing the alignment and lining (which are the preliminary and final aspects of construction of a tunnel) comprise the under mentioned intermediate operations:

(a) Excavation, picking or blasting, the first being applicable to soft soils, the second for medium soils and very soft rocks and the last applicable to medium and hard rocks.
(b) Fixing temporary supports, as necessary.
(c) Removal of blasted/excavated material (mucking).
(d) Dressing, fixing permanent supports, followed by
(e) Final operation of lining, wherever necessary.

6.2 PRELIMINARY WORK

The face from which tunnelling work is to commence has to be decided first. This is done with reference to the type of material and the rock cover. The minimum cover with which the tunnel can be started depends on the type and structure of the rock or soil material and also the shape of the tunnel. Also to be taken into consideration are the geological structure and fault line of the strata, as already discussed in Chapter 2. The relative economics of adopting open cut in preference to a tunnel must also be determined. In some cases the cost of protective works of open cuts and protection of the Portal may be so high that commencement of the tunnelling operation even with a smaller cover is preferable.

6.3 TUNNELLING METHODOLOGIES

6.3.1 Alternative Tunnelling Methods

The adoption of any one of a number of possible methods for tunnelling depends on the nature of the soil profile. Soils can be broadly grouped under soft strata or rock. Soft strata may be cohesive soil (clay), or granular soil or a mix. Tunnelling through soft strata is done by using one of the traditional methods of driving, i.e., excavation by digging or with a tunnelling machine. The machine is generally used in very soft strata and clay and prevails in subway construction. Wherever seepage flow is heavy, the operation has to be supplemented by use of compressed air. Occasionally some drilling and blasting is also done, especially in the case of mix with soft rock and in laterite types of soil.

A number of methods have been developed for tunnelling by machine and the choice will generally depend on the type of soil, type of structure etc.

Broadly speaking, there are seven main methods of constructing tunnels, as listed below. Shape of tunnel suitable for each are indicated in brackets1

- Drill and blast (Circular, Horseshoe, Segmental)
- Sequential excavation (Horseshoe)
- Shield driven (Circular)
- Bored (Circular)
- Immersed tube (Circular, Rectangular)
- Cut and Cover (Rectangular)
- Jacking (Rectangular, Circular)

Drill and Blast:

This method has been the most often used methodology for older tunnels and is still used for tunnels in rocky soils and in difficult conditions. In such cases, they may be considered cost- effective and fast.

Sequential Excavation Method (SEM)

This method is suitable for soils which have sufficient strength to stand by itself when done in small increments with no direct support but the exposed soil face has to be supported as soon as possible by shotcrete (with or without mesh as additional support before excavation is continued on next segment.

Cohesion of soil/ rock is increased by injecting grout and providing rock bolts as required.

Shield driven tunnel

This method involves building a short length shield, pushing it to cut through the soil, excavation of the soil in face in small lengths, and providing a lining inside the shield before it is pushed forward. It is suitable for soft soils.

Bored tunnels

This is an extension of the shield tunnelling technology. A TBM (Tunnel Boring Machine) is used, which can work on full face of the tunnel at one time, by varying the type of tools to suit the type of soil, slush to rock. The TBM is designed to support the surrounding soil after excavation, till the lining is provided behind. Generally CI or precast segmental linings are provided.

Immersed Tunnels

This method is used to provide sub-aqueous tunnels, i.e., tunnels to be constructed across, canals, rivers or under-sea, using precast, floating and sinking technology. A trench is cut at the bottom of the water and prefabricated tunnel units or precast segments are taken to the site, made watertight and lowered in position butting against each other tightly. Joints are grouted to make them leak proof. Trench may be backfilled to provide the cover to the tunnel and waterway clear for water borne traffic.

Cut and Cover

As the name suggests, this method involves excavating an open trench, in which the tunnel section is built at desired level. The trench is filled back, compacted and restored to original level and other pre-existing facilities provided, as required. There are different methods used for cutting the trench and its protection, in form of piles, diaphragm walls etc.' and same tied back as required. This is the commonly used for subway construction.

Jacked tunnels

This method, alternatively known as pipe pushing method, is used for only short lengths in case of large sections and adopted for tunnelling below existing utilities like highways, railway lines, buildings, and other utilities spread over an area, and where cutting in open and constructing shallow tunnels is not possible, since the obstructions cannot be removed temporarily and replaced. In this, first a jacking pit is prepared in which the tunnel section in short length is constructed. After it gets sufficient strength, it is pushed into the soft ground with powerful hydraulic jacks, at the same time the soil in front is systematically removed. As the built unit advances, further lengths are built up behind till full length is covered. In case soil is too soft, it should be stabililised by means of grouting etc. in advance.

6.3.2 Traditional Methods[2]

The traditional methods of sequencing operations to achieve full profile may be classified as follows:

(a) full face method;
(b) top heading and benching method;
(c) bottom heading and stoping method;
(d) 'Drift' method, subclassified into wallplate drift method, side drift method and multiple drift method.

The first three methods are generally used for rocks (aided by blasting) and medium type soils, while the last (d) is used for soft rock and disintegrated rock (requiring ground support). Clayey soils are generally dealt with by the shield method (see below), a method gradually refined over the years with the advent of tunnelling machines.

(a) *Full Face:* This method is used wherever the strata are such that they can stand long enough to allow erection of supports after excavation. It is also suitable for tunnels of small cross-sections in which the mucking quantity is small and the support needed is not high. In this method the full face is bored through in one continuous operation, followed by mucking, supporting and lining, as necessary.

(b) *Top Heading and Benching:* As the name suggests, this method is adopted in two stages. The top part, generally the arch portion, is first excavated and supported. This may be done for the full length or in parts ahead by boring the top portion advanced first. This is followed by benching, i.e., removal of material for the bottom portion (in parts step by step), each line supporting the soil to the required profile and extending the supports below. While doing benching, the top heading supports (continuously supported over a wallplate) are generally underpinned. This operation is followed by providing necessary permanent rib supports and lining.

(c) *Bottom Heading and Stoping:* This is the reverse of the top heading and 'benching method. In this method tunnelling generally, commences from the top face, while in most other methods it is started from the bottom face or both faces concomitantly. As this method requires special treatment of the soil for stabilising and securing, it is more suitable for hard rock which can stand by itself by arch action longer.

(d) *Drift Method:* in the case of disintegrated rocks and fault zones, tunnelling is done in phases by driving small-size tunnels as pilot tunnels. Alternatively, the top drift may be driven first, followed by the side drifts to provide side supports. Subsequently, the material between and on the sides is removed up to the required profile. The drift method can be executed in a number of ways:

(i) *Wallplate drift:* This method is used when both heading and benching or top heading methods have to be supplemented by drifts on each side because the rock is so bad that only a short advance can be made per 'pull' in the heading.

(ii) *Side drift method:* This is employed in large-size tunnels through bedrock which requires support before mucking. A drift is driven ahead on either side at subgrade revel and posts and wallplates are erected. If the strata permit, a full-face operation can then be carried out and roof ribs erected quickly over the wallplates already erected and mucking initiated.

(iii) *Multiple drift method:* This method is used for soft ground conditions not suitable for free operation. It is a combination of crown drift and side drift. It is generally adopted for large-size tunnels and for going through crushed rock in fault zones requiring light blasting. The sequence of operation depends on the type of rock. In one type the top drift is driven first and the side drifts driven later; in other rock types the modus operandi is reversed.

6.4 TUNNELLING IN ROCK

6.4.1 Drill and Blast Method

In the case of tunnelling through rock, blasting has to be invariably resorted to. Support (temporary or permanent, may depend on type of rock. In this case also all four alternative methods for sequencing described above can be used, depending on type of rock and size of tunnel.

(i) *Full-face method*: Suitable for sound rock and up to medium-size tunnels.

(ii) *Top heading and benching:* Recommended for large-size tunnels and/or for rocks not structurally sound. Length of heading to be excavated or advanced forward depends on strata, site condition and length of tunnel.

(iii) *Bottom heading and stoping:* Recommended wherever tunnel size is very large and rock is consistent and sound.

(iv) *Drift method:* Used for large tunnels wherever economical to drive one or more small tunnels called drifts or pilot tunnels. Drifts can be 'top', 'centre', 'bottom' or 'side' drifts according to their position with reference to the main bore.

6.4.2 Codes Applicable for Major Operations

Tunnelling in rock is essentially a process comprising drilling and blasting of rock and removal of muck. These two are major operations and hence is covered in detail in Chapter 7. Bureau of Indian Standards code IS: 5878-1970 (Part II/Sec. 1) also details them. The operation should conform to the safety codes IS: 4089-1967 concerning blasting and drilling operations and IS: 4137-1967 for working in compressed air. The general safety precautions to be followed in all tunnelling works are laid down in IS: 4756-1968.

6.5 TUNNEL SUPPORTS[3]

6.5.1 Purpose of Tunnel Support

Supporting the tunnel section is required temporarily or permanently for the purpose of preventing collapse or cave-in of the top or the sides of the bored portion. Support may be part of a permanent lining or temporary until such time the permanent lining which has to take the rock or soil pressures is ultimately provided. In solid rock structures tunnelling can be done with no support However, in other types of rocks and soils supporting is essential. 'Guidelines for Tunnel Supports' are covered in IS: 5878-(Part IV)-1971[3].

The main components of tunnel supports are ribs, posts, invert struts, wallplates, crownbars, truss panels, bracings and Iaggings. In addition, in some places a certain amount of packing may be required between the aforementioned components and the exposed surface. In the early days these tunnel supports were mostly made of timber. The tunnelling itself was done by making a pilot tunnel or a drift which was properly supported and different methods were used for extending the section and simultaneously extending the supporting system. One typical arrangement is shown in Figure 6.1. known as the English Crownbar method, it is a form of the drift method. The Figure is self-explanatory. In case of Tunnelling in rock, where necessary steel posts and arch ribs were used with or without wooden plank lagging to suit nature of rock, soft/ fissured or solid/ hard. Fig 6.2 shows different types of steel supports used.

Stages and sequencing of excavation varied from country to country. The different practices are briefly discussed in Para 6.6.

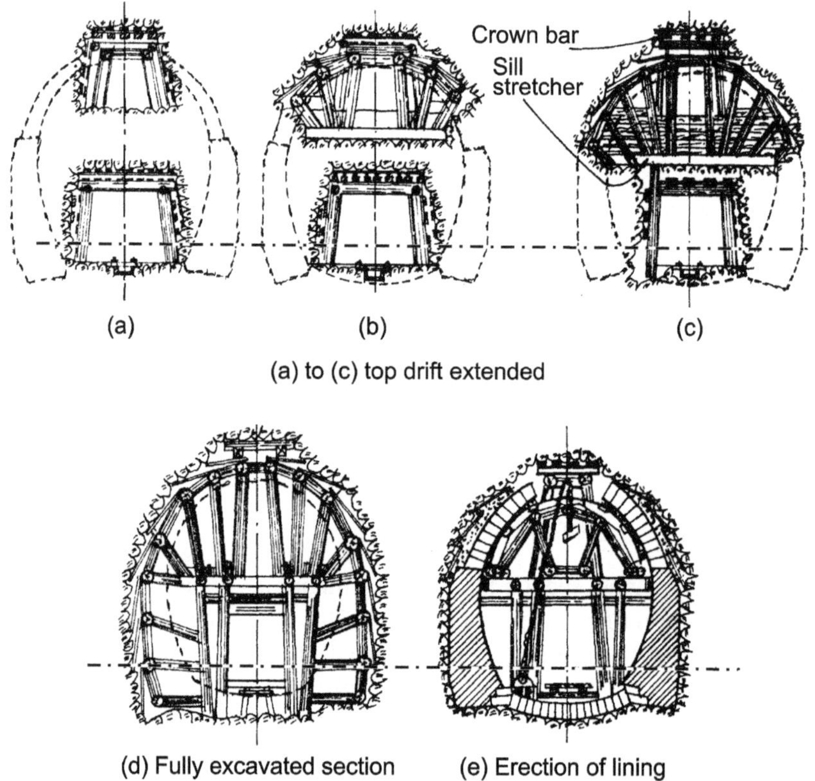

Figure 6.1 Tunnel Support Stages at Different Stages of Tunnelling in Old methods[3]

6.5.2 Types of Steel Supports

The present practice is to generally use steel members either for temporary or permanent supports as they can be fabricated in standard sizes and can be easily transported and assembled as necessary. Where necessary, rock bolting and or pressure grouting of exposed surface is done to reduce the roof settlement. Full face or top heading and benching methods are adopted.

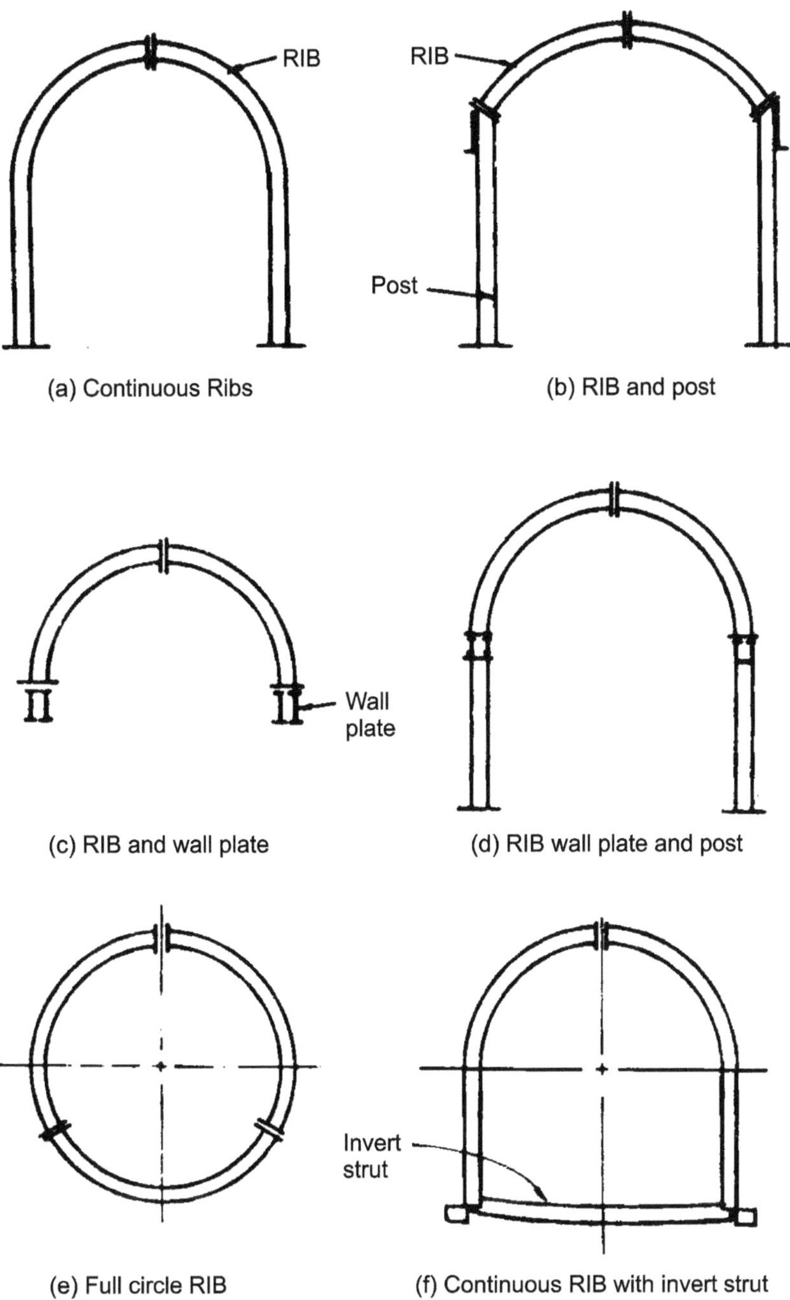

Figure 6.2 Typical Steel Support Systems[3].

Different types of supports provided, as illustrated above, are:

(a) *Continuous ribs*: These are made in two pieces and are used when full-face tunnelling is being done. They are recommended for use in rocks whose bridge action period is long enough to permit removal of gases and mucking.

(b) *Rib and post*: Also used for full-face working and for tunnels whose roof joins the side walls at an angle instead of a smooth curve. These are in three pieces.

(c) *Rib and wallplate:* As already mentioned, this is used for supporting the top heading and wherever sections with high straight sides through good rock or large circular tunnels are to be bored. This is recommended for spalling and soft rocks.

(d) *Rib, wallplate and post:* This becomes necessary for a tunnel with high vertical sides and is especially mandatory whenever spacing of the ribs and posts differ.

(e) *Full-circle rib:* Used for providing a circular section or near-circular section and is necessary for use in tunnels in squeezing and crushed type of rock or in rock that imposes considerable side pressure.

The ribs are the transverse supports used for supporting the roof. They are in arched form. They may be made of bent I beams or braced frames. Posts are provided to support the ribs while extending the tunnel section below by benching; they thus transmit the roof load to the base and also provide a framework for providing support for the sides. The invert strut, as the term implies, is provided at the invert level and it struts between the feet of the posts to-hold them in position. Wallplates are longitudinal members provided between the posts and ribs. These are generally provided when tunnelling is done by the top heading and benching method. In this system the ribs are first erected over the wallplates located on the floor of the heading. These wallplates are in turn underpinned and supported by the posts during benching. Crownbars are provided between the ribs and the exposed surface of the tunnel roof as it is difficult to make the ribs fit correctly against the exposed surface. These crownbars are placed longitudinally, i.e., parallel to the axis of the tunnel. See Figure 6.1(c).

Truss panels may be provided for lateral support to the ribs so that the latter are kept in correct spacing; they also provide lateral restraint to the same while doing underpinning. Bracings are also provided longitudinally between the ribs and posts so as to increase their resistance to buckling about their minor axis and to prevent their displacement while blasting ahead is done. This purpose can alternatively be served if lagging is provided. Lagging serves one or more of the following functions:

(i) to provide protection from falling rock or spalls;
(ii) to receive and transfer loads to the rib sets;
(iii) to provide a convenient surface against which to block, in case it is not convenient to block directly against the rib because of irregular over break;
(iv) to provide a surface against which to place backpacking;
(v) to serve as an outside form for concrete lining, if concrete is not required to be poured against the rock surface.

Spacing of various components depends on the type of rock or soil structure. Laggings are sometimes used, as already mentioned, in such a way that they form the shuttering for the lining also. As far as possible, use of timber should be avoided as it is difficult to remove safely and furthermore is likely to deteriorate and prove a source of weakness. However, timber lagging is economical if it is to serve as inside shuttering in tectonic type rocks as the regular face need not necessarily be fully and directly supported by the lining.

6.5.3 Tolerances

Various tolerances and methods of fabrication and erection of supports are indicated in IS: 5878-1970, Part Ill. The ribs shall generally be bent cold. For small jobs, ribs may be fabricated as polygons. The accuracy of bending shall be such that after bending, each segment shall conform to a true template at the ends; intermediate portions may depart from a true template by not more than plus or minus 10 mm. The web shall be true and wrinkles or buckles shall not exceed 5 mm when measured from a straight edge held flush against either side or web on a radial plane. To ensure proper space at the crown after wedging, the ribs shall be fabricated to a slightly larger radius than theoretically computed. In tunnels with small cross-section, the crown joints may be 'kicked' up or the ribs set high to accommodate the concrete delivery pipe.

Welding for fabrication of support is governed by the provisions of IS. 816-1969 (Welding Code).

Erection of supports is done within the following tolerances[2]:

Spacing of ribs: Average of three measurements taken around periphery shall not differ from the spacing shown on the drawings by more than ±30 mm. Internal dimensions: Dimensions between the inner flange of ribs shall be checked at 2 or 3 points in the horizontal plane and such dimensions shall not vary from the theoretical dimensions by more than +30 mm.

Level at crown: Level of the crown shall be within –20 mm and + 40 mm of the required level. Deviation in vertical plane: The deviation from vertical or the required inclination shall not be more than ±20 mm when measured at the lowest point. Spacing between tie rods: This shall be as shown on the drawings with a tolerance of ±30 mm.

Gap between joints: The gap between the joints of the butt plates of the ribs shall not be more than 5 mm.

6.6 TUNNELLING IN SOFT ROCK AND SOILS

6.6.1 Methods used in Different Countries

Traditional methods of tunnelling through medium rock, earth and stiff soils is to drive in drifts or smaller headings and successively enlarging them by excavation and lining. They are also called Classical or Mining methods. There are a number of variations in sequencing the headings/ drifts to suit different soils and evolved in different countries. With the shortage of timber for providing supports and with development of better technology they are not much in vogue now. NATM with use of steel and shotcrete supports is more in use. However, these methods are detailed here to understand how old tunnels were done. The major ones are[4]:

- English Method or Crown bar method;
- Classical methods- Belgian flying arch method; German core leaving method;
- Austrian or Cross bar method;
- Centre cut method; and Alternate ring method.
- Italian- invert arch method
- Combined methods developed as a combination of two or more methods mentioned combining selected elements of same.

As the names suggest, the methods vary mainly in respect of location and sequencing of drifts, their widening process and timing the construction of permanent supports. A few major ones are briefly described below. For more details, Szechy (1970) and Merriman and Wiggin (1943)[4] may be referred to.

(a) The English method

This is a full face method and is suitable for very firm ground and heavy and wet grounds i.e., soils which can stand on its own for some time for full face to be opened up before the lining can be built up from below. It is done in short lengths of 3 to 6 m. Work starts with driving a small drift (at bottom (which would help in haulage of muck and also drainage) This is followed by a similar drift just below crown, the roof being supported by providing roof bars supported on props at far end. Transverse polings are provided over these. The top heading is then widened for that width by placing breast boards against face and propping them. The heading is then widened and supported till full arch portion is opened. Similarly the bottom heading is widened and supported all-round, till full section is opened for selected length and roof and sides are firmly supported. Once this is done, masons take over and build the side walls and arch followed by the invert. The process is repeated for further lengths. Figures 6.1 and 6.3 (a) show the different stages and how timber supports are provided at various stages.

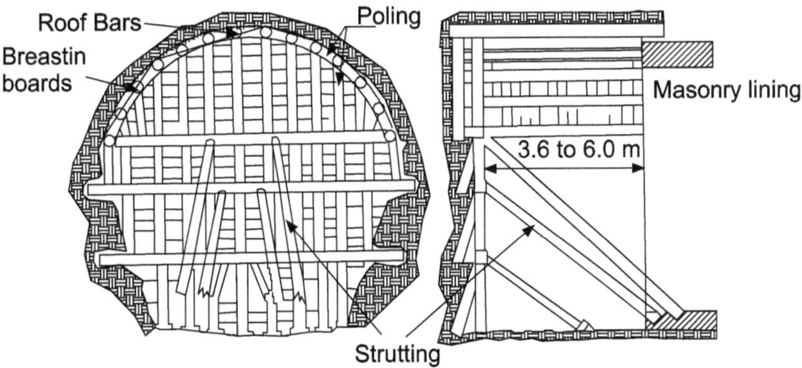

Figure 6.3(a) English Method[4].

(b) Belgian method:

This method is used in firm ground, where generally loads would be primarily vertical with minimal side pressures. In this method, top excavation is continued over longer length ahead of the lower part excavation. Transverse poling is not used over crown bars to support roof, but props are set against the earth in between roof bars. Once arch portion is excavated, the arch masonry is built up and strutted and supported at springing. Then central trench is cut below for full depth at intervals. The arch masonry is underpinned at intervals. Cut is then widened to extend underpinning to facilitate building of one side wall. The process is repeated on the side also and side wall built. Floor is then cleared and invert is built. Stages are conceptually shown longitudinally in sketch at Figure 6.3(b).

The method can be extended to heavier soils, but with considerable risk, by doing same keeping the timbered length in shorter stretches. Main disadvantage of the method is the likely uneven settlement of the arch which is first supported on earth and subsequently on timber at intervals.

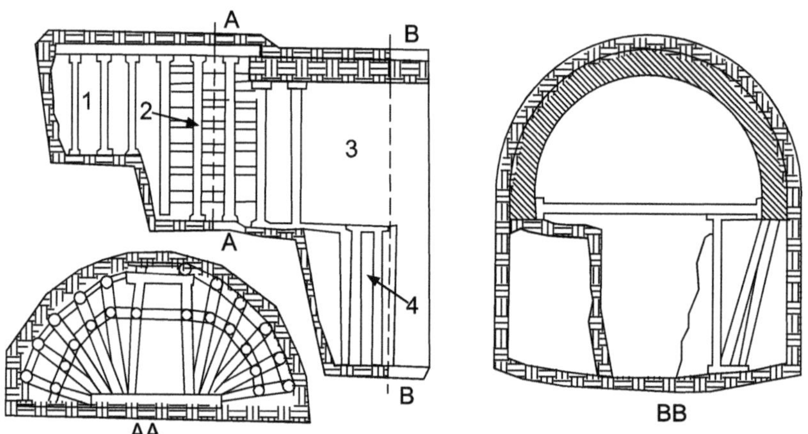

Figure 6.3(b) Belgian Method[4].

(c) German Method

In German Method, three drifts are run, one larger one at top centre just below roof and two at bottom at foot on either side. In the side drifts, the side walls are built as high as the supports would permit. On top of these, two more side headings are done and side walls extended up. The top drift is widened by timbering. Then, in stages the arch ring is built over the side walls on both sides, till they join at centre. Core is then removed and the invert concrete laid or built up. Figure 6.3(c)

Figure 6.3(c) German Method[4].

(d) Austrian System

It is similar to German system with the difference that, first the soil in centre is removed by driving two central headings, one below the other as shown in Figure 6.3(d). They are widened, the bottom heading on one side first. Over this, poling boards are driven for the arch roof above and widening out with arch the top drift done in short lengths (supported with propos and struts). Side wall is built in short lengths and extended over. The same process is repeated on the other side. Finally the invert is built, as in other cases.

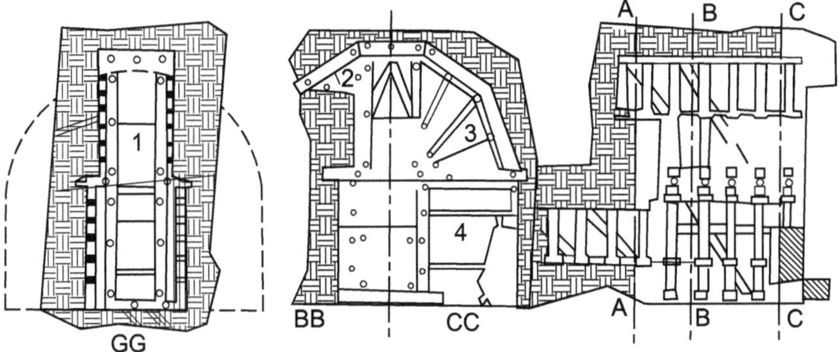

Figure 6.3(d) Austrian Method[4]

(e) Italian Method

This was developed for very soft and treacherous soils. In this, only short lengths (2 to 3m) and small areas are opened at a time. First a centre drift is driven for about one eigth height of tunnel. It is extended on either side, and supported all-round with heavy timbering. Within this, the invert and as much height as possible of side walls can be built forming a bench. The open area between is filled back with earth. A centre top heading is then driven and enlarged for half height of tunnel and for full profile, and opening is well supported by props and braces. On the sides, after trenching through the bench and clearing earth upto top of side walls already built, the side walls are extended along with the arch above in one operation. Figure 6.3(e)

6.6.2 Sequential Excavation Method (SEM)[5]

'SEM (Sequential Excavation Method is defined as a method where surrounding rock or soil formation of a tunnel or underground opening are

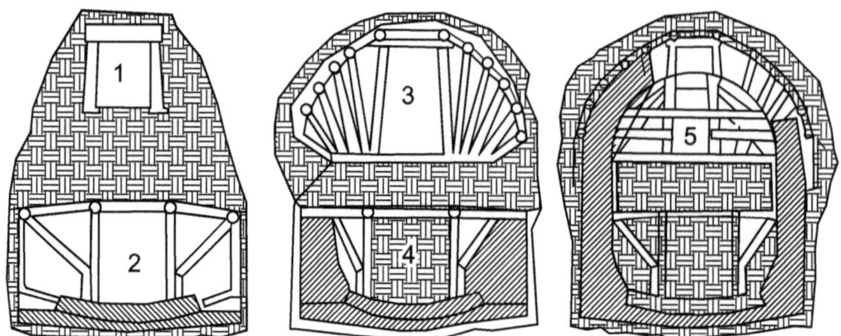

Figure 6.3(e) Italian Method[4]

integrated into the overall ring-like support structure and following principles are observed.:

- 'The geotechnical behaviour of the ground must be taken into account
- Adverse states of stresses and deformations must be avoided by applying adequate support system
- Completion of the invert will give the above mentioned ring like structure, static properties of a 'tube'
- Support system should be optimized according to the advancing deformation
- General control, geotechnical measurements and constant checks on (the functioning and) optimization of the pre-established support means must be performed.'

This method as evolved was considered suitable for Short tunnels, large openings such as stations, unusual shapes or complex structure, Intersections and Enlargements and may not be competitive enough for long tunnels. With the advent of NATM, which is typically an SEM, it is being used for long tunnels also now.

6.6.3 New Austrian Tunnelling Method (NATM)

Popularly known as NATM, this method has been used extensively in many countries. This has been developed over the Modified Austrian Method and was first tried in large scale in 1950s.[5] It is in fact based on an 'accumulation of expertise from all over the world' It is a concept developed taking into account the self carrying capacity of the rock surrounding the tunnel by allowing deformations within limits'. It owes its refinement Prof Rabcowitz (1965). When the opening is made, there will be some movement from all around the opening and there will be some adjustment in load sharing by surrounding rock due to elastic nature of the rock and due to elastic nature of rock, there will be some yielding in immediately surrounding rock layer and the rock beyond will start sharing the load, thus relieving the rock nearer tunnel face of some stresses. This can result in their cracking and becoming loose. If not held in place, it may crumble in places and fall. Provision of temporary supports partake some of these forces and reduce this effect. Instead, if the surface of exposed rock in tunnel is provided a shotcrete layer and some elastic support in form of grids or steel mesh and rock bolts, such layer would form a protective layer and also in integration with the surrounding rock resist the forces from the surrounding rock mass. By observing the movement of this surface and measuring the stresses in them and rock around, it will be possible to assess the forces to be resisted over time and design the permanent lining and provide the same as soon as possible.

The method itself is quite simple and can be used in weak rock and cohesive type soils. It was employed early in construction of the Vienna Metro, for example. In developing countries, apart from detailed soil investigations, it requires measurement of strength of surrounding rock and its movement soon after the cavity is made and till it stabilizes. This process requires import of certain expertise from abroad. In some stretches depressing the water table by well point pumping may also be necessary. NATM may be cheaper than shield tunnelling and can be used in varying geological conditions, such as rock outcrops coming in-between brown silty clay, and progress will be faster than shield tunnelling. Restrictions such as minimum cover on top and minimum centre-to-centre distances between tunnels, as required for compressed air shield tunnelling, could be considerably reduced. This would in turn allow the tunnels to pass at a comparatively lower depth and reduce inconvenience to commuters. In this method, as soon as the excavation is done to required profile the exposed surface is gunited to stabilise it. This is followed by proper lining work.

This method is now widely used in India in mixed soil conditions, especially in the Himalayan region. It has been used along with Drill and Blast for boring the 11 km long Pir Panjal Tunnel, details of which are covered in a Case Study at Annexure 6.2

6.7 EXCAVATION AND ADVANCING FACE[2,4]

6.7.1 This section covers details of how excavation at tunnel face is carried out. The excavation operations would differ according to type of soil. Five different methods have long been used for tunnelling through soft soils and to some extent medium strata, using manual labour:

(a) Forepoling method
(b) Tunnelling with liner plates
(c) Needle beam method
(d) Flying arch method
(e) Shield tunnelling.

The first four methods were developed some hundreds of years ago and the last in the early nineteenth century for full face tunnelling in soft ground.

6.7.2 Forepoling Method

Forepoling is used particularly in running ground for small tunnels or drifts, as well as for driving headings in some other cases. Forepoles as originally used were timber boards (about 2m long) or scantlings with wedged ends. (Now steel pipes of longer lengths with wedged end are used for driving headings in cohesive or mixed soils). They can provide support to the ground above when driven in. Figure 6.4 (a) shows schematic of how they work.

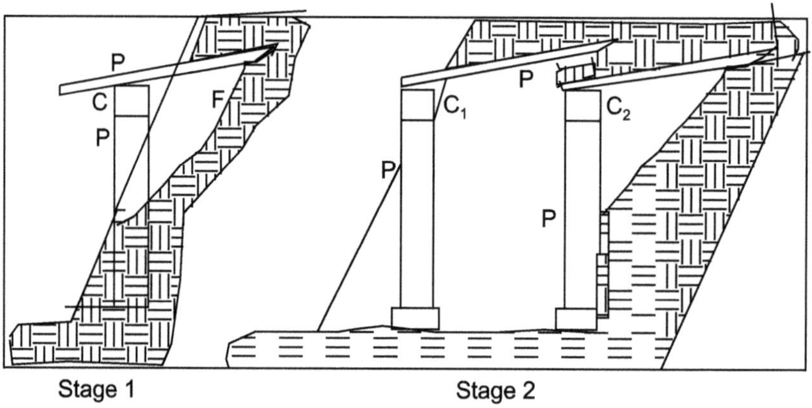

P- Post; C- Capping beam p- Poling board/ Poles;
b- Breasting board; F- Tunnel face

Figure 6.4(a) Forepoling for Mining/ Tunnelling in earth.[4]

Work is started with fixing posts and a capping beam set against tunnel face. Using it as a bent for levering poling boards or forepoles are driven through the soil at an angle (about $\tan^{-1} 0.15$) and adjacent to each other so that they would form a roof over and earth can be excavated below. With this protection excavation is done to extent possible for erecting next set of bent (posts and cap). If, however, the soil cannot be held for duration required to erect next set of posts and cap, some short vertical boards are driven into ground (as shown in rear of C_2) and similarly the bent erected. Next set of poles can be driven into the ground next and process repeated.

For more details of stages of fore poling in running soils, Reference 2 may be seen. With the development of shield tunnelling, latter is preferred for running soils.

In partially cohesive soils, the forepoles can be driven at the roof level in an overlapping manner as shown in Figure 6.4(b). Transverse ribs or lattice girder normally used for supporting roof level near the face of the tunnel is used as the bent mentioned above for driving in and supporting rear end of forepoles. (Face is normally supported with breast boards below the girder, to suit the nature of soil). A row of forepoles are driven into the soil with the girder providing rear support. With this row of poles providing support soil below for some small depth is cleared for some distance and another transverse girder is laid across. The intervening distance can be about 0.600 to 0.900 m. Another set of forepoles can be driven in to extend the support for further excavation. In this manner forepoles are driven in an overlapping manner so that there is no gap between them and in the coverage as shown in Figure 6.4(b). Excavation below this umbrella can be done for proposed pull length. As it is done, face support in form of breast boards is provided beyond the pull length, shifting from the front. Part of the pole will be cut off and the remaining will be left buried behind the lining.

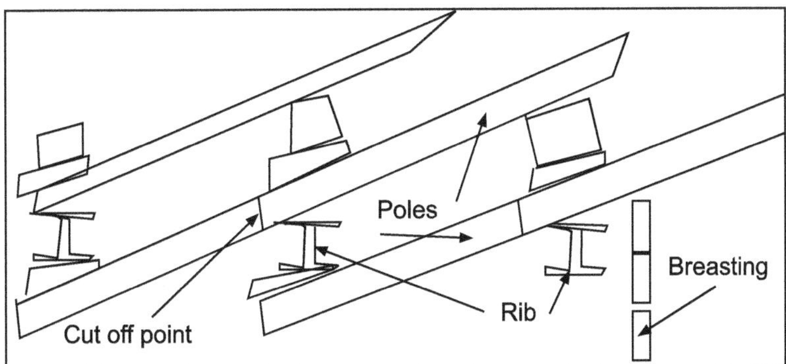

Source: Manual for Design of Tunnels and Bridges, FHWA

Figure 6.4(b) Forepoling Method of Supporting Running Ground.

Forepoling with longer pipes and larger overlap is now used extensively for driving headings of tunnels in clayey earth, with full or partial breast board protection for holding the tunnel face during the time taken for driving a set of forepoles.

6.7.3 Tunnelling with Liner Plates

This is generally used for either forming drifts or headings on medium soft ground. It can also be adopted for small cross-section drifts in running ground, combined with compressed air. The first liner plate is placed at the crown segment in a pre-excavated cavity at the top. After the hole has been sufficiently widened, adjacent liner plates are bolted to it, one on either side. These plates are temporarily supported by trench jacks or props placed below and tightened carefully. The arch section can be extended down to the springing line in a similar manner. The liner plate ring for the section so formed should be wedged outward from the wallplates or wall beams placed in the grooves. This process can be continued below also if a full-face operation is intended.

6.7.4 Needle Beam Method

This method is a modification of the liner plate method. It is applicable wherever the roof can stand for a few minutes and the sides for an hour or two as in stiff clay. The full section of the tunnel is broken up into successive portions, as indicated in Figure 6.5. Work is started in the top heading portion and then extended below.

To start with, a monkey type of drift is driven into the face of tunnel. The roof of the drift is supported with laggings carried over wooden segments. These are supported by trench jacks which should leave the central portion clear. After a sufficient length is cleared, including space along the axis for inserting the needle beam, the latter is inserted with its front portion resting on planks placed on the floor of the drift. The rear end will be supported on a post resting on the finished floor of the tunnel. Further lagging plates are set up one by one and are supported radially using trench jacks or props from the centrally placed (longitudinal girder) needle beam.

The needle beam consists of two IRSJs bolted together or a timber beam. If the needle beam is made up with joists, the space between the webs of steel joists is filled in with hard wood. The length of the needle beam chosen will depend on the length of daily progress, generally 1.5 to 2 m.

The remaining part of the entire section is cleared and the exposed surface is supported by plates and trench jacks and the tunnel advanced, as shown in Figure 6.6, stage 3. This method causes difficulty in placing concrete lining due to the obstruction posed by the needle beams. Brick lining is easier to place. This method is suitable for small- and medium-size tunnels (2.5 to 4.0 m dia).

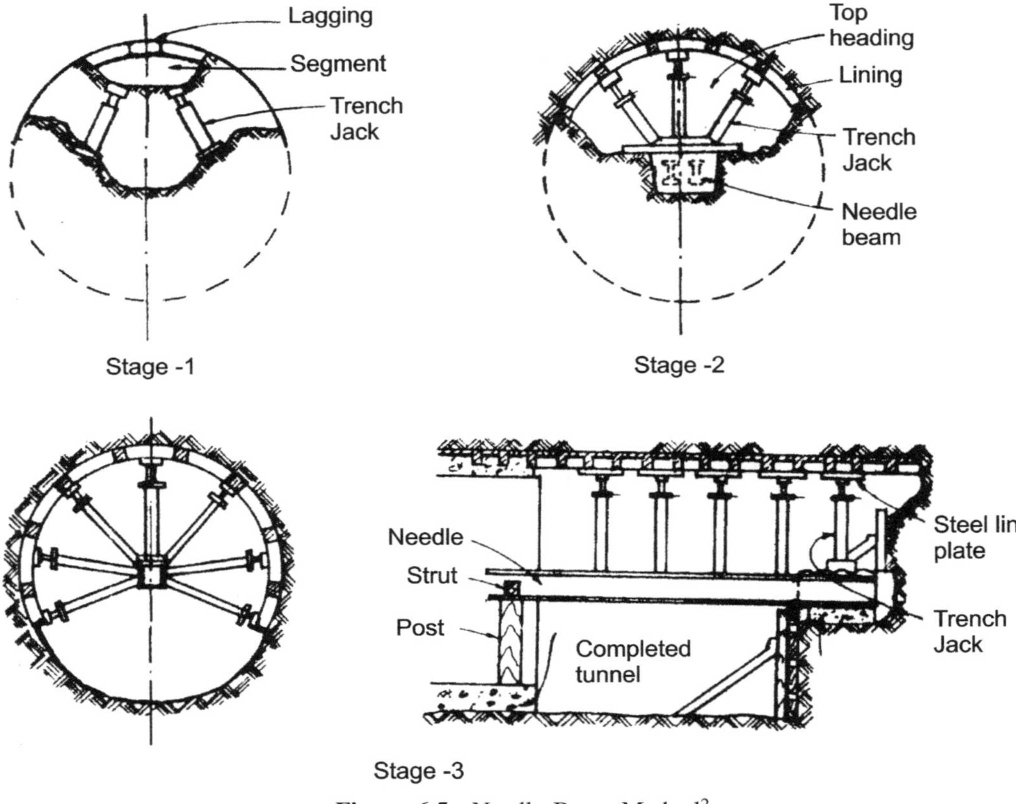

Figure 6.5 Needle Beam Method[2].

6.7.5 Flying Arch Method

The flying arch method is similar to the needle beam method except that no beam is used for supporting the liner plates. As the top heading is driven, the liner plates of the arch are supported by trench jacks resting on the bench itself. Each day's drive is concreted with half-round arch forms, handled and filled by hand with a plank used as footing under the concrete. After the heading has been driven about 20 to 25 m, the benching is done for extending the section below, after which side and invert concrete are placed (Figure 6.6).

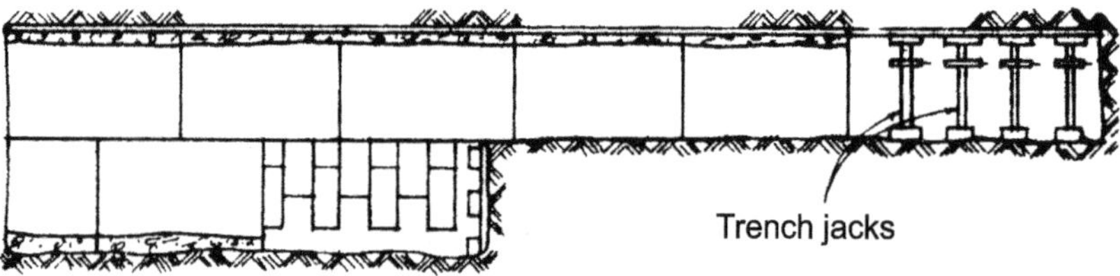

Figure 6.6 Flying Arch Method.

6.8 SOFT-GROUND TUNNELLING[7]

6.8.1 Tunnelling in soft ground has been recognised as one of the most challenging and complicated construction operations. The requirements for satisfactory tunnelling in soft ground are:

(a) The method should be such that it is possible to advance the tunnel consistent with safety and to maintain the integrity of the opening temporarily until the permanent lining is built/ erected and gains strength to support the pressures exerted on it.
(b) Construction of the tunnel should not result in any damage to adjacent or overlying buildings, streets or utilities.
(c) The tunnel after construction should be capable of withstanding the influences to which it may be subjected during its lifetime, such as earth pressure due to the surrounding soil, overburden, presence of stations, ventilation shafts, cross-tunnels etc., including disruptions caused by sub-sequent tunnelling alongside completed tunnels.

Different methods used in soft ground Tunnelling are listed below with brief details.

6.8.2 Shield Tunnelling[6]

In most subway constructions shield tunnelling has to be carried out with the use of compressed air since the level at which it is done will be below the subsoil water level, or under water bodies such as streams and rivers. Where depth is not large or in highly cohesive soils where subsoil water ingress can be easily managed, use of compressed air can be dispensed with. Technology in shield tunnelling has advanced so much that there are machines, Tunnel Boring Machines (TBM), mounted at the face of the tunnel which can cut the soil using rotary cutters; excavated soil is collected and conveyed through a conveyor-belt system mounted on the equipment. This equipment advances automatically like a mole as face soil is cut and cleared and lining in the rear is erected or poured. Such equipment, known as 'moles', is more popular in subway tunnel construction. Moles are successful wherever soil is fairly uniform, cohesive and not too hard and can achieve a progress rate of as much as 0.9 m per hour. The methodology involved in shield tunnelling is discussed in more detail in *Section* 6.9.

6.8.3 Tunnelling by Freezing

Tunnelling in soft ground below the water table can also be done by the freezing method. Its essential feature is solidification by freezing of ground-water in the soil through which the tunnel is to be constructed, by circulating brine cooled to low temperatures. Circulation of brine results in the formation of an ice wall due to freezing of the water inside the soil pores. Even though freezing of vertical shafts by this means has been found successful, freezing a horizontal tunnel has rarely been tried, because of the complexity and expense of the operation. This method has been resorted to for some lengths in a recent London tube construction. Even though tunnelling by freezing is technically feasible in water-bearing sandy soils, the technique would be uneconomical for construction of a tunnel of the size required for a metro system, especially in tropical countries. The technical know-how and equipment for the process are also not indigenously available in developing countries, for example India.

6.8.4 Tunnelling with Chemical Grouting

Soil stabilisation to facilitate tunnelling is possible also by a special chemical grouting system. In this method sodium silicate, calcium chloride or silica gel is injected under pressure. Chemical grouting is not practicable, however, in soils with considerable variation in permeability of different formations which call for different grouting materials of different strengths, thus leaving several doubts about the effectiveness of the injection. The sequence of tunnelling using chemical grouting is indicated in Figure 6.7.

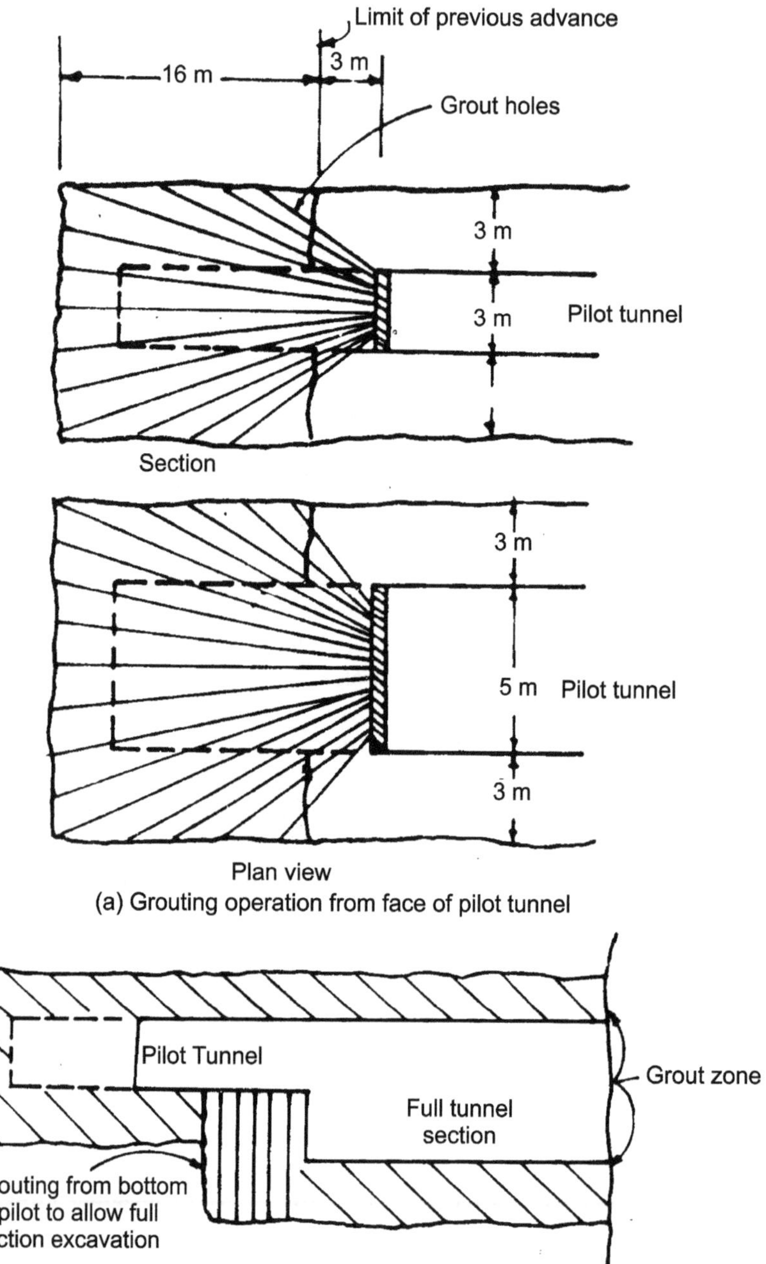

Figure 6.7 Sequence of Grouting Operations Using Pilot Tunnel to advance Tunnel Face.

6.8.5 Slung Wall Construction

This method is used in cut-and-cover construction without additional piling or diaphragm wall. It also avoids need for underpinning adjacent buildings.

The side walls are constructed for full depth in trenches as for diaphragm wall construction, the difference being that the trenched wall itself forms the side walls of the subway. Once the walls are completed, the earth between them is excavated to the underside of the proposed roof level of the subway (using struts to hold the walls laterally). Using ground as the form, the roof slab is cast and keyed into the side wall. After this, the earth below the roof slab and side walls is mined by working from the ends. The earth above the roof slab can be simultaneously filled (after the roof slab has gained sufficient strength) and the road surface restored. Once the excavation by 'mining' reaches base level, the floor stab is cast and well keyed into the side wall. This method was experimentally implemented for a short length in the USA but proved costly and hence has not been pursued. Its main advantage is reduction in interruption of road traffic.

6.8.6 Cut-and-Cover Tunnels

Wherever the permanent protective arrangements for continuing an open cut are expensive and fixing soil-supporting arrangements for boring the tunnel complicated and/or costly, that part of the length can be made an artificial tunnel by using the cut-and-cover method. This method was used for entire lengths of some tunnels on the Budni-Barkhera section (on Nagpur-Delhi line) of the Central Railway, where the soil was neither dependable for stabilisation in open cut nor considered easy to support and tunnel through. Similarly, in the loose sandy strata area on the Assam Rail Link Project the cut-and-cover method was used for a number of tunnels in 1947 and recently on a rail link project in Tripura and on Konkan Railway. This type of tunnelling is more often used for metro tunnels.

While establishing a tunnel face, the following items have to be attended to:

(a) Open excavation in overburden and rock or excavation of shaft from the bottom of which tunnel excavation can start;
(b) Arrangement for collection of surface water and its drainage by gravity or pumping;
(c) Access roads or rail tracks to mucking areas;
(d) Erection of winching and hauling equipment;
(e) Establishment of a field workshop, compressors, pumps, water-lines, ventilation fans and ducts.

In tunnels in shallow depth, where cut-and-cover method is adopted, diaphragm walls on either side are constructed using bentonite slurry. After these set, excavation from above can be done duly strutting the walls. This method is adopted more in subway tunnels and is discussed in more detail in Chapter 8.

6.9 SHIELD TUNNELLING[5,6]

6.9.1 Where used[5]

The shield tunnelling method is used in loose, non-cohesive or soft ground. It is mostly circular in shape. It can be driven either in free air or under compressed air. The former method is used when the depth of the tunnel is shallow and/or there is not much likelihood of ingress of water, as in clayey soils. The advantage of the shield tunnel is that it permits excavation of soil and erection of primary lining under safe conditions. It also provides better control of ground settlement from above as well as on the sides.

The shield tunnel was invented by Marc Brunel and his son who completed a 360-m long tunnel between 1825 and 1845. It was 11.5 m wide and 6.6 m high and was made in free air under the Thames River. The first circular shield was designed by Greathead who used it in 1896 to build the second tunnel in London stiff clay.

Use of compressed air was first proposed by Cochran. The first shield tunnel driven using compressed air appears to be the St. Clair tunnel between Port Huron in Michigan and Sarnia in Canada in 1898 using a 'Beach' hydraulic shield. This tunnel is 6.3 m dia in section and nearly 2 km long. Air pressure, as high as 2 atmospheres, was used.

6.9.2 Shield[6]

The equipment comprises a shield made of an outer envelope of steel plate, slightly larger than the outside dia of the tunnel lining. This is known as the 'skin' (Figure 6.8 (a)). In small shields the skin is stiffened internally by means of a diaphragm made of a steel plate with a large central aperture. In large shields it is braced by heavy framework of structural steel. A fixed cutting edge made of cast steel is mounted in front of the skin. This is pushed into the face and cuts into the soil and trims the excavation to the shape of the tunnel. A series of hydraulic jack (rams) mounted around the inside of the skin in the rear portion of the shield are used for pushing the shield forward as each successive length of the tunnel is completed. To aid in cutting harder ground, sometimes the shield cutting edge is coated with abrasion-resistant welding material.

In some shields a hood is provided in the upper half projecting one shove ahead of the shield. Under the protection of this hood, the excavators can advance and fix breast boards to prevent the soil in front from collapsing, i.e., to hold the face prior to the next shove.

During use in clayey soils the front is not closed and hence no breast boards are necessary. While using the shield in very soft clay, silt or fine running sand, the front face of the shield is generally closed by a steel bulkhead provided with portholes through which the soil can be excavated. These ports can also be closed with steel doors for control when necessary.

6.9.3 Accessories

The other accessories of the shield are:

(a) Shove jacks,
(b) Breast jacks,
(c) Table jacks,
(d) Erector arm,
(e) Tail seat and
(f) Hydraulic equipment.

Shove jacks: These are a series of hydraulic jacks mounted around the periphery at the tail end of the shield. Their capacity is generally about 70 t/m^2 of the face and they have a stroke 150 mm longer than the shove proposed. These jacks are set in stiffening rings of the body of the shield and bear against the last set of primary lining erected for the completed tunnel in the rear. They bear against jacking shoes resting on the face of the lining. The number of jacks used may vary from 6 for small shields to as many as 40 for large openings. Each jack may exert a force ranging from 10 to 100 t. Total shove effort may reach 1000 *t*.

Breast jacks: These jacks are mounted on the upper face of the shield to hold the breast boards (where used) bearing against the soil. They are held by and bear against suitable vertical posts fixed on the diaphragm. They retract gradually when the shield is advancing and concomitantly maintain pressure on the breast stands. Table jacks: These are used to support steel plates or platforms which are used as the work deck for giving the excavators a working platform or table and sometimes are used for holding the breast boards.

Erector arm: This is used for picking up the liner segments from the delivery trolleys or from the bottom of the tunnel in the rear and placing/

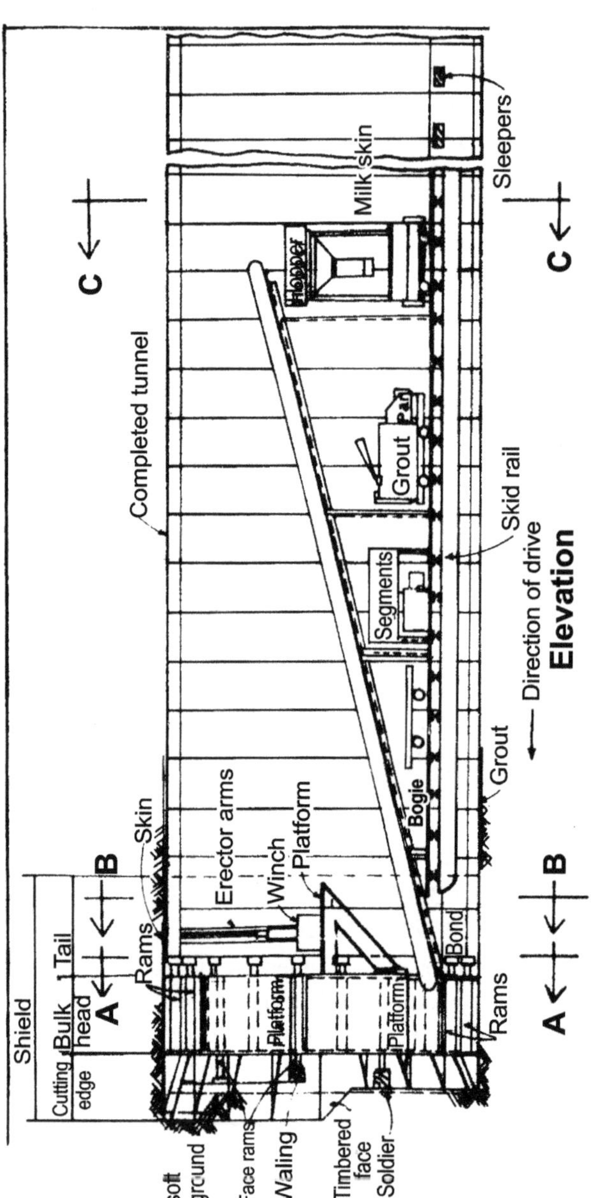

Figure 6.8(a) Shield Tunnelling Equipment.

(contd.)

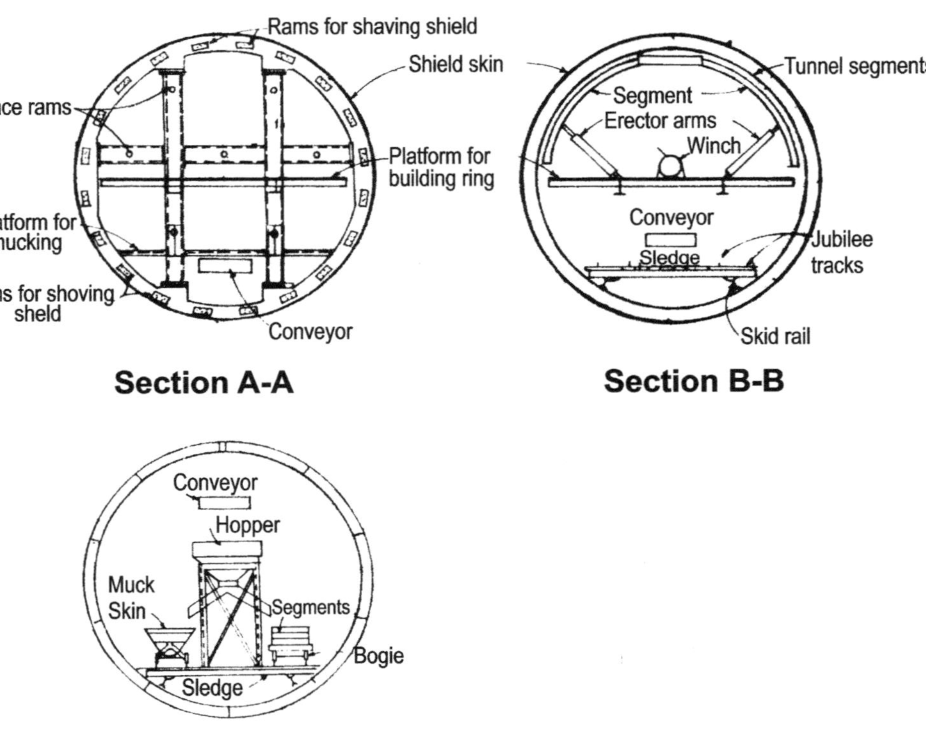

Figure 6.8(a) Shield Tunnelling Equipment.

pushing them into position. The arm holds the liners until they are bolted in place. These are mounted at the rear diaphragm of the shield just above the horizontal axis.

Tail seal: Rubber seals are bolted onto the inside of the tail at the rear. They prevent soil, gravel, grout or water flowing into the back of the shield. Precaution has to be taken against their getting 'frozen' into the grout.

Hydraulic equipment: All the jacks and the erector arm are operated from a central hydraulic system. The shove jacks are controlled by valves placed in the four quadrants, namely upper, lower, left and right. The breast jacks and table jacks as well as the erector arms have individual controls. The fluid used for the hydraulic system should be fire-resistant. The hydraulic system is generally operated by an electrically driven pump which has remote controls for starting, stopping and varying the pressure. The normal maximum working pressure is about 28 N/mm^2 but the system is designed for 2.5 times that pressure. In fact, when the shield does not move at 28 N/mm^2, a check has to be made to ascertain whether some major obstruction in the form of rock, boulder etc. is impeding the cutting edge.

6.9.4 Work Cycle[8]

The cycle of working the shields without the use of compressed air is:

(i) The ground is excavated ahead of the shield to a standard distance equal to the width of one section of the tunnel lining (usually between 0.5 m to 0.8 m). The lining is erected and properly fixed onto the excavated face.

(ii) The shield is jacked bodily forward, the thrust of the rams being taken by the ring of lining last erected.

(iii) The rams are withdrawn, leaving a space inside the tail of the skin for the erection of a new ring of lining (at no time should any ground be exposed at the rear of the shield).

(iv) The space between the excavated surface and the outer surface of the previous ring of lining left by the withdrawal of the skin of the shield is filled with cement grout injected through special holes in the lining.

(v) The next length of ground is excavated in the face and the process is repeated.

Figure 6.8(a) shows different stages of work with the shield. Figure 6.8(b) shows how in-situ concreting is done, in conjunction with shield work. This procedure will be equally applicable for use in other methods of doing tunnelling full face.

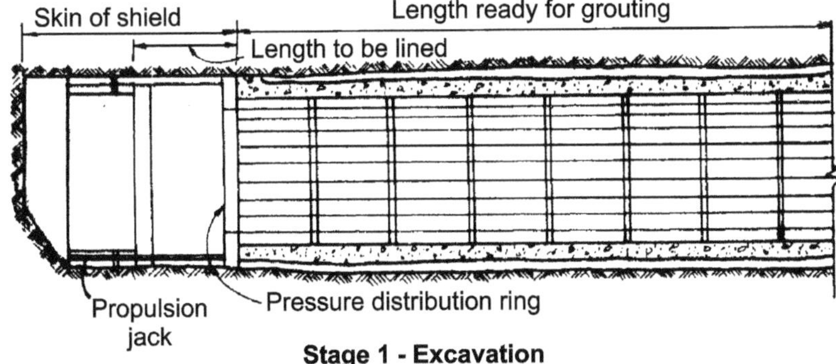

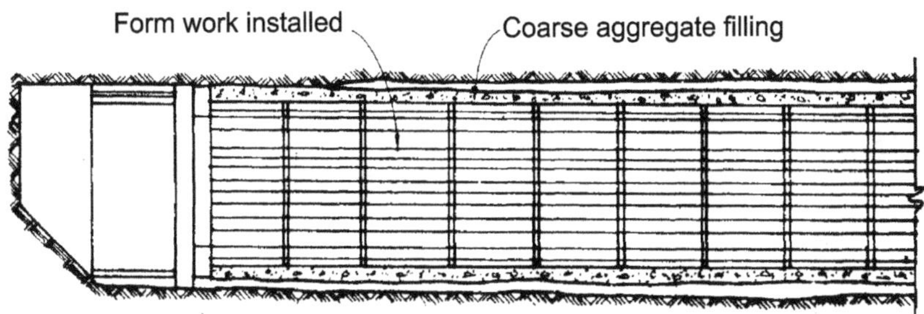

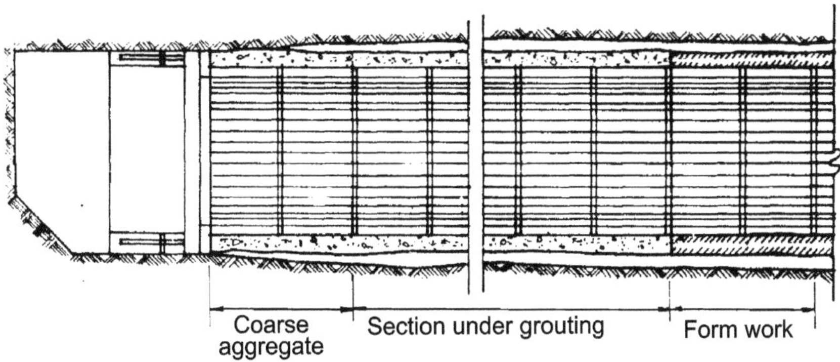

Figure 6.8(b) Construction Stages of Cast-in-situ Concrete Lining in the Shield Method[6].

6.9.5 Special Precautions

If at the beginning of the shove, one side or the other is advanced a small amount more by operating all the jacks on that side, there will be a tendency for the shield to turn slightly, either squeezing the ground on one side or leaving a void on the other. This can be avoided by advancing the shield using all the shove jacks uniformly and by keeping a surveyor on each side of the tunnel to closely monitor the lead distances as the shield advances and to ensure that it is moving uniformly in the desired direction.

Shield 'diving': Since most shields are nose-heavy, they have a tendency to dip, i.e., 'dive'. This has to be closely watched and controlled by using only the lower group of jacks when such a tendency is observed. Sometimes an extra jack is added at the bottom to correct this problem.

Curves are very difficult to drive with shields and hence the minimum radius of curvature of alignment while using a shield is kept as 175 m (10 degrees). The segments supplied for this will be in the form of tapered rings. Shield rolling: All circular shields have a tendency to roll; to prevent this, some are designed with removable fins on the sides. A simple way of getting over this difficulty is to fix a few of the shove jacks in slotted holes. In that case the cover plates can be removed and the jack forced to the end of the slot by wooden wedges. When a shove is made with this arrangement, it gives a spiral effect to the shove and prevents rolling.

Running sand and gravel: When driving a tunnel through (loose) sandy and gravelly soils, hoods must be provided for the shield and the face must always be breasted. While doing the excavation, the upper breast board will first be removed carefully and muck drawn into the tunnel. This board will then be advanced for one shove length and held temporarily by two trench jacks. Work is then similarly continued below by advancing the remaining breast boards one at a time. This operation is continued until the soldier beams are free and can be reset and held by breast jacks. The trench jacks can then be removed and the shield shoved ahead.

Tunnel Boring Machines have been developed, which work as sealed units, with facility for fixing Segmental CI or Precast segments for lining as the machine advances. Annexure 6.1 gives a brief description of working of a TBM.

In fine water-bearing sand, advancement using this method is slow and costly. Even if the overburden is not much, it is advantageous to use compressed air or to lower the water table by external pumping in such cases. Alternatively, the surrounding ground can be stiffened by chemical grouting. Such problems arise, as a matter of course, in subway tunnels, particularly in cites, and in coastal areas. A case study is given in Para 6.10 to illustrate how such problems can arise in a hill tunnel and how they have been dealt with.

6.9.6 Tunnelling by Bentonite Slurry Shield

The need for a tunnelling method to bore through non-cohesive soils without disturbing overlying property and preferably without exposing men to work in compressed air, has been felt for long. With the development of Shield Tunnelling, it has been possible to use bentonite slurry for advancing the shield for stabilizing the cut and clearing the cut soil as slurry mixed with bentonite and pumping it out of the tunnel. Some have been tried, such as the continental 'compressed air in the face alone' with variable openings in a full face with rotating cutter head etc., Such machines do not offer constant and regulated support to the working face. Safety of the tunnel and the overlying property relies heavily on the judgement of the machine operator if such machines are used. An approach to the problem of supporting the face while tunnelling in non-cohesive ground was recently developed in the United Kingdom and Japan. This new tunnelling technique is the result of the realisation that bentonite slurry which supports surfaces of an open trench in non-cohesive ground can be equally effective in supporting a tunnel face. The essential features of this methodology are presented in Figure 6.9.

6.10 CASE STUDIES

6.10.1 Challenges in Soft Ground Tunnelling on Konkan Railway Line[8]

This 760 km long Broad Gauge line on the West Coast of India, was constructed in the last decade of the last millennium. It passes along the west coast of India and for most part passes through the hilly terrain of Western Ghats. This long overdue link between Roha (near Mumbai) and Managalore in south substantially reduces the travel distances and travel times between many destinations in the north to south of the country, apart from opening up the area for economic development. It gives a more direct rail link to Goa, a major tourist spot, from both north and south. This happens to be the first BOT Rail project to have been completed in record time.

In the 760 km the northern 600 km length passes through a formidable hilly territory. In the total length of 760 km, there are 92 tunnels totaling a length of 83.6 km; the longest tunnel being 6.506 km long (Karbude) all in rock. It has 178 numbers major bridges (total length 20.5 km), longest Sharavati Bridge being 2065 m long.

Nearly 74 km of the tunnels is through hard rock, balance being in strata varying from soft rock to lithomargic clay. They generally were with hard porous laterite crust on top for 4 to 12 m depth followed by silty clay over

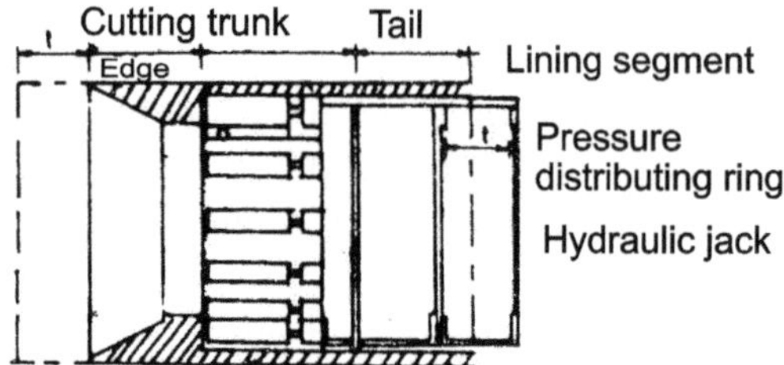

Stage I Excavation with Temporary Support of the front Face

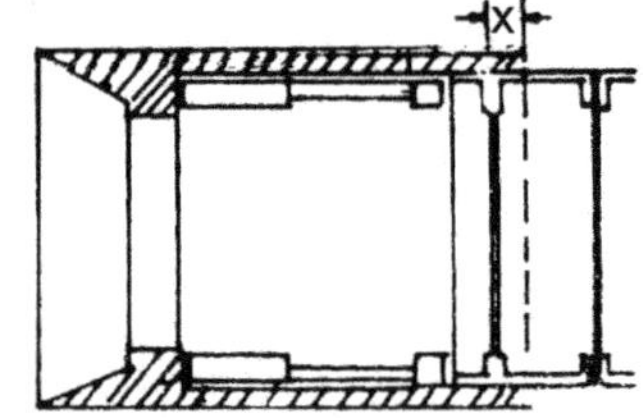

Stage II Advancing the shield

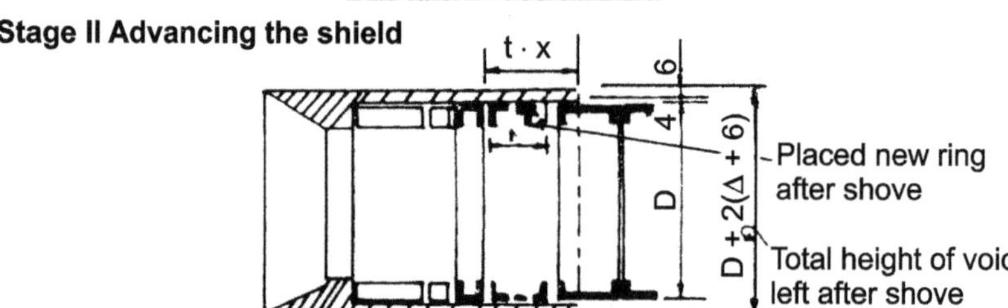

Stage III Placing another ring of the lining

(a) Shield tunnelling

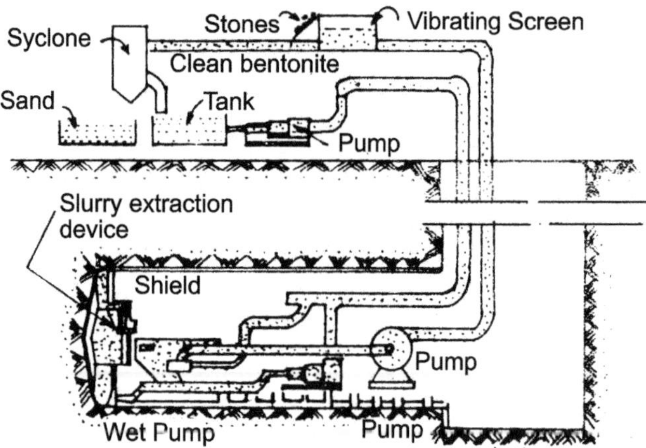

(b) Bentonite Shield tunnel

Figure 6.9 Bentonite Slurry Method.

lithomargic clay. During rains (which used to be heavy in that area), water would percolate down and the saturated clayey layer used to cause mud flow, resulting in creation of cavity extending upto the crust, which would cave in causing craters in many places. Of the hard rock tunnels 9 were longer than 2.25 km (being the longest of then existing tunnels on Indian railways).

6.10.2 Honavar Tunnel[8,9]

Of the soft strata tunnels, 7 tunnels aggregating to 3.8 km length were fully through soft soil. All except about 1500 m of a tunnel at Honavar were tackled by conventional heading and benching process. For the 1500 m mentioned, two 7.9m OD Shields were used to start with. The tunnel consisted of two stretches, one 300 m long and the other 1200 m long.

Work was started from both ends with two shields and had been planned to bore this tunnel through with the two shields. The type of shield used is known as 'Blade shield'. The ones used here were for providing tunnels with 7.2 m (inner diameter). Allowing for lining thickness and a play of 100mm, the inner diameter of shield was chosen as 7.9m. The hood of the shield had two circular bearing rings of 6.1 m ID and 7.7 m OD and fabricated with 20 mm and 25 mm plates welded to stiffeners inside. The rings are spaced at 3.448 m. 6.8 m long blades with sharpened front ends formed front blades and tail blades were 1.2 m long. Total number of blades was 29. Sixty ton capacity jacks with 600mm stroke length, one each attached to front shield and bearing against rear bearing ring, were used for propulsion into the tunnel face.

A number of problems were met with while carrying out the work. During the boring, large boulders were encountered, which had to be blasted and removed. In the process, there were some collapses resulting in cavity formation on top of shields. Such problems had to be overcome by grouting the soil from surface. Since such interruptions caused delay, in order to expedite work, it was decided to open up more faces in between and do work by conventional 'heading and benching' method, in addition.

Additional faces were created, one pair by additional Adit and one pair by sinking a 10m dia shield 20 m deep as shown in Fig 6.10 (a). Fig 6.10 (b) shows the schematic of an overhead collapse and how it was dealt with. Due to such problems and slow rate of progress, shield Tunnelling method was restricted and conventional heading and benching method adopted as indicated in Figure 6.10(a)

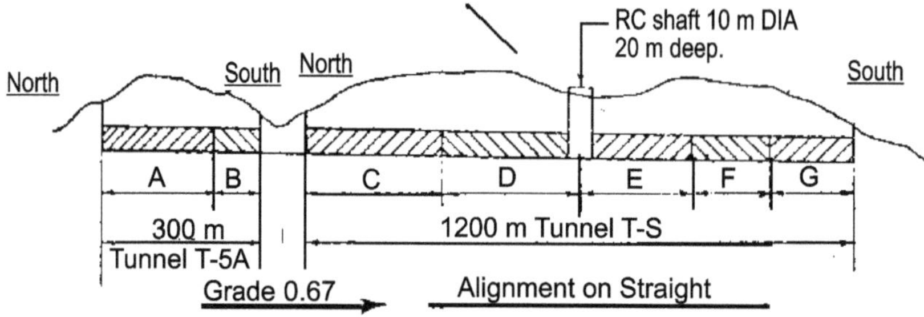

A, G-By Shield Method–524 m; B, C, Conventional Heading, Benching –306 m; D, E – Conventional Method from Shaft 511m; F-Conventional from South – 260m

(a) Tunnelling Strategies for Honavar Tunnel 5 and 5A Grouting Holes.

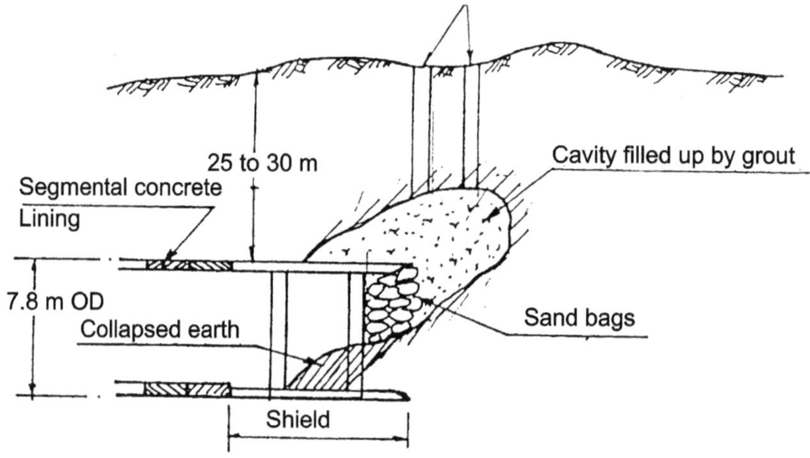

(b) Honavar Tunnel- Problem in Progress of Shield.

Figure 6.10 Tunnelling through Soft Strata at Hannover on Konkan Railway[9]

6.10.3 Alternative Strategies for Soft Ground Tunnelling[9]

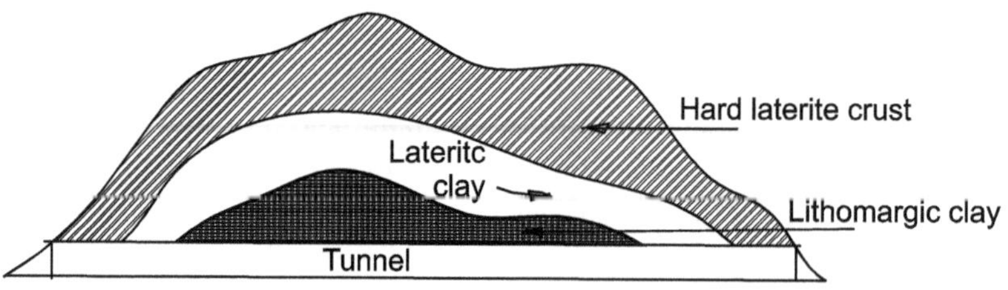

Figure 6.11(a) General Structure of Soil with Lateritic Crust[9].

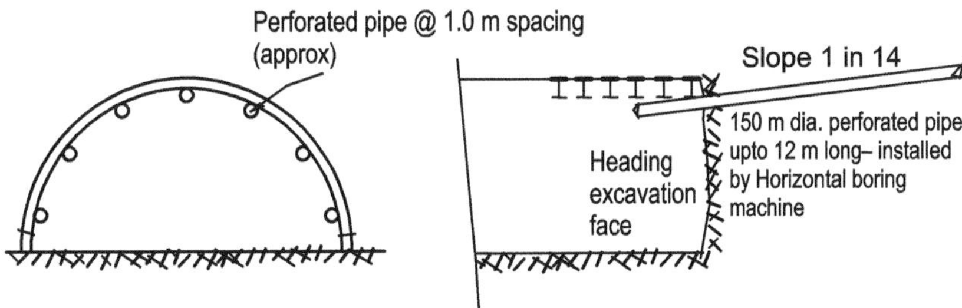

(b) Poling with perforated pipeswith horizontal boring machine

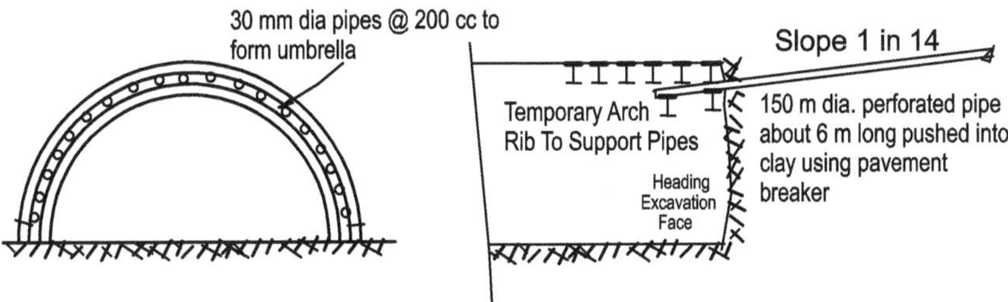

(c) Provision of Pipe Umbrella For Heading

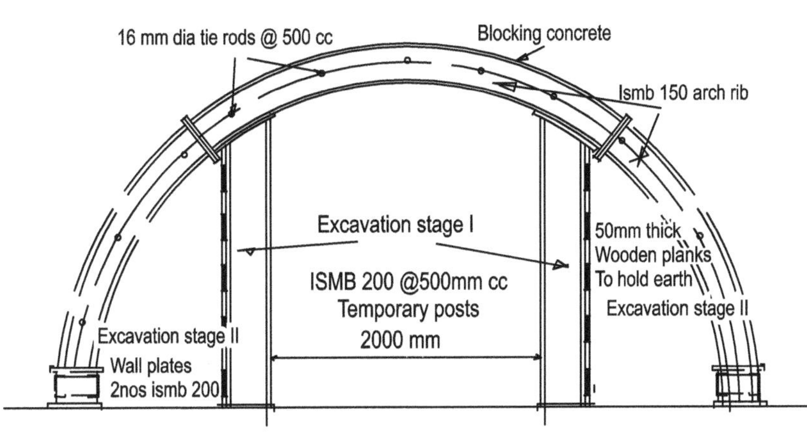

(d) Multi drift method of tunneling

Figure 6.11 Different Strategies adopted in dealing with Lateritic Soil and Clay[9]

Figure 6.11(a) shows the general strata structure in locations with lateritic rock crust. The problem of seepage of water, mud flow and/ or cavity formation and caving mentioned above were dealt with by four different strategies adopted as listed below:

(a) Installation of perforated pipes (150 mm dia) at the heading face for advancing it as shown in Figure 6.11(b). These pipes were driven through using horizontal boring machines. These suited in medium hard strata.
(b) Providing pipe umbrella to advance hold the heading using smaller (30mm dia) pipes at closer spacing (200mm), by pushing them through the clay with jack hammer. The rear ends were temporarily supported on temporary arch ribs. This is illustrated in Figure 6.11(c)
(c) Using multi-drifts as shown in Figure 6.11(d).

The cavities met with, shown in Figure 6.10(b) were filled by placing concrete in them. For passing through cavities, nine blades of the shield were extended by upto 2.5 m. For transition between the extended and non-extended portions, four more blades were extended partially. Conventional method of filling rock spalls in cavity did not work. In order to combat this problem, a number of 100mm dia holes (upto 4) were drilled through the overburden and a grout formed with a mixture of stone chips, sand, bentonite and cement was poured through the bores. As such mix was found to be costly, later a grout consisting of cement: clay: fine sand in ratio of 1:8:8 was used successfully[8].

6.11 REFERENCES

1. Agarwal, M.M. and Miglani, K.K., (2014) 'Global experience of design, construction & maintenance of Railway Tunnels with special reference to India', Proceedings of National Technical Seminar, on Management of P. Way Works through need based Outsourcing and Design, Construction and Maintenance of Railway Tunnels, Jaipur 2014 pp315-356.
2. IS: 5878- (Part III)- 1972 ,'Code of Practice for Construction of Tunnels: Part Ill Underground Excavation in Soft Strata', Bureau of Indian Standards, New Delhi.
3. IS: 5878- (PART IV)- 1971- ' Construction of Tunnels- Part IV- 'Tunnel Supports' , Bureau of Indian Standards, New Delhi.
4. Silas H. Woodard (1947), 'Dams, Aqueducts, Canals, Shafts, Tunnels' in Ed. Thaddeus Merriman and Thomas H.Wiggin, American Civil Engineers handbook, John Wiley Inc. New York, 5th Edn. 1947, pp 1614-1652.
5. FHWA Technical Manual for Design and Construction of Road Tunnels and Bridges, Civil Elements Chapter 7., USDOT, Federal Highway Administration
6. Bickel and Kuesel, (1982) Tunnel Engineering Handbook , Von Nostrand Reinhold Company, New York ,1942 pp 93-120, 417-444

7. Pequinot, C.A., (1963) "Tunnels and Tunnelling'- Hutchinson, Scientific and Technical, London
8. Raju, C., Narayanan, G., and Kurien, A.P., 'Soft soil tunnelling using shield for Honavar tunnels'. Indian Concrete Journal, February 1994, Associated Cement Companies, Mumbai
9. Limaye, S.D., (1997), 'Konkan Railway Project: Tunnelling -Important Aspects', Structural Engineering Convention, Mumbai. (Unpublished)
10. Vipul Kumar (2014), 'Managing Surprises in Tunnel Projects', Proceedings of National Technical Seminar, on Management of P.Way Works through need based Outsourcing and Design, Construction and Maintenance of Railway Tunnels , Jaipur 2014 pp 247-257.
11. Ankur Jain, (2013), 'Pir Panjal Tunnel', Paper presented in Course 824, Indian Railway Institute of Civil engineering, Pune, IRICEN web site.
12. Rastogi, V.K., 'Instrumentation and monitoring of underground structures and metro railway', World Tunnel Congress 2008 on Underground Facilities for Better Environment and Safety- India

ANNEXURE 6.1

TUNNEL BORING MACHINE

General

Tunnel Boring Machine is a compact system with the Cutter head with or without short shield leading and the mechanical thrust system (jacks, driving unit etc.,) conveyors for the muck removal and control unit linked to it in rear. They move as a unit. The cutting is done by the rotation of the cutter head. Thrust is provided by hydraulic jacks acting against gripper shoes mechanism provided against the tunnel walls of the completed portion in rear in hard rock or against the concrete rings. TBM is also called a 'mole',

TBMs of different types are available for use in all types of soils varying from soft strata to hard rock fro tunnels of circular cross section. Their main advantage is that they limit the disturbance to the surrounding ground. It, thus is eminently suitable for tunnelling through heavily built up urban area. It saves the cost of lining in hard strata as it leaves a smooth tunnel wall behind. The segmental lining installation is quick resulting in saving of time. The main disadvantage is its cost (as for example TBMs being used in Chennai cost about Rs. 90 crores each, as in 2013). First such boring machine is reported to have been built in 1846 for a rail tunnel across Alps but not used due to a Revolution. One such was built in USA in 1853 with Cast Iron as a Stone Cutting Machine but failed after drilling about 3m length. First successful use of tunnel boring machine is reported to be in 1952 in Oahe dam by James Robinson. Since then they have been improved upon and are now available for boring holes of diameter varying from 1 metre to 17.5 m (earth balancer type) developed in 2013 for Highway 99 tunnel project in Seattle. Largest hard rock TBM is for 15.62 m diameter, (weighing 4500 tons, made for Sparvo gallery of Italian Motorway Pass A1. (Wikipedia-March 2014).

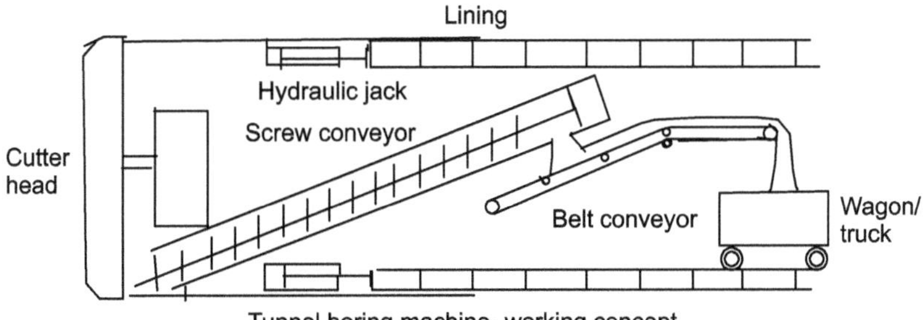

Figure A 6.1.1 Schematic View of a Tunnel Boring Machine in Soft Ground.

Types of TBM

There are basically five types of TBM, two for Rock and three for Soft Ground Tunnelling as indicated below:

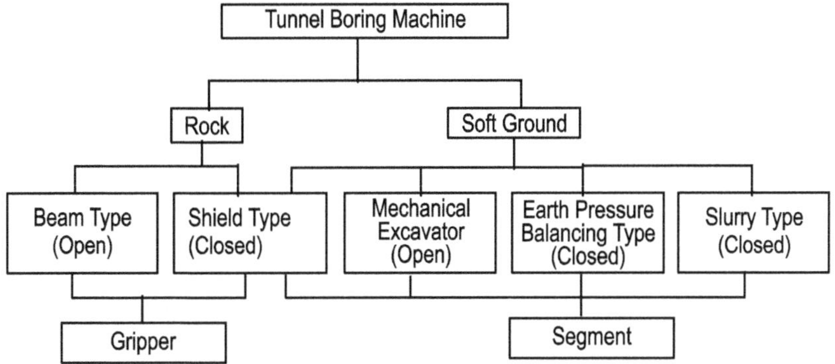

Source : Google Image

Figure A 6.1.2 Types of Tunnel Boring Machines

Working of TBM

Hard Rock

Hard rock TBMs can be open or shielded type. In either case, the excavation is done by 'disc cutters' mounted on the cutter head. The thrust applied by the machine on the cutter head makes the 'disc cutters create compressive stress fractures in the rock' at the 'tunnel face' resulting in chipping away of the rock and same fall through the openings in the cutter head. It is collected and is carried through an overhead conveyor system to the muck cars for removal from the tunnel. In hard rock, open type is used, in which the rear of the cutter head is left open for the exposed rock to provide the support. Advancing of the machine is done by a gripper system, in which a pair of gripper shoes is pushed on side walls of excavated tunnel for them to get support. During forward stroke, the jacks acting against the grip bars cause the compression required for cutting the rock and for the cutter head to move. At the end of the stroke, 'rear legs of the machine is lowered and the grippers and propel cylinders retracted'. Then the gripper assembly is moved forward and repositioned for the next cycle. Erection of concrete segments is not done in the open/ beam type TBMs. If required, the exposed rock can be supported using ground support systems like, ring beams or Ring steel, rock bolts, shotcrete with or without steel straps and wire mesh. See Figure A 6.1.1.

Shielded (closed) TBMs are used in fractured or softer rocks, which require installation of concrete segments to support the unstable tunnel walls. In this case, the pushing and advancing is done using thrust cylinders acting against the installed tunnel segments behind. In this there are two types, single shield TBM and Double shield TBM. The former can push forward only by thrust cylinders acting against the lining segments and hence are used in fractured ground only. The Double shield type has both capabilities viz., use of grip against tunnel walls or use of thrust cylinders. Hence it can be used where there is a mix of the two types of rocks met with.

Soft Ground

There are three main types of TBM for soft ground Tunnelling viz., open face excavation type; Earth Pressure Balancing type; and Slurry type (Figure A 6.1.2). The latter two types work as closed Single Shield TBMs and the cutter head is advanced by pushing off against segments. The open cut one also uses a shield for maintaining the profile and advancing but does not use a cutter head since these machines are used in ground, face of which can stand without support for a short period of time during excavation. Excavation is done with a back actor arm (or a cutter head) upto about 150 cm of the front edge of the shield. As the shield is advanced forward, the cutter in front cuts any remaining ground to obtain the circular profile. Ground support to the excavated tunnel is provided by erection of precast concrete or Spheroidal (Graphite Iron) segments. This type of TBM can be used for boring through softer rocks of upto 10 MPa strength.

Earth Pressure Balanced type is the most commonly used TBM. In this the cutter head uses a combination of 'tungsten carbide cutting bits and carbide disc cutters. As the name suggests, as the cutting head is jacked against the earth face, the earth and any other pressure coming on the face is balanced by applied pressure and additional thrust and rotation of the cutter head causes the excavation of the face and excavated soil flows down through the openings in the cutter behind it. That soil is scooped and conveyed by a screw conveyor to the overhead conveyor for ultimate removal from the tunnel. As soil is excavated, the cutter head is advanced

There is an automated system through which the operator can maintain the rate of machine advance with the rate of soil removal. The soil in front is stabilised, as necessary, by injecting chemical grouts, polymers, bentonite or foam to suit the type of soil, through hole provided in the cutter head for the purpose. Any pre-existing bore holes in the area of influence of the tunnel will have to be sealed well in advance. Suitable precautions have to be taken for avoiding or minimizing any settlement of ground in the area of influence.

In soft grounds with high sub-soil water and very high water pressure, it will be necessary to use Slurry Shield Type TBM. Working in these has to be in a completely enclosed environment. The soil in front is stabilised by mixing it with bentonite and as the shield advances, the soil mixed with bentonite will flow through the cutter head. It is collected and sucked/ pumped and removed through a system of slurry tubes to the outside of tunnel. There, large slurry separation plants have to be available on surface for separation and removal of muck and the reclaimed bentonite slurry is recycled back into tunnel. Where the water pressure is very high, pneumatic pressure may have to be applied ahead of cutting head and system of locks for men and for muck removal will have to be provided behind cutter head. In this case, various stipulations and precautions will have to be taken in selection of workers, their health condition, reduced duration of work etc..

Special Types of TBM

Of late, site specific TBMs are being developed with the advancement of the technique. For example, recently a special type of TBM was developed in Malaysia for use in two different kinds of ground. At one end of tunnel, normal soft ground is met with and the tunnel has to pass through a lime stone deposit, in which it has to cut through lengths with hollows in the rock formation.

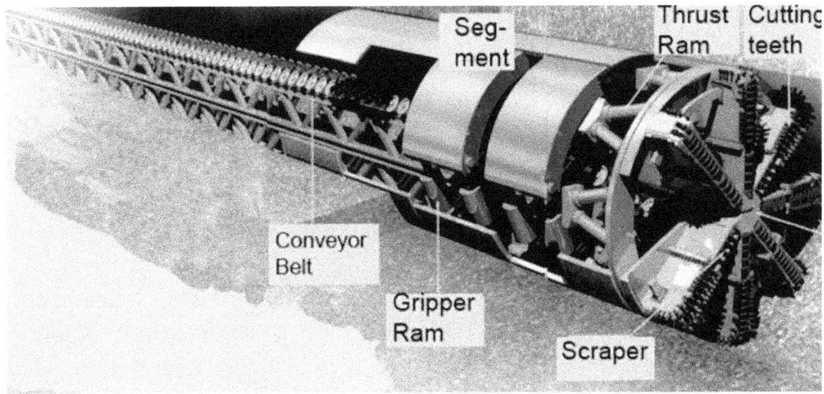

Source : Google Images

Figure A 6.3 View of a TBM and Parts.

ANNEXURE 6.2

CASE STUDY : Example of a Railway Tunnel With Service Road done using New Austrian Method

The 345 km Udampur- Baramulla rail line project in the Jammu-Kashmir region on the Himalayas can be considered the most difficult rail line project in mountain range carried out by Indian Railways since Independence. Of this, the most and challenging length is the 129 km line, cutting through Pir Panjal Mountain range, including 62 bridges and 35 tunnels (aggregate tunnel length being 103 km). Pir Panjal Tunnel is the longest Vehicular Tunnel in India as of 2014. It is 10.96 km long on a straight alignment. Situated at an elevation of about 1700m above MSL, it is about 440m below the Highway Tunnel passing through the same range. It is less likely to go under snow. The work was started in 2006 and took about six years to complete; the tunnel was commissioned in June 2013.

Geological Features[10]

Geological investigations consisted of consultations with the expert geologists and supported by drilling exploration bore holes as follows: 6 bore in 2003 in RITES investigations; 15 bores upto 225 m in 2004 RITES detailed investigations; and 7 bores in 2004 upto 640 m depth as part of MECL investigations. It can be seen that the deepest drill hole was 640m as against normal practice of about 300m. A layout of the bores can be seen in Figure A.6.2.1

It has been found that the tunnel passes through different types of rock and soil, consisting mainly of 'silicified lime stone, andesite, quartzite, sandstone and lime stone'. There are shale intercalations, agglomerates, shale and tuffs are also met with. The portal areas are in soft ground, situated in 'fluvioglacial sediments'. Trap and Quartzite is met with in the middle length.

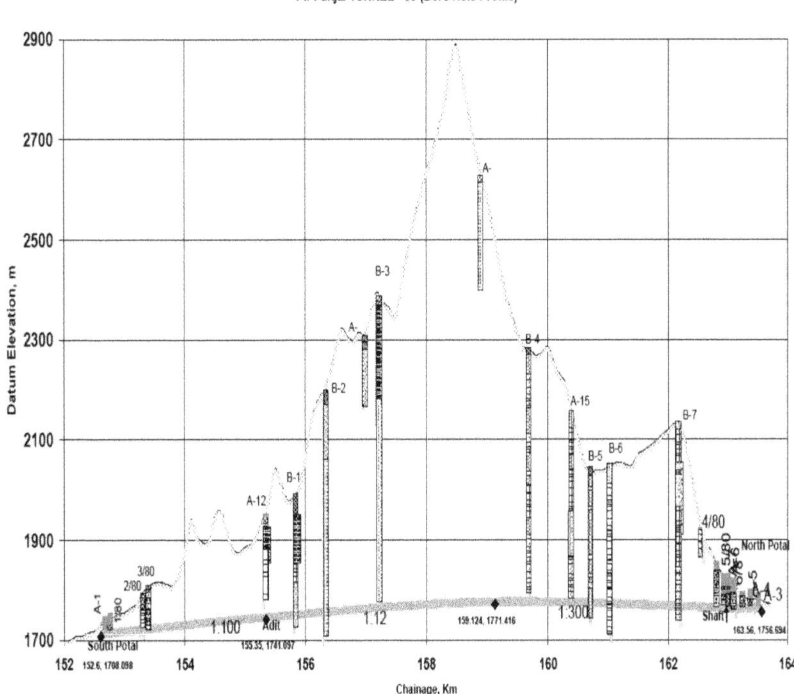

Layout of Invstigation Bore Holes for Pir Panjal Tunnel

Source: Reference 10

Figure A.6.2.1 Layout of Bore Holes for Pir Panjal Tunnel Construction.

The central areas of the range have a 'distinctive folding' and a number of fault zones have been encountered. General direction of strike in mountain range is NW- SE while the tunnel runs almost North- South. With an overburden of upto 1100m, the tunnel bore was anticipated to be subjected to high squeezing, apart from heavy water incursion, in the lime stone area. These conditions have had a major impact on the choice of tunnelling methodology.

Salient Features of the Tunnel[11]

The tunnel section has been designed to accommodate a Single line BG track and a 3m wide road by the side running for the full length for the purpose of maintenance, attention to emergency rescue and relief measures. It is lined. It is in arch form with maximum width of about 8.40 m and height of 7.39 m. Figure A 7.2.1 shows a typical section with the lining. Finished area is 48 sqm, while excavation area varied from 67 sqm to 78 sqm.

Proceeding from south, it has a rising gradient of 1% for about 6720 m from the South end and a falling gradient of 0.5% beyond to the North end, (summit being at Km 159.134).

Apart from the two portals, it has a 774 m long Adit (meeting tunnel at 2750m, isolating the soft ground) and a 12m dia 55 m deep shaft at 600m from the North end (isolating the north end tunnel). They provided additional working faces and now serve for ventilation. The Adit gives an alternative approach for use in emergencies and for maintenance. A total of about 1 million cum of underground excavation was involved. Provision has been made for ventilation, fire fighting and safety monitoring and lighting for the full length. Extensive instrumentation was done during execution for monitoring and also updating design parameters and revision as required.

Tunnelling Methodology:

Use of TBM was ruled out for this tunnel for following reasons, even though it would have been faster:

- The section being non-circular, with the use of TBM, expansion of section requiring enlargement of the section calling for use of other means.
- The heterogeneous geology varying from the portals to the middle and the need to go through a number of fault zones, requiring changes in regulation of applied pressure and types of cutting tools.
- High squeezing of rock anticipated and also heavy water flow likely in the limestone area
- Retrieval of TBMs working from two ends in middle poses problems like providing caverns etc.
- TBMs having to be designed to suit varying geological conditions, import and transport up the hill would have required a very long mobilisation time.

NATM (New Austrian Tunnelling Methodology) was the next best choice for the section chosen and for the varying geological conditions met with. Soil excavation has been done using a road header and 'drill and blast' methods.

The sequence of tunnelling work was as follows:

Excavation: Eight classes of rock mass have been found. Where the rock mass class was found excellent, very good and good, full face excavation was done; depending on the composition, excavation was done by heading and benching. For further inferior classes considered poor and very poor ones, excavation was done by segmentation. At the excavation face, the face was 'divided into small cells that would help the ground to stand until completion of the lining'. This primary lining comprised of shotcreting with steel wire mesh embedded in it. The opening was gradually widened in steps. Depending on the geometry and size of the tunnel section, number of such steps was decided. Figure A.6.2.2(a) shows how they were segmented and sequenced when work was done in soft ground. Work sequence can be understood from the table of cycle time (in hours) for different pull lengths given below:

Description of task	**Cycle time for 2 m pull (in hours)**	**Cycle time for 1.25 m pull (in hours)**
Drilling and excavation	1.7	0.9
Charging, blasting, defusing	2.7	1.5
Diffusing and mucking	1.1	0.8
Fix wire mesh and lattice girder supports	2.4	2.3
Shotcreting	0.9	0.8
Rock bolting	1.8	1.0
Preparation for next face	0.3	0.3
Maintenance	1.6	1.6
Forepoling	1.6	1.7
	14.1	10.4

Source: Rahul Sinhal from 'Academia' web site

Figure A.6.2.3 (b) shows the work of primary lining as the arch is widened upto spring, the roof being supported with a lattice girder type of rib and A.6.2.4 shows fixing of a rock bolt.

Design of the Support System

The support system design is integrated with the deformation characteristics of the ground around. It is done using 'rock support interaction' diagrams. Geotechnical classification of rock mass obtained initially and those evolved by observations during construction and continuous analysis of the rock pressure and movement of primary lining were used to obtain these . For this purpose, instruments like load cells on rock bolts, strain gauges and extensometers were fixed at intervals and continuous monitoring and measurement of actual stress levels and movements were noted. (Details of these instruments and their locations are briefly discussed in chapter 11.). The support system has two components, viz., primary which is done as soon as possible after excavation and the permanent one in form of final lining. Details of latter are not covered here. The former is designed based on preliminary observation of the rock mass and the final one after assessing the stress levels and movement pattern. It is done after the movements of the

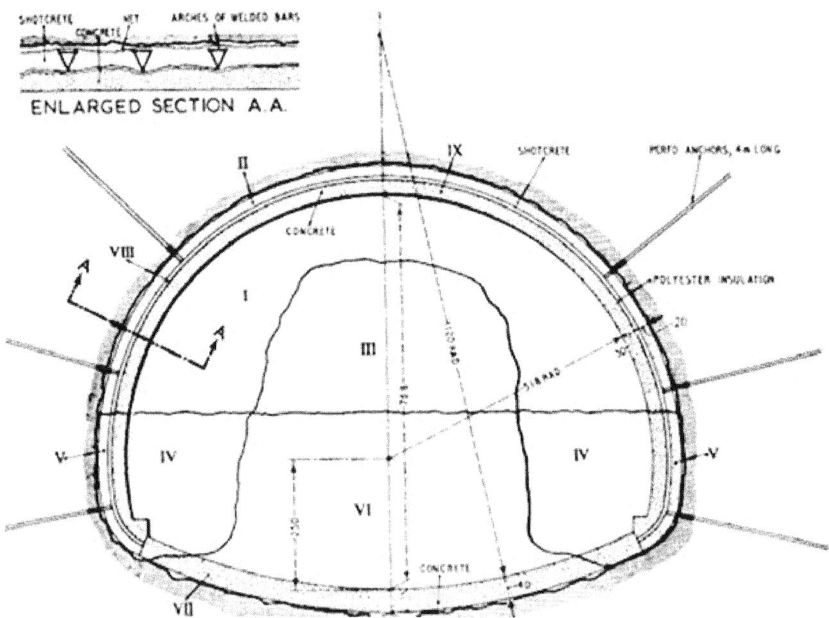

Figure A.6.2.2 Stages of Excavation[11]

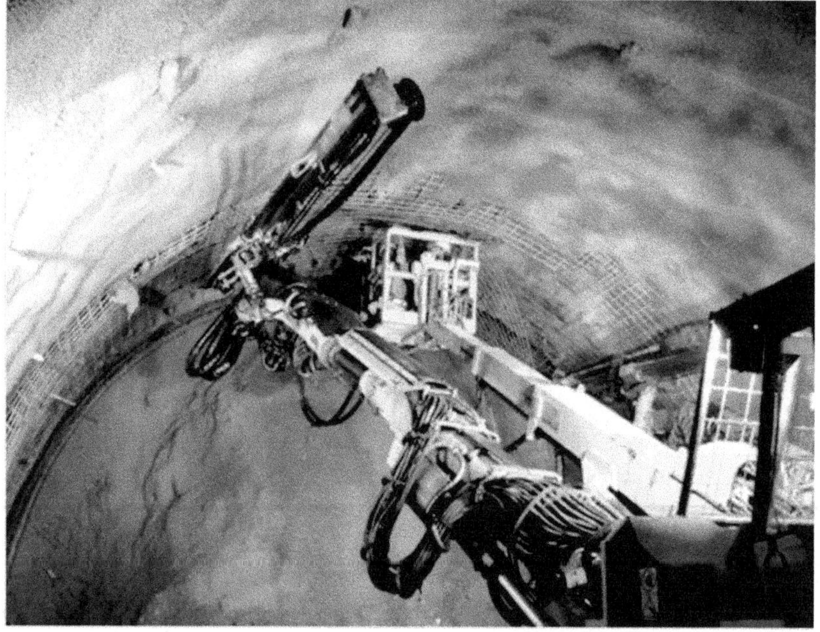

Figure A.6.2.3 Shotcrete and Lattice girder Support[11]

Source: Reference 11

Figure A.6.2.4 Drilling for Rock bolt

excavated surface stabilises and reaches a steady state or is within prescribed limits. Table 6.1 shows details of Primary support system provided for different classes of rock / soil met with.

Table 6.1 Elements of Support system and Tunnel advancing used for different Rock mass classes

Rock Class	Sealing Shotcrete	Fore pole	Wire mesh	Shotcrete mm	Lattice girder	Rock bolt
I	√	–	–	50	–	–
II	√	–	–	100	–	√
III	√	–	2 layers	150	–	√
IV	√	√	2 layers	250	√	√
V	√	√	2 layers	300	√	√
VI	√	√	2 layers	300	√	√
VII	√	√	2 layers	300	√	√
VIII	√	√	2 layers	300	√	√

√- Yes provided/ used.

Some important actions suggested to be taken as part of NATM system by different authorities[5] are:

Inherent strength of rock or soil around (tunnel domain) has to be preserved and mobilised to the maximum extent possible; Such mobilisation is achieved by controlled deformation of the ground; it is achieved by the initial and primary supporting systems by way of shotcrete and systematic rock bolting/ anchoring plus semi flexible sprayed concrete.

The closure of ring (i.e., invert) has to be appropriately timed, which would vary with rock and soil conditions

Tests (and measurements) and monitoring of deformation of supports and ground and forces have to be done

Length of unsupported span should be as short as possible

All people involved in the work of design, execution and supervision will have to clearly understand the approach to this system, accept the same and react to and act in close co-ordination anticipating any problem arising.

Method of Excavation and Machines used for different Lengths.

Table 6.2 Shows the split of mode of tunnelling done in different lengths of the tunnel

Description of Segment	Length in metres	Main method
Main Tunnel Zone VA	6130	Drill & blast plus soft ground by Hydraulic excavator in SEM
Main Tunnel Zone VB	4245	Drill & blast
Main Tunnel Zone VB to North end	585	Hydraulic excavator
Adit, 12m dia Shaft 55 m deep and cross passage	863 m (approx)	Drill & blast.

Tunnel Instrumentation

The tunnel has been provided with RCC permanent lining. The initial investigations and study of the soil/ rock properties helped in decision on profile of the tunnel, tunnelling methodology to be used and design of primary supporting arrangements (shotcreting, mesh fixing , rock bolting, and provision of trussed girder arch and vertical supports). The stability of the tunnel and forces exerted by the enveloping soil/ rock had to be monitored for not only taking any additional measures to maintain integrity of the profile, but also for designing the permanent support and also to decide when the movements of the profile is such that the permanent lining can be provided. '3-D optical deformation sections at spacing of 5 to 25 m with prism targets around the perimeter needed to be installed for the purpose.'

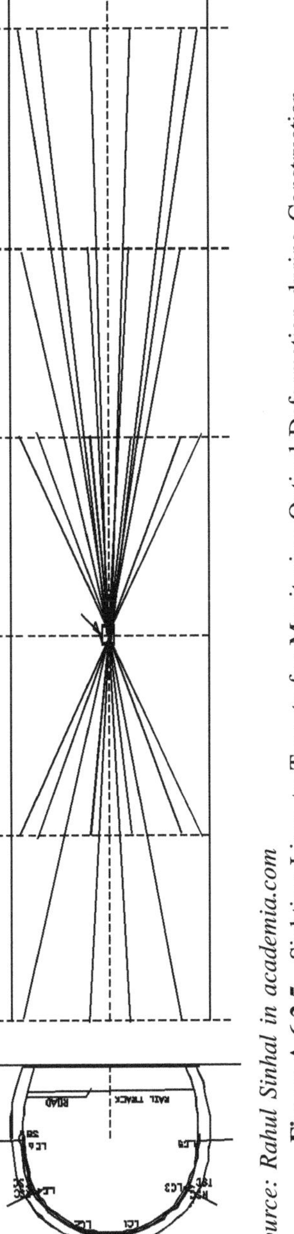

Source: Rahul Sinhal in academia.com

Figure A.6.2.5 Sighting Lines to Targets for Monitoring Optical Deformation during Construction

A typical arrangement how monitoring of optical deformation system was done is shown in Figure A.6.2.5.

In addition, other instruments used were convergence measuring devices, shotcrete pressure cells and strain gauges, load cells on anchors and extensometers as required. This was a continuous process. Typical arrangements are discussed in the Chapter 11 on 'Instrumentation of Tunnels'. On completion of the tunnel, the behaviour of the tunnel and components are monitored by installing load cells, strain gauges etc., at sections in regular intervals. The spacing may be 500 to 1000 m for a long tunnel like Pir Panjal tunnel. A typical arrangement is shown in Figure A.6.2.6

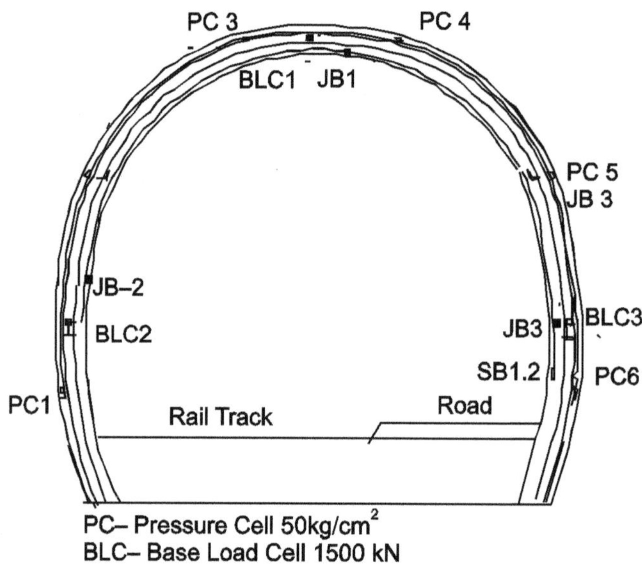

Source: Rastogi, Reference 12

Figure A.6.2.6 Typical Arrangements of Instrumentation for Tunnel Monitoring

Tunnel Ventilation[3]

The tunnel ventilation system has been provided for this tunnel after considering a number of alternatives and detailed studies. The design provides for quick clearance of emissions that would arise in either direction (North or South bound) and at different gradients for a 5000 ton train. Based on the calculations, 25 numbers of jet fans (each having 1500 mm diameter) have been provided in the main tunnel positioned in five groups of 5 fans each. Minimum distance between the machines has been kept at 150 m. Three numbers 1500 mm dia jet fans have been proposed in the access tunnel. 2 numbers axial fans are proposed to be installed in the Shaft

Emissions during train moving uphill would control the design. It was found that the emission level of NOx and particles for the system provided would cross the threshold limit in some segments of the tunnel. This would require more time for the tunnel to be cleared of gas before the next train can enter. In such a situation, the 25 fans would have to be operated simultaneously in the Northern direction generating an air velocity 5.7 m / sec. so as to accelerate clearance and to bring down waiting time for a train. To meet with complex situations in operation, a stable SCADA has to be used for control.

Figure A6.7 Shows the layout of the tunnel, adit and shaft.

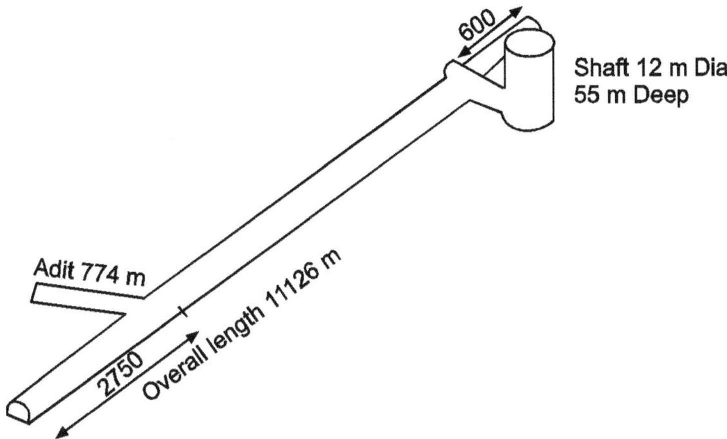

Figure A 6.7 Layout of Piar Panjal Tunnel, Adit and Shaft

CHAPTER

7

Drilling, Blasting and Mucking

7.1 GENERAL

Blasting is resorted to while dealing with hard rock and soft rock of compact nature. Special care has to be taken while blasting. The blasting operation has to cover just the profile and the tunnel has to advance progressively. It has to be done in such a way as to minimise shock, vibration and any other disturbance to the rock structure around the profile of the tunnel. It should be such as to ensure minimum 'overbreak', i.e., the extra section over and above the minimum section required to be removed (see Figure 7.1). This is particularly important when the tunnel is to be lined. Otherwise, extra lining material, an expensive item, will have to be used. The basic principle is to loosen the volume of virgin rock in such a way that when it is removed, the length of the tunnel has advanced in the correct direction, and conforming as nearly as possible to the correct cross-section.

Hence preparation for the blast is a vital factor that affects the speed and economy of tunnel driving. It is therefore meticulously planned with the fullest regard for the characteristics of the rock to be removed. Details of drilling pattern, depth of holes, quantities of explosives to be used and sequence of detonation are to be suitably worked out. They should also be amenable to revision, based on the performance and changes in the structure as driving proceeds. Thus it is apparent that blasting for the entire section is not done in one go; rather, a sequence is adopted in detonating various holes, enabling delayed detonation arrangements.

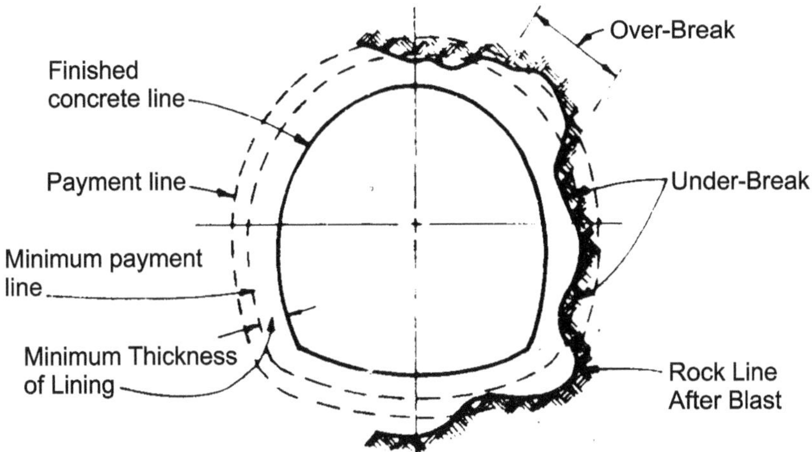

Figure 7.1 Trim and Payment Lines.

7.2 THE ROUND

The work of drilling and blasting rock is done in cycles known as 'the round'. It comprises of: (a) drilling, (b) blasting, (c) smoke time, (d) mucking and (e) supporting[1].

To maximise use of time, while drilling is in progress, water lines, air lines, ventilation and power lines should be extended. Normally the number of men required for the drilling operation will be more than that for the mucking operation. If it is proposed to utilise the same men for both operations, it is advantageous to use the extra men for transporting the utilities during the mucking period. If the same men cannot be used for both operations, it is economical to work in two rounds in adjacent lengths at a time, so that drilling and mucking work can be done alternately. Smoke time is the time needed for clearing/reducing smoke to a safe level for men to work. This clearing is done by blowers for exhausting gas through ventilation lines of the heading or by exhausters. The latter are more effective.

7.3 DRILLING

7.3.1 Drilling Pattern

The drilling pattern is very significantly important. It has to be designed such that the holes are easy to drill; the total quantity of explosive consumed should be held to the minimum (taking into consideration the overall work and not individual holes); and the periphery of the space left after the blast should conform as nearly as possible to the required tunnel section after allowing for supports and lining wherever necessary.

The terms 'underbreak' and 'overbreak' refer to deficient section and extra section and can be understood by reference to Figure 7.1, which indicates also the 'trim line' (finished concrete line) and 'payment line'.

7.3.2 Classification of Drill Holes

Holes are drilled by using pneumatically operated rock drills with wet drilling. The drill holes are divided into three classes: (a) cut holes, (b) easers and (c) trimmers. The section given in Figure 7.2 indicates direction of drilling the various holes. In the case of larger sections, further refinement can be made by adding secondary easers, secondary trimmers, lifters and blank holes.

Cut holes are also known as 'breaking-in' shot holes. Cut holes are usually in the centre of the drilling pattern, 150 to 300 mm deeper than the other holes and converge towards the centre of the section away from the face. They are to be charged first in the sequence of charging. While detonating they will remove a cone or wedge of material in the centre initially, which in turn will serve as a free surface for 'breaking off' when successive holes are detonated.

The location of cut holes is determined by the following considerations:

(a) They should be as low as possible in order to restrict the throw of blasted Rock

(b) They should be placed in the centre when the choice of different delay intervals is limited;

(c) In narrow tunnels one may try cut holes placed on the side to gain maximum advance.

The types of cuts are described below

(i) *Horizontal wedge cut:* The holes are placed symmetrically with respect to the vertical centre line. They are all horizontal and inclined towards the centre. They are easy to drill but it is not possible with wedge cuts to advance more than half the width of the tunnel (Figure 7.2(a)).

(ii) *Pyramid cut:* Three or four holes are so inclined both vertically and horizontally that they are directed to converge towards a point further on towards the direction of the tunnelling (Figure 7.2(b)). This cut is suitable for small-size tunnels.

(iii) *Fan cut:* This type is used when the holes are directed towards one side of the heading of the bore and laid out in a fan shape (Figure 7.2(c))

(iv) *V-cut:* Similar to the horizontal wedge cut and suitable for tunnels of very large size (30 to 100 m^2 area) (Figure 7.2(d)).

(v) *Michigan or cylinder cut:* Refers to the pattern of a hole of large dia (70 to 100 mm) drilled in the centre and surrounded by a series of smaller holes forming a geometric pattern (two pentagons or two triangles). All holes are drilled horizontally. The larger holes are not charged while the smaller ones are loaded and blasted in such a way that the holes close to the central ones are charged first (Figure 7.2(e)).

(vi) *Burn cut:* Similar to the Michigan cut. All holes are normal to the tunnel face and the charged holes are located inside the cut (larger) holes and broken towards the surrounding uncharged holes. This is suitable for homogeneous rocks. Blasted stones are also not thrown so far from the working face as with wedge cuts (Figure 7.2(f)).

Easers: The function of the 'easers' is to blast the area around the central cone or wedge created by the cut holes. The charges in these will normally be less than in the cut holes.

Trimmers: Provided along the periphery of the profile to give the required final shape and section of the tunnel bore. These will require still lighter charges in order to avoid excessive 'overbreak'.

7.3.3 Depth of Holes

The 'pull' or advance per round or cycle of holes drilled and fired is controlled by three factors:

(a) the cross-sectional area and shape of the tunnel;
(b) the nature of the rock; and
(c) the degree of accuracy required in the excavation.

It also depends, to a certain extent, on the drilling pattern chosen and the type of explosive used. These determine the depth of holes to be drilled.

Drilling shot holes, charging and detonating them and leading out debris, i.e., mucking, are considered the productive activities and those like moving plant and equipment away from the blast zone before blasting and returning it afterwards both for blasting and mucking are ancillary (non-productive) activities. The length of the pull affects the later operations considerably. For better progress, the number of latter operations has to be kept to the

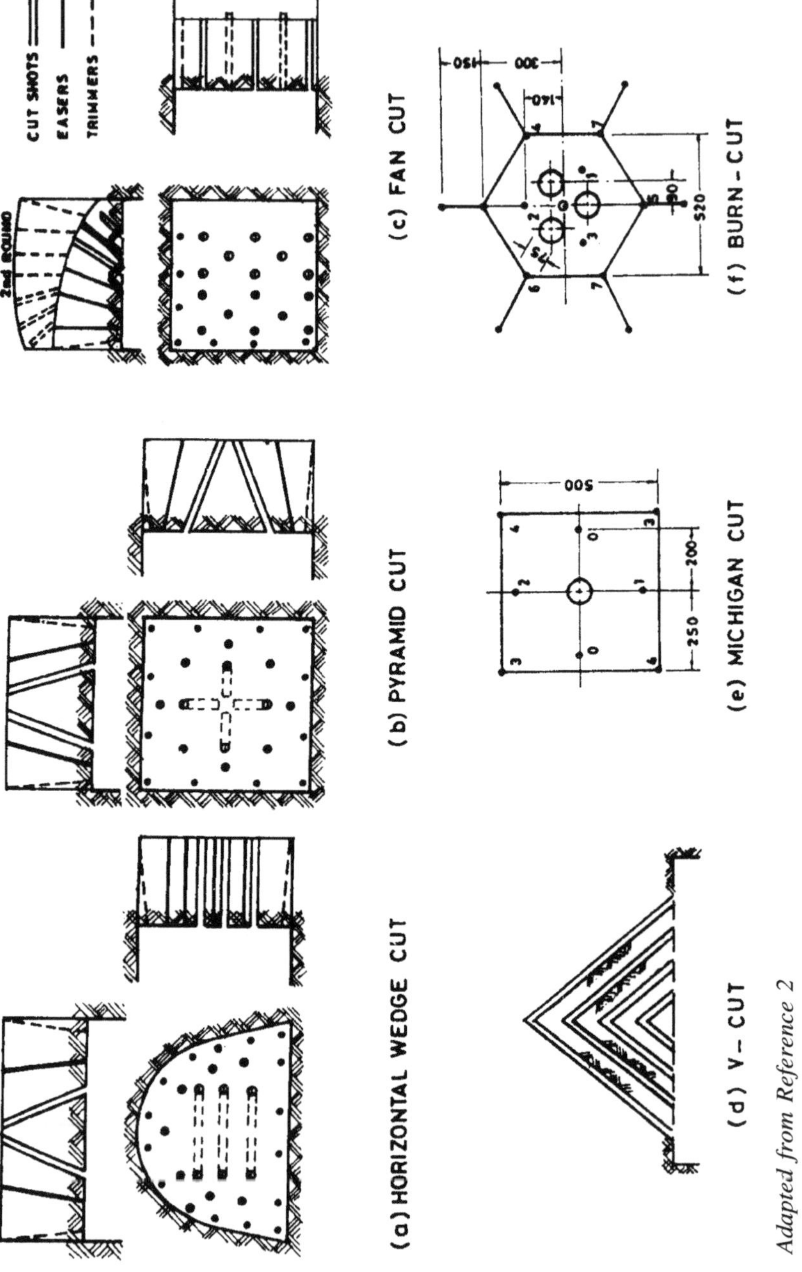

Figure 7.2 Typical Drilling Pattern.

Adapted from Reference 2

minimum. A rough guide for determining the maximum economical depth of pull is to image a square neatly inside the tunnel contour and to make the holes 300 mm shorter than the length of one side of the square' (Pequignot, 1963)[1]. However, there is a physical limitation to the depths that can be drilled and charged. In soft rocks the maximum is about 2.4 to 3.0 m and in harder rocks, is generally restricted to about 1.5 to 1.8 m. Presently with use of pushers and Tunnelling Jumbos and larger diameter rotary percussion drills, drilling holes upto 4m depth is possible.

7.3.4 Diameter of Hole

The diameter of the blast hole depends on the size of the charge material that is generally available. For example, in the U.K. the standard charge used is 32 mm dia. In the USA and France, charges of more than 22 mm dia are in use. The dia of the holes should be at least 3 mm more than the dia of explosive charge. In India the hole dia. generally used is 30 to 35 mm to suit normal cartridges of 25 to 30 mm dia. (Cartridges up to 63 mm dia are also available.) A suitable number of cartridges of 200 to 245 mm in length have to be loaded, with the primer cartridge placed next to the farther end, i.e., blind end of the hole.

7.4 EXPLOSIVES

The quantity of explosives used for loading the individual holes depends on hole spacing, burden of the rock to be loosened and removed, and type of rock. Many empirical rules for calculating this requirement have been evolved.

With explosives having a density of 28 g per 25 mm length of cartridge, the length of drilling required is taken as 2.2 m per kg of explosive. This figure divided by the drilling length gives the approximate number of holes needed in the area of the tunnel face. Another method for determining the holes is to ascertain the area of tunnel face per hole based on Table 7.1. Table 7.2 indicates the tunnel size and the quantity of explosive per m^3 excavated, as adapted from British practice (Pequignot, 1963). For example, for a tunnel face of 4.2 m x 3 m in medium hard strata to be blasted, the depth per charge is generally kept at 1.8 m. This will require 21.1 kg of explosives, which will give the length of all holes to be drilled as 46.4 m. The probable number of holes will be 26. The velocity of detonation is dependent on the type of explosives used, as shown in Table 7.3.

Table 7.1 Blast Hole Requirements

Area of tunnel lace (m^2)	Tunnel face per hole (m^2)	
	Soft or highly Fractured	Hard or massive
9.0	0.41	0.20
18.0	048	0.30
22.5	0.55	0.36
36.0	0.60	0.41
45.0	0.64	0.45

Table7.2 Explosive Requirements for Blasting

	Quantity of explosive required per cum.excavated		
Tunnel area (m^2)	Hard strata Kg	Medium strata Kg	Soft strata Kg
5.6 to 7.5	1.64	1.15	0.98
7.5 to 10.0	1.45	1.02	0.88
10.0 to 13.0	1.32	0.93	0.79
13.0 to 15.8	1.14	0.80	0.68

Table7.3 Properties of Explosives

	Weight/strength ratio (%)	Velocity of detonation (m/s)	Density (g/cc)
Polar ammon. gelignite	78	2500	1.5
Polar ammon. gelatin dynamite	90	2500	1.5
Submarine blasting gelatin	95	7500	1.6
Gelignol	82	3000	1.2

IS: 5878 (Part II)[2] recommends the following types of explosives for use in tunnelling operations:

(a) *Blasting gelatin:* To be used for hard and tough rocks as it has the highest concentrated power. It is fully waterproof and suitable for wet work also

(b) *Special gelatin 90% to 40%:* Requisite strength has to be chosen to suit actual rock conditions. At high altitudes, low freezing types should be used.

(c) *Ammonia dynamites:* These types contain an equal amount of nitroglycerine and nitrate of ammonia and are made in strengths of 15 to 60%. They are suitable for soft rock.

(d) *Semi-gelatin:* These contain ammonia, gelatinised nitroglycerine and nitro cotton and come in two strengths, 45 and 60% They are cheaper and water resistant and emit less harmful fumes. They are used for soft rock and limestone.

As already indicated, the firing of shots has to be done in a controlled sequence, neither singly nor simultaneously, as otherwise surrounding rock will be disturbed and uniform removal of materials inhibited. This technique was evolved in 1831 consequent to Bickgord's invention of the safety fuse. But major developments of controlled delayed detonation have taken place since 1939. Initially, the delay period between consecutive blasts was 1 second but later a series of detonations with 0.5s delay intervals was introduced. A range of still shorter delay detonation has recently gained popularity. Delay numbers and delay period used with 0.5s delay detonations and with the present short-delay detonators are indicated in Table 7.4.

Table 7.4 Normal Delay Detonation Periods

Delay no (s)	Normal delay period for half-second delay detonations	Normal delay period for short delay detonations (ms)
0	0	0
1	0.5	25
2	1.0	50
3	1.5	75
4	2.0	100
5	2.5	150
6	3.0	184
7	3.5	225
8	4.0	265
9	4.5	305
10	5.0	345
11	–	400
12	–	465
13	–	535
14	–	615
15	–	700

Delay detonators are available in delay numbers 0 to 6 with nominal delay intervals of 25 milliseconds each. They are widely used in tunnelling and open excavations as they give much better and more uniform fragmentation, reduce the need for secondary blasting to a minimum, and the amount and direction of pull can be controlled better. With these, more holes can be fired in a single blast with less vibration, concussion and noise. Alternatively, long-delay or half-second delay detonators are also available with delay series of 0 to 10 numbers with nominal delay intervals of 300 milliseconds (500 milliseconds in the case of 0.5s detonators). The typical drilling pattern, sequence of blasting and quantum of charge used recently for heading and benching operations in a tunnel worked through hard rock on the Bombay-Nasik section of the Central Railway are indicated in Figures 7.3(a) and 7.3(b)[3].

7.5 DRILLING OPERATIONS

7.5.1 Drilling by Wet Method

The work of drilling has to be done very carefully and is to be done by the normal wet method. The wet method is favoured mainly for suppressing dust as the dust nuisance in a confined space has an adverse effect on the men working inside. Water also cools the drill bit at the bottom of the slot hole, which reduces breakage of bits. On the other hand, dry drilling is more efficient. In some places use of water may not be practicable or desirable. In such cases some mode of dry-dust suppression is needed, which can be achieved by either of two methods:

(i) Dust is withdrawn through drilling hoses and through a swivel external to the drilling machine.
(ii) Alternatively, dust is withdrawn through the drilling tools and the machine itself. Dust is collected in filter units.

7.5.2 Types of Drills

(i) *Percussion drill with rifle bar rotation:* The most common and used most satisfactorily for hard rock.
(ii) *Percussion drill with separate positive method of drill rotations:* used on soft rock to achieve greater percussion than that given by the above percussion drill. if used for hard rock, the cost of bits may become excessive
(iii) *Rotary drill:* This gives high penetration in extremely soft rock, shale or mud.
(iv) *Augur drill:* Suitable for very soft rock of hardness similar to coal (cohesive type).

Drills are commonly identified by size of cylinder bore. Common sizes now used vary from 50 to 125 mm. All underground drilling is done wet to

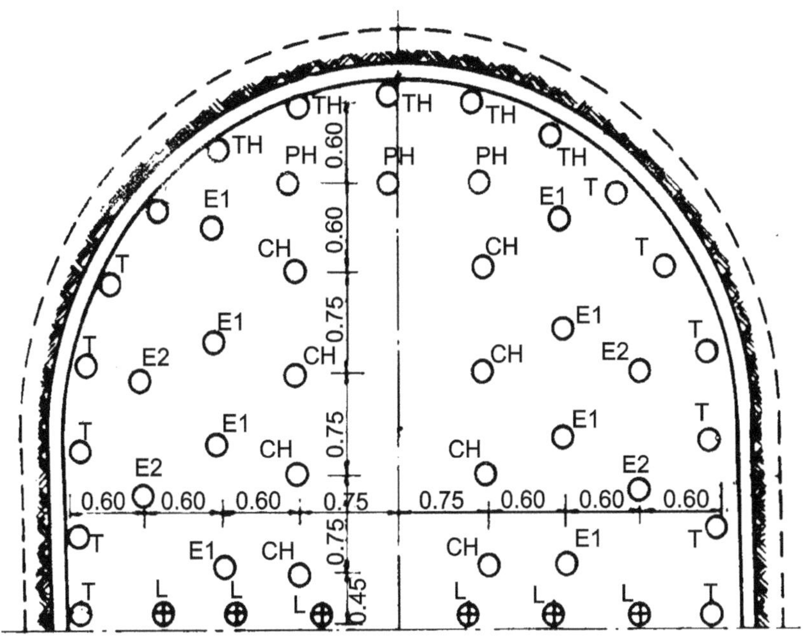

Mark	Position	No of Holes	Rating of Detonator	Length of Holes m	Charge in each hole kg	Total Charge kg
○CH	Cut Hole	8	0	2.743	1.20	9.6
○E1	Easer 1	8	1	2.438	1.20	9.6
○E2	Easer 2	4	2	2.438	0.80	3.2
○PH	Plate Hole	3	3	2.438	0.70	2.1
○T	Trimmer	12	4	2.438	0.75	9.0
○L	Lifter	8	5	2.438	0.75	4.5
○TH	Top Hole	5	5	2.438	0.80	4.0
	Total	48				42.0
	Average Pull Per Blast = 2 m					

Figure 7.3(a) Typical Blasting Pattern for Heading of Tunnels[3].

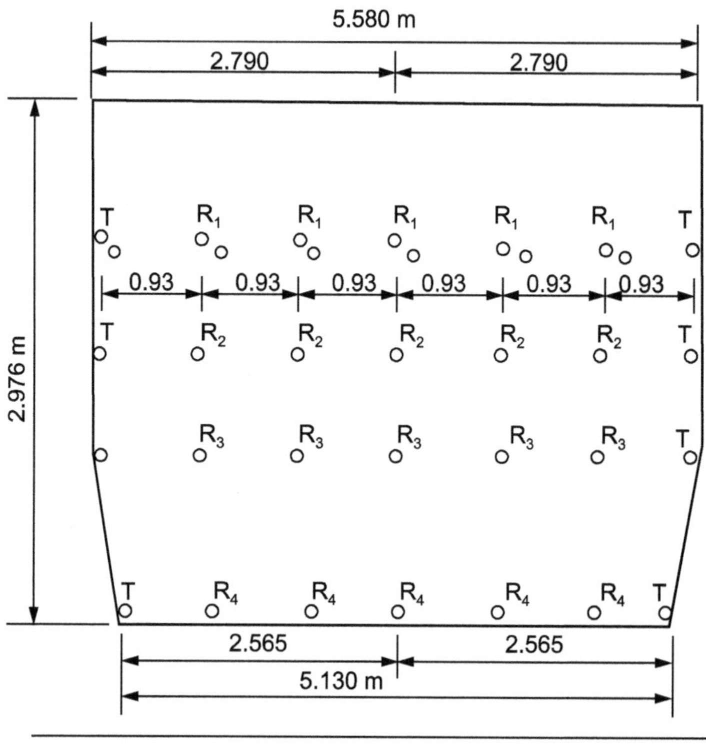

Schedule of Charging					
Mark	Length of Hole m	No. of Holes	Charge in each hole kg	Total Charge kg	Delay Detonator No.
R_1	30.5	5	0.9	4.5	0
R_2	3.05	5	1.1	5.5	1
R_3	3.05	5	1.2	6.0	2
R_4	3.05	5	1.4	7.0	3
T	3.05	8	1.0	8.0	4
	Total	28		31.0	

Area of Face = 16.6 m^2 Average Pull = 2.0 m

Quantity of Excavation of Each Blast = 30.2 m^3

Rate of Gelating 1 m^3 = 0.98 kg

Figure 7.3(b) Typical Pattern of Drilling and Blasting for Benching[3].

reduce rock dust breathed by miners. In percussion drilling water clears the cuttings out of the hole.

It also has a secondary benefit less consumption of compressed air. Economy means lower total cost, which in drilling may be obtained by less capital cost plus increased direct cost in medium jobs. Hence it is advisable to choose light hand tools or feed leg supported jack hammer.

For soft rock, a drill giving good penetration is advised. Penetration rates of 85 to 110 mm in various types of soft rock are quite similar in result. A smaller size drill is used to reduce capital cost.

In hard rock the progress of excavation is directly proportional to time to drill out a round cycle. Maximum output is achieved by using the longest drill that can be used with standard size bits without excessive steel breakage. Percussion drills are used with sinkers (140 to 320 N by weight) and jack hammers (designed for hand holding) classified by weight 140 to 300 N or more. Feed legs and jack legs are used for drilling lateral and overhead holes.

In hard rock, which can stand on its own the drilling and blasting is done on full face and apart from speed of work, the number of holes are reduced and some economy is achieved. A typical section of such work and type and sequence of blasting as done on many tunnels on Konkan Railway is shown in Figure 7.4.

7.6 SAFETY[2]

Tunnelling work is of a specialised nature and is hazardous due to cramped working space in the heading, wet and slippery flooring wherever seepage exists, artificial lighting, inadequate ventilation and presence of obnoxious gases. Unseen weaknesses in the rock and any lack of care in handling of explosives can cause accidents. All possible precautions have to be taken from the moment of procurement and storage of explosives. A number of safety precautions are imperative while drilling and charging the holes and blasting. The main precautions generally advised in drilling and blasting operations are indicated below.

(i) The explosives must be stored only in a magazine that is clean, dry, well ventilated and reasonably cool. The magazine should also be correctly located, substantially constructed with bulletproof and fire-resistant material and securely locked.

(ii) Detonators must be stored separately. Suitable fire extinguishers should be provided at places of storage of explosives and detonators and explosives periodically examined. Figure 7.5 shows the typical arrangements for magazine, detonator and fuse store. A typical site

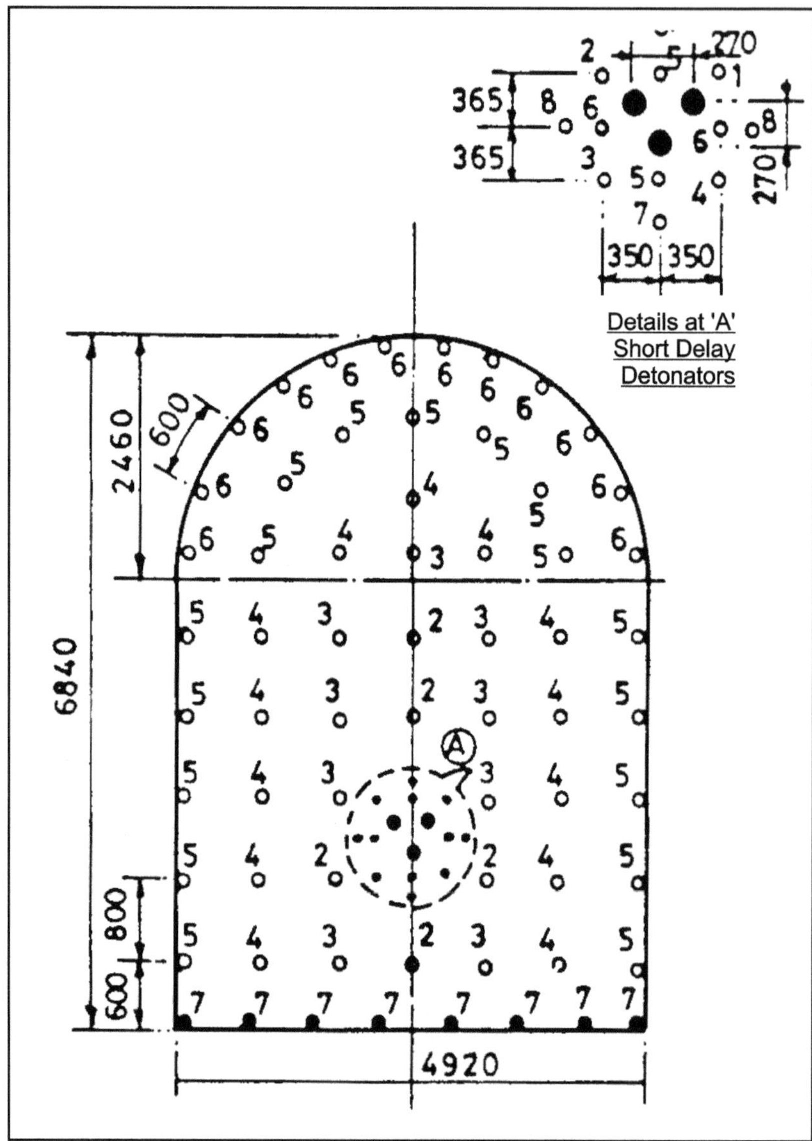

Source: Indian Concrete Journal, February, 1994

Figure 7.4 Blasting Pattern on Full face in Karbude Tunnel[4].

plan for magazine location with minimum safety distances from other structures is shown in Figure 7.6.

(iii) Vehicles used for transporting explosives must be in good working condition and have a floor made of light wood or non-sparking metal

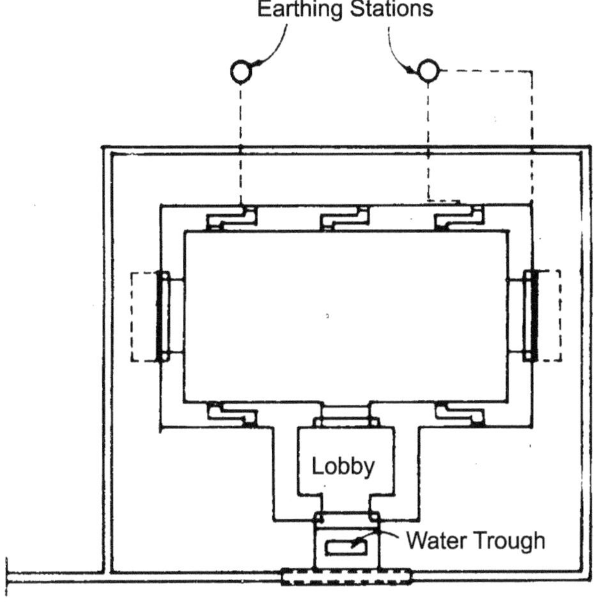

(a) Plan for magazine building

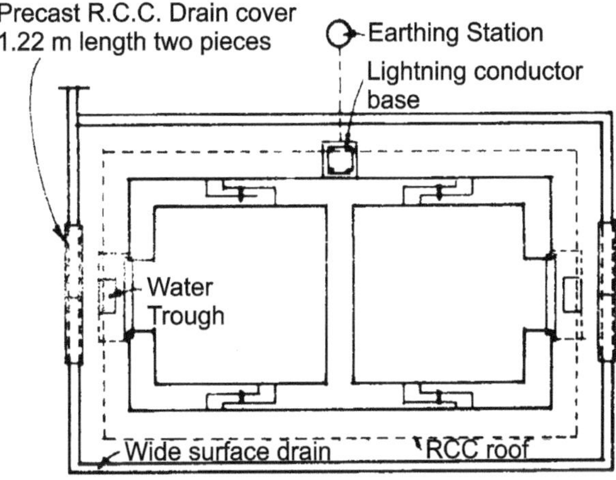

(b) Plan for detonator and fuse store

Figure 7.5 Typical Magazine Building, Detonator and Fuse Store.

such as copper, brass etc., with sides and ends sufficiently high. The electrical wiring of the vehicle must be fully insulated to preclude the danger of any short circuiting or sparks. At least two fire extinguishers of carbon tetrachloride type must be carried in the vehicle. Explosives and detonators must not be transported in the same vehicle.

(iv) Explosives and detonators should be brought to the working places in separate well-insulated containers.

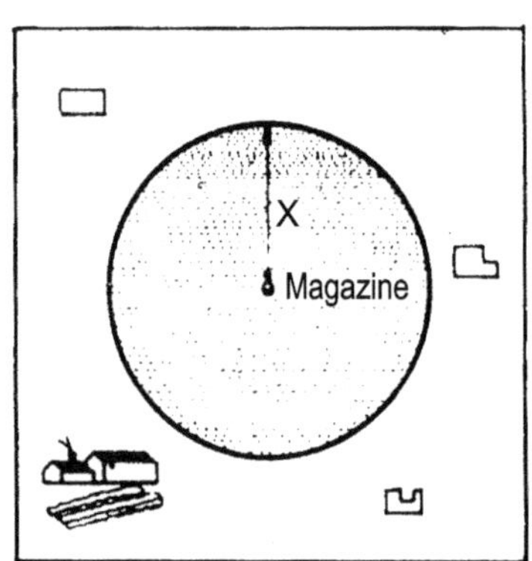

Safety Distance

Magazine capacity kN	Distance 'X' m
10	140
20	215
30	270
100	345

Figure 7.6 Typical Plan for Magazine Location with Respect to Safety Distances.

(v) Blasting must be carried out using a suitable exploder with 25% excess capacity. The revolving handle of the exploder must be in the custody of the blasting foreman to prevent anyone else firing the shot by mistake when the blasting foreman and/or other persons are inside.

(vi) All explosives are dangerous and must be handled and used with care under close supervision and direction of competent, experienced and licensed persons. The provisions of the Indian Explosives Act, 1984 and the Explosives Rules, 1940 must be scrupulously observed.

(vii) Before undertaking drilling operations for blasting, the nature of the strata and the overburden should be carefully examined to avoid possible leaks through fissures and landslides after blasting.

(viii) Before firing, sufficient warning must be given to enable people working in the blasting area to get clear of the danger zone, which should be suitably demarcated with flagmen posted at important points to prevent strangers from straying into the area. It is better to schedule firing at a fixed time of the day.

(ix) None of the fittings, including locks of rooms or containers, should be of steel as it is likely to cause sparks. Brass fittings are to be provided.

(x) Adequate provision should be readily available for rendering prompt and adequate first aid to any person injured at the work site.

(xi) All persons entering the tunnel or a shaft should be provided with protective clothing, such as helmets, gumboots etc. They should also have torches.

(xii) In view of the limited and confined space inside the tunnel, only the minimum equipment and materials required should be allowed to be kept near and at the face of work.

(xiii) The entire electrical installation must be carried out according to the provisions of the Indian Electricity Rules and wiring should have good-quality insulation. Exposed wires should be scrupulously avoided. The voltage of the supply line for lighting purposes should be reduced in tunnels from 230 V to 110 V. For machinery requiring 440 V supply, reliable moisture proof cables must be used.

(xiv) As stray currents can cause accidents while loading, utmost care must be taken in removing all faults from electrical circuits. Electric power, light and other circuits within 70 m of the loading points must be switched off after charging the explosives.

(xv) Where high-voltage electric traction is in use on railway lines close by, special care for prevention of induction and sparking is necessary.

(xvi) After every cycle of blasting, the blasting foreman must make a careful inspection of the face to determine whether all charges have exploded. No other person must be allowed near the area of blast unless an all-clear signal is given by the blasting foreman after checking.

(xvii) Drilling should not be resumed after blasting until a thorough examination is made by the blasting foreman to ensure that there are no misfires and no unexploded charges left.

7.7 ROCK MASS CHARACTERISTICS AND ALTERNATE STRATEGIES

The Drill Blast strategy with nominal steel rib supporting discussed above are applicable to competent rock, which generally are self supporting over fairly long periods. But, in nature, the type of rocks met with will be with varying static strengths, joint structure, contact behaviour at joints and seepage or intrusion of water. They will require not only alternative strategies for supporting but also in choice of section and even tunneling methodology. Similar to the development on NATM for soft ground tunnelling, a methodology known as NMT, Norwegian Method of Tunnelling is considered more suitable in such cases[5].

7.7.1 Norwegian Method of Tunnelling

This method is considered by many as the most suitable tunnelling method for hard rock masses, where jointing in rock structure and over breaks are dominant, and in which Drill & Blast or use of Hard rock TBMs are the usual methods used for excavation. In case of NMT, (i) firstly rock mass charaterisation is done to predict the rock mass quality and support needs. The information needs to be updated during tunnelling, so that suitable modification can be made in supporting. (ii) Rock bolting forms the dominant form of 'rock support' for mobilizing the strength of the surrounding rock and same is done systematically. (iii) The support provided by bolts/ anchors is supplemented by Shotcrete/ steel fibre reinforced shotcrete (SFRS) in potentially unstable rock masses like ones with clay filled joints and discontinuities. (iv) A thick load bearing ring (formed by reinforced rib in shotcrete) is formed, as required to match the uneven profile finally.

This method has been found more suitable for the type of geologic structure met with in the tunnels on USBRL, in addition to Drill and Blast used for harder rocks. The main components of the support system used here are: a) Installation of shotcrete lining with rock bolts, to allow limited deformations but preventing loosening of rock mass. (b) Providing steel supports to avoid several layers of shotcrete that would have to be provided otherwise infringing tunnel space. As mentioned above, the design of support system first requires 'Rock Mass Classification' as they are dependent on Rock mass rating RMR. RMR itself is related to Q the Rock Mass Classification.

7.7.2 Rock Mass Classification Determination[6]

As mentioned above, Q is a relationship combining different characteristics of rock like RQD, the main characteristic denoting the core strength of rock as listed below:

$$Q = \frac{RQD}{J_n} \times \frac{J_r}{J_a} \times \frac{J_w}{SRF}$$

- Where, J_n = Rating for the number of joint sets (9 for 3 sets, 4 for 2 sets etc) in the same domain

- J_r = Rating for the roughness of the least favourable of these joint sets or filled discontinuities
- J_a = Rating for the degree of alteration or clay filling of the least favourable joint set or filled discontinuity
- J_a = rating for the water inflow and pressure effects , which may cause outwash of discontinuity infillings, and
- SRF = The rating for faulting, for strength/ stress ratios in hard massive rocks, for squeezing or for swelling

The three different terms represent characteristics/ factors as follow:

- RQD/J_n indicates the relative block size (useful for detecting massive, rock- burst prone rock)
- J_r / J_a indicates relative frictional strength (least favourable joint or filled discontinuity)
- J_w/ SRF indicates effects of water, faulting, strength/ stress ratio, squeezing or swelling type. This is a stress term.

Tables at Annexure 7.1 give the quantitative ranges for the rock structure at a location in respective characteristics Jn ,etc., used for quantitatively determining the Rock Mass. Rocks are grouped under different classes based on (tunnelling quality) Index Q as per N.G.I as given in Table 7.5 below[6];

Table 7.5 Rock Mass Classification as per N.G.I- Based on Index Q

Sl. No	Class	Q	Rock mass description
1	Class I	➢ 40	Very good rock
2	Class II	10–40	Good rock
3	Class III	4–10	Fair rock
4	Class IV	4	Poor Rock
5	Class V	0.1–1	Very poor rock
6	Class VI	< 0.1	Squeezing rock

Bieniawski has given the following relationship between Q and RMR

$$RMR = 9 \log_e Q + 44$$

Bieniawski has classified the rocks based on their cohesion, average stand up time, average internal friction and RMR as given in Table 7.6 below[5]:

7.7.3 Rock Mass Index and Tunnel Support

Knowing the RMR value and stand up time, the manner of excavation and type of support and lining can be determined. The following table gives some guidance in that respect.

Table 7.6 Classification of Rock Mass (Bieniawski (1979)[5]

Sl. No	Rock Classification	Cohesion in Rock Mass	Average Stand up Time	Angle of Friction internal	RMR
1	Very good	0.4	10 yrs for 15 m span	➢ 45°	100–81
2	Good	0.3–0.4	6 months for 8 m span	35°–45°	80–61
3	Fair	0.2–0.3	1 week for 5 m span	25°–35°	60–41
4	Poor	0.1–0.2	10 hrs for 2.5 m span	15°–25°	40–21
5	Very poor	< 0.1	30 min for 1 m span	< 15°	< 20

While drilling holes for blasting, probe holes for a depth equal to about 2 to 3 times the length proposed to be excavated should be done to see if there is intrusion of water and also to know the type of soil likely to be met with ahead. It will help to suitably modify the methodology and take precautionary measures. For excavation, alternatively road headers with rock cutter provision can be used for lighter rocks and soils. Hard Rock TBMs are also used as an alternative tool for tunnelling through rocks.

In case of rocks, Singh et al (1995) have suggested different support pressures and some additional requirements of support. Their recommendations for Type V and VI rocks, are given in Table 7.8. These are the types of rock often met with in the tunnels considered by the designers for tunnels done in Himalayan Region.

Rock pressures for structural analysis of the support system in these cases have been done based on Q values as indicated in Table 7.9.

As against these value, actually the values as per Modified Terzaghi loads, Pv = 3kg/ sqcm and Ph = 1kg/sqcm as against 2.82 kg/sqcm and 2.08 kg/sqcm respectively, e.g., for comparison of D shaped and E shaped (egg shaped) tunnels are used.

7.7.4 Suggested Support System for different Rocks

To sum up, the tunnel system proposed at crown for different types of rocks are:

- Rock class I - Spot bolting with local shotcrete
- Rock class II- 50 mm thick shotcrete + 25 dia Rock bolts 3.5 m long @ 1.5 m crs.

Table 7.7 Rock Mass rating vs Excavation and Support[5].

Rock Mass Class	Excavation	Support		
		Rock bolts	Shotcrete	Steel Ribs
Very good rock- RMR 100- 81	Full face 3m advance	Generally no support required		
Good rock RMR 80- 61	Full face 1.0 to 1.5 m advance. Support 20 mm from face	Locally- 3m bolts in crown spaced 2.5 m- Occasional mesh	50 m in crown when required	None
Fair rock RMR 60- 41	Top heading and bench; 1.5 - 3 m advance; Support after blast	Systematic 4 m bolts @ 1.5- 2 m at crown and walls with mesh in crown	50- 100 mm in crown; 30 mm in side walls	None
Poor rock RMR 40- 21	Top heading and bench; 1 - 1.5 m advance; Support concurrently with excavation 10 m from face.	Systematic 4.5 m bolts @ 1-1.5 m in crown and walls with mesh in crown	100- 150 mm in crown; and 100 mm on sides	Light ribs spaced @ 1.5 m where required
Very poor rock RMR 40- 21	Multiple drifts 0.5- 1 m advance; Support concurrently with excavation. Shotcrete soon after blast	Systematic 5- 6 m long bolts @ 1- 1.5 m in crown and walls with mesh; Bolts in invert	150- 200 mm in crown; and 150 mm on sides; 50 mm on face	Medium/ heavy ribs spaced @ 0.75 m with steel lagging; Forepoling if required; Close inverts

Table 7.8 Recommendations of Singh et al (1995) on Support Pressure for Rock Tunnels and Caverns[5]

Terzaghi Classification			Classification by Singh et al (Modified Terzaghi's Classification				Remarks
Category	Rock condition	Rock load	Category factor (Hp)	Rock condition support pressure	Recommended		
					Pv	Ph	
V	Very blocky, seamy, shattered, arched	0.35 to 1.1 (B+ Ht)	V	Very blocky, seamy, shatterd, highly jointed, thin shear zone of fault	0.2	$(0\text{-}0.5)p_h$	Inverts may be required, arched roof preferred
VI	Completely crushed but chemically in tact	1.0 (B +Ht)	VI	Completely crushed but chemically unaltered, thick shear and fault line	0.3	$(0.3\text{-}1.0)p_v$	Inverts essential, arched roof essential

Terzaghi's classification modified by B.Singh has been considered, since earlier Classification by Terzaghi gives loads on higher side.

Table 7.9 Rock Pressure based on 'Q' system

Rock Type	Q value	J_r value	Average 'Q'	P- Roof Kg/ sqcm	P- wall Kg/ sqcm
Good	10-40	1.5	25	0.528	0.308
Fair	4-10	1.5	7	0.507	0.594
Poor	1-4	1.5	2.5	1.13	0.628
Very poor	1.0	1.0	0.55	2.82	2.08

- Rock class III- 50 mm thick shotcrete + 25 mm dia Rock bolts 3.5 m long @ 1.2 m crs.
- Rock class IV- 100 mm thick shotcrete with wire mesh + 25 mm dia Rock bolts 4.5 m long @ 1.0 m crs.
- Rock class V- 100 mm thick shotcrete with wire mesh + 25 mm dia Rock bolts 4.5 m long @ 1.0 m crs + steel ribs ISHB 150 @ 50- 100 cm crs with back fill concrete.

Figure 7.7 shows the typical arrangement of steel ribs and back fill concrete.

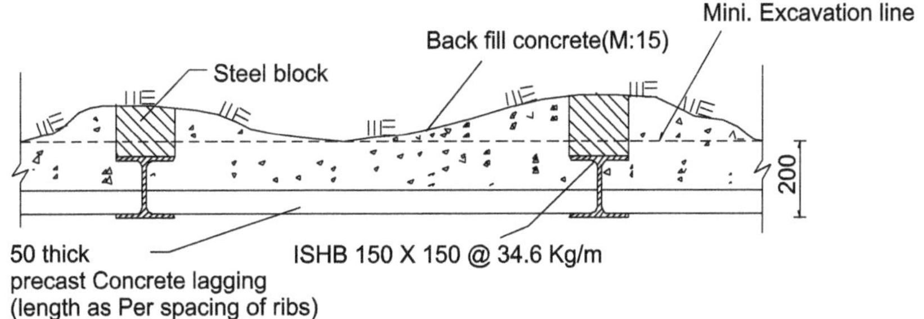

Figure 7.7 Arrangement of Ribs, blocks and Backfill concrete (Primary) Support system for tunnels[5]

Support System for Very Poor Rock

In very poor rock, the tunnel shape and supporting system described below has been found suitable:

(i) Semi circular arch roof, curved walls and curved invert to make a continuous supporting ring (near to an ellipse), which will help in controlling rock deformations and resist both vertical loads and side pressure;

(ii) Support system provided during construction is a combination of shotcrete, rock bolts and steel rib including invert strut;

(iii) Backfill concrete encasing the steel ribs forming an essential element of initial support, both 'overt and invert';

(iv) Heading and benching system of excavation should be followed;
(v) A haunch system (for heading) is provided, aiming at following advantages-
 — Footing (for arch ribs) is not disturbed during benching, thus making it safe, and
 — Most of the roof thrust being conveyed to abutments, side wall segments are made safer
(vi) In the heading stage also a curved invert is provided with shotcrete/ concrete.

A typical section of the support system proposed in such cases is shown in Figure 7.8.

7.7.5 Squeezing and Swelling Soils[7]

Behaviour of squeezing soils- By squeezing, one refers to the soil when it 'squeezes or extrudes plastically into the tunnel, without visible fracturing or loss of continuity and without any perceptible increase in water content.' This can be considered as ductile, plastic yield of soil due to overstress.
Typical soils that cause such squeezing are:-

Ground with low flexural strength subject to overstress- the rate of squeezing of soil depends on degree of overstress. It may occur at shallow to medium depths in clay of very soft to medium consistency. Stiff to hard clay under high stress also may move, in combination with raveling at excavation surface and squeezing at depths behind surface of excavation.

Rocks with low geo-structural quality (RMR) also can lead to squeezing when it is subjected to high stress in relation to its strength[7]. The behaviour of rock as its geo-structural quality deteriorates is explained diagrammatically by some as indicated in Figure 7.9. Rock with most unfavourable RMR and low strength can lead to squeezing when subjected to high stress.

The type of rocks with low strength and high deformability include, for example, 'phillytes, schists, serpentines, mudstones, tuffs, certain kinds of flysch and chemically alterated igneous rocks' (Kovari, 1998). In such cases, caving is also possible. In India, this type of condition has been met with extensively in the Himalayan Region as has been experienced during large scale tunnelling on the USBRL project. New methodologies like NMT and NAMT have been adopted in them extensively and the structural forms evolved to suit, as indicated in Para 7.7.3. In such cases, recommended support pressures and support system as per modification by B.Singh is given below, which can be considered a continuation of Table 7.8.

As examples, Annexure 7.2 gives some case studies of typical problems met with while tunnelling on some tunnels on Katra- Dharam section of USBRL project.

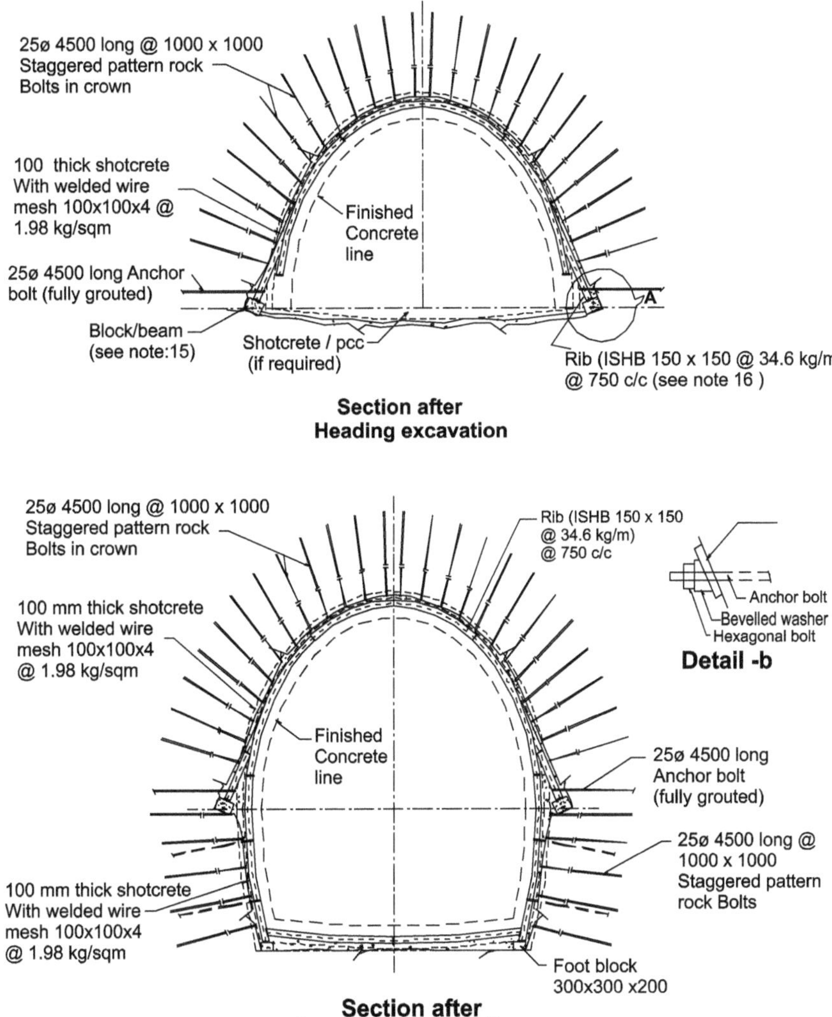

Source : KRCL

Figure 7.8 Typical Support and Lining details for Tunnel in Very Poor Rock[5].

Continuation of Table 7.8 (See Para 7.7.5)

Terzaghi Classification			**Classification by Singh et al (Modified Terzaghi's Classification**				**Remarks**
Category	**Rock condition**	**Rock load factor (Hp)**	**Category**	**Rock condition**	**Recommended support pressure**		
					Pv	**Ph**	
IX	Swelling Rock	Upto 80 m	VIII	A. Mild swelling	0.3–0.8	Depends upon type and content of swelling clays, May exceed Pv	Inverts essential in excavation, arched roof essential Inverts essential, arched roof essential
				B. Moderate swelling	0.8 to 1.4	-Do- -Do-	
				C. High swelling	1.4 to 2.0	-Do-	-do-

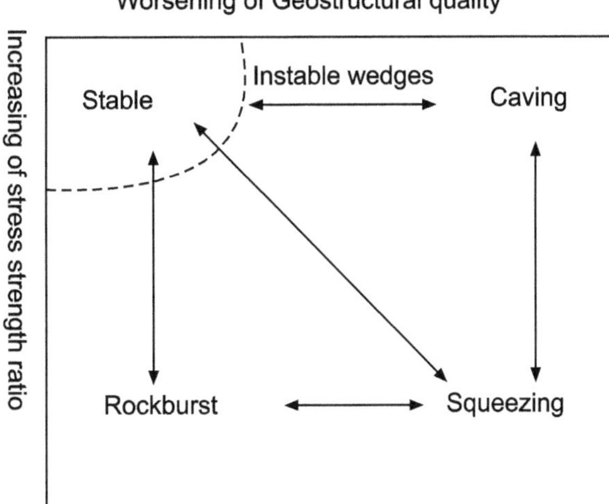

Source : Russo

Figure 7.9 Conceptual behaviour of Ground on excavation[7]

7.8 MUCKING PROCEDURE[1]

7.8.1 General

'Mucking' is the term used for removal of the spoil, i.e., debris that has resulted after each blasting or excavation operation. This should be done as soon as the blasting is over and It is considered safe for men and machines to go into the tunnel up to its face. In the case of rocky soils mucking can be commenced immediately. The first task is to spray the debris with water to settle the dust. This operation is repeated during the mucking process whenever considered necessary.

In the early days mucking was done by running small trolley trucks or open top wagons over narrow tracks laid on the floor. They were manually pushed in small jobs and for short leads. Nowadays mucking is alternatively done using individual motorised tippers and trucks. A wide range of mechanical equipment is available, from which one or more can be chosen to suit the particular circumstances, tunnel size, type of material etc. The time taken in mucking represents about one-third to one-half of the total cycle time of tunnelling operations and any quicker means adopted will help to hasten progress of the work.

In the case of unstable soils not considered safe, or there is danger of roof falls and side slips, mucking operations should be commenced only after the sides and roof of the tunnel are temporarily supported properly. The supporting can be done by means of timber struts and lagging in the case of smaller tunnels and headings. In larger tunnels and headings it may be necessary to install a collapsible steel framework conforming to the shape of the tunnel. This framework would cover the entire section in the case of full-face tunnels. Individual frames are to be properly tied to one another longitudinally. In tunnels bored with the use of headings, they should be supported with longitudinal runners known as 'wallplates', which are to be kept at the base on edges of the heading. They will bear against the steel frames mentioned above. Typical supporting arrangements are shown in Figure 5.6. Supporting and mucking should progress gradually to cause minimal disturbance by way of vibration inside the tunnel.

7.8.2 Use of Rail Trucks

There are two methods of doing mucking in long tunnels. One method uses wagons or open trucks pushed by a locomotive over a track laid on the floor. This track has to be extended in small lengths. The latter are generally supported on steel plate sleepers or wooden sleepers suitably packed to keep them level and aligned. Where the tunnel is wide enough, in order to expedite the work a number of wagons/trucks may be taken in tandem and an intermediate truck-passing arrangement (loop or parallel track) laid so that it is possible to transfer and move the loaded trucks back towards the entry/opening while the next batch of empty trucks is moved forward to be loaded.

Where the width of the tunnel is not large enough and only a single track can be laid inside, such transfer within the tunnel may not be possible. In that case there should be two lines/spurs outside, one for holding the wagons to be taken in for loading and the other for bringing out and stabling the wagons loaded with material to be offloaded or dumped. These wagons are moved on to the haul line extending into the tunnel, through a pair of railway switches. These switches are light in weight and portable and hence can be shifted from location to location without much difficulty.

Alternatively, motorised vehicles either with rubber tyres or with tracked wheels are used. They can be reversed into the tunnel one by one and after loading driven out for dumping the spoil outside.

7.8.3 Loading

Loading inside can be done either manually or by mechanical bucket loaders selected to suit the limited head room available inside. These loaders may be of the rocker-arm type, which can directly load onto the car or onto a conveyor behind the machine (Pokrovsky, 1977)[8]. Rocker-arm type shovel loaders are usually driven by compressed air and are mounted on a track. The bucket is filled by pushing the machine forward into the spoil heap, the bucket lifted, and the machine reversed for discharging into a truck or over the conveyor belt. The bucket returns to its initial position by gravity when machine recedes. Compressed air drive makes it easy to connect the machine to the power supply at the face and, with reasonable maintenance; the shovel will give reliable and economic service. Machines with a bucket capacity of 0.4 m^3 are generally used. Machines are available, however, with bucket capacities varying from 0.2 to 0.5 M^3 and working capacity of 0.5 to 2.0 m^3/min. Some are provided with turntables for swinging sideways. Alternatively, the bucket assembly can be swung by a swivel mechanism. The working heights vary from 2.2 to 2.8 m. The machines may be crawler mounted or mounted over wheels moving over a narrow gauge (600 mm to 900 mm) rail track.

Alternatively, electrically operated machines may also be used. In one such, the gathering arm loader is mounted on crawler/tracked wheels. It collects spoil by means of two eccentrically operated arms or claws. A flight chain conveyor carries the muck backwards and dumps it into cars or trucks. Another type of loader is the shaking-pan or vibratory type which consists of a trough with a shovel-shaped front end which can be vibrated into the spoil heap. The vibration carries the material up the slope of the trough and into skips waiting at the rear of the loader. This type can work only inside tunnels with even floors. Yet another method of loading employs a slusher or scraper loader. In this case the sloping steel chute at the front of the machine is used as a ramp over which a double-drum winch hauls a scraping bucket to deposit its load into a waiting skip. A typical loading arrangement is shown in Figure 7.10.

The method of loading adopted at a particular location depends on the availability of men and machinery. In India, in short tunnels and tunnels of difficult access, loading is done manually. The stones are broken and the debris shoveled or loaded onto trucks/dumpers manually.

7.8.4 Cycle Time

Where compressed air is used for drilling and loading, air pressure is increased to 0.8 N/ mm^2. Such a pressure can work three drills at a time or operate one loader at a time. With a typical shovel loader, the operation time for mucking and blasting a typical job would be as shown in Table 7.10 (Pokrovsky, 1977)[8].

Figure 7.10 Muck Loading to Train with a Conveyor Belt.

Table 7.10 A Cycle of Mucking Operation[5]

Operation	Time (normal) min	Time (shuffle car) min
Clear smoke	15.0	12.0
Prepare to muck	11.5	10.0
Muck	80.0	42.0
Prepare to drill	5.0	5.0
Drill	34.0	35.0
Charge, stern and fire	20.0	20.0
Lay rails	9.5	8.0
	Total 175.0	132.0

7.8.5 Haulage

No improvement in cycle time can be achieved without adequate provision for getting the loaded spoil out of the tunnel quickly. Hence, for this purpose, where rail mounted trucks/wagons are used, diesel operated locomotives of suitable haulage capacity are employed. Use of small diesel shunters reduces the smoke problem, so less ventilation air is needed, and also increases efficiency. Use of such locomotives requires proper track laying, with rails of suitable weight and sleepers adequately spaced and packed. A poorly laid track will mean slower travel and may also involve expensive derailment. Due to the nature of the work and unevenness of the base of a tunnel, the general tendency is to hurry and carelessly lay the track on an uneven, undressed bed. Any time saved in such laying without adequate precautions will be far exceeded by the time spent in slower movements and in rerailing/ removal of derailed vehicles. The gauge used is generally narrow, 610 mm or 641 mm. In wider tunnels, tracks of 1000 or 1067 mm gauge have also been used. A few typical arrangements for layout of tracks are shown in Figure 7.11.

The use of railway locomotives requires that the tunnel be well ventilated and the engine in good working order. Even though diesel exhaust contains no carbon monoxide, the fumes emitted from a badly adjusted engine are nauseating to breathe, heat the surrounding air and hence reduce the working efficiency of the men unless quickly cleared out of the tunnel.

While rail-mounted skips or trucks can be moved even in small-size tunnels, motorised or individually hauled units of trucks or dumpers can be used only when the tunnel is large enough to accommodate them. However, use of these dispenses with the need for laying a track inside. The general practice in the U.K. is to use (tracked wheel) crawler type trucks while in America and Europe pneumatic tyred vehicles are used more often. The same practice had been adopted in short tunnels and in areas of difficult access in India.

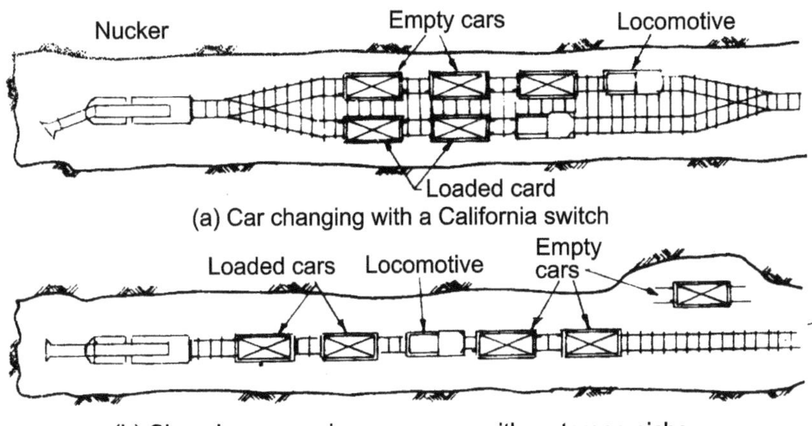

Figure 7.11 Typical Arrangements for Switching Wagon[9].

The relative advantages and disadvantages of using rubber tyred or diesel-powered individual haul units and rail haul units are detailed below (Parker, 1970)[9]:

(a) ***Individual haul units:***

The main advantages of using individual haul units such as tippers, trucks etc. are that they

- — can be mobilised quickly,
- — will operate easily on grades up to 10%,
- — need less capital expenditure,
- — are flexible in operation,
- — can readily move from one heading to another when different heading crews are working,
- — can be easily moved from one place to another over even rough roads, and

— can dispose of muck at any chosen location, instead of the rear mouth of the tunnel.

The disadvantages are:

— Small tunnels need small units and the latter cannot operate on steep grades due to low power.
— Units wider than rail units of the same capacity need more moving room and wider bed/invert
— Individual units are relatively heavier and so are less efficient and need more horsepower per payload.
— They need more ventilation per unit on long hauls and a large ventilation requirement makes them impractical.
— The road units need a well-graded firm surface and dry road-bed inside the tunnel, both of which are difficult to obtain and maintain in a wet tunnel.

(b) ***Rail haul units:***

Advantages from rail haul units result from their work in train formation.

— They are more efficient in long tunnels and yield more haul with less horsepower and fewer operators.
— The ventilation problem is less critical.
— Being compact and relatively narrow, they can operate in small-diameter tunnels.
— They are more suitable in wet tunnels and can operate in flooded tunnels also.
— They are easily adaptable for concrete (lining) placing operations after mucking is over.
— When operated with cables and hoists, they can be used on any grade.
— Being of robust construction with only the locomotive or winch to be powered, they require less maintenance.

Their disadvantages are:

— Longer period of mobilisation required.
— In short tunnels, more expensive due to larger crew requirement.
— In large tunnels, one half has to be mucked out at a time or two muckers operated abreast of each other.
— Lack the flexibility necessary for use in excavation of underground metro tunnels.
— Work well normally on grades up to 2%, application restricted with grades of 2% to 4% and must be winched up/down for grades over 4%.

However, pneumatic-tyred individual vehicles are now used more in view of easy availability of tippers, flexibility and especially lower cost for short- and medium-length tunnels. Para 7.9.2 gives more details.

7.8.6 Mucking by Pumping

Where water seepage occurs, the mucking operation becomes more difficult. Adequate trench arrangements have to be provided to clear the water so that it does not interfere with the various operations nor hamper the laying and maintenance of track or movement of the dumpers. In short, soil mucking is done by loading the spoil through large-diameter pipes, mixing with compressed air and pumping outside the tunnel.

In some countries belt conveyors have also been used in recent years for conveying muck in long tunnels. They are yet to be used in India on any large scale.

7.9 TUNNELLING MACHINERY[10]

In order to speed up the major operations involved in tunnelling, specially boring long tunnels in rock, mechanization of major operations is necessary. Two major operations involved in rock tunneling are Drilling holes and Muck loading and disposal.

7.9.1 Drilling Operation

Traditionally, drilling holes in rock was done manually by hammer men using jumper. But progress was hardly two metres of hole per hour. Rotary-cum-percussion drilling with hand held or otherwise supported tools followed. Steam power was used to aid in this. Pneumatic operation to aid for drilling was the first major development since the beginning of 19th century. This was followed by the invention of 'Pusher leg', which helps the operator of drill to exert less. It is learnt that in construction of St Gothard Railway tunnel, hydraulic drilling tools were used (as early as 1876), but this technique got boost only with development of different hydraulic tools in 1960s. One main problem in use of hydraulic tools was non- availability of hydraulic hose pipes which could transmit the high pressure fluid without leak at joints. Manufacture of pneumatic tools and steam driven or diesel compressors were found cheaper. Later development of hydraulic tools and transmission system gave a new impetus to use of hydraulic drills. Initially it used pneumatic percussion mechanism. The development of rotary drilling mechanism and use of 'compressed nitrogen accumulators in the circuit to impart energy to piston' and use of indexing for avoiding continuous rotation helped making the system more economical. Figure 7.12 shows how the drilling technology has been developed and refined in past over a hundred years. (Anand, 1994)[10]. Present day drilling systems with use of water at tool end for lubrication and washing helps in reducing dust and over-heating of the tool.

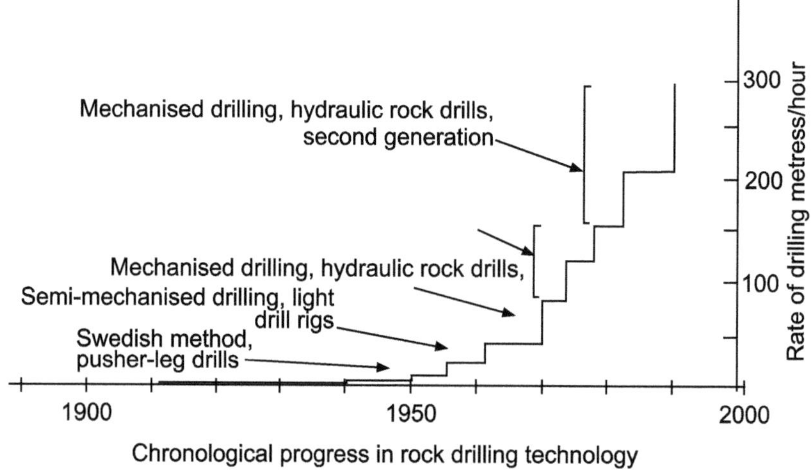

Adapted from Reference 10

Figure 7.12 Development of Drilling Technology and Rate of drilling.

Earlier, single tool was mounted on each unit, but jumbos have been developed with two or three such drills mounted on each mobile unit, capable of dealing with two or three tools in parallel. Feed lengths of such drills now are from 3.5 to 6 m. Jumbos suitable for use on tunnel sizes of 4.3 × 7.1 m to 13.2 x19.2 m are available. Their operation can be hydraulic or electro-hydraulic. For movement from portal/ niche to working face and back, diesel power is used. In India, Konkan Railway Corporation had as early as 1992 imported nine such jumbos (having two booms) from Sweden for expediting work on a number of long tunnels. They had to tackle a total 78 tunnels (totaling a length of 79 km), out of which there were 9 tunnels varying in length from 2033 m to 6520m. They found that, by using conventional equipment about 70m length only could be done per month, while with the Jumbos, they aimed at a progress of about 150 m per face per month. In fact, they achieved a record progress of 180.3 m in a month on the 4376 m long Nathuwadi tunnel and in general average of over 120 m. On the longest Karbude tunnel they used one such from each end and conventional method on 6 more faces created by sinking three shafts. On the second longest Nathuwadi tunnel, due to heavy overburden (about 350m), sinking shaft for developing additional faces was not considered practicable and they worked from two ends only with one jumbo at each end.

- 'KRCL adopted various mechanized methods for excavation and loading in their tunnels in order to expedite progress, such as-
- use of clay diggers for excavation
- Deployment of electrically operated conveyor belts for loading muck/ spoils in tippers / trolleys
- Use of JCBs for excavation in enlarged sections of heading
- Deployment of Westfalia Road Header with soil cutting tools for excavation -with conveyor belts
- Use of Alpine Miner AM50 for excavation'

Figure 7.13 (a) shows a modern Jumbo (tunnel Running 2L Jumbo) which is designed to work on a face with maximum height of 7.98m and width of 12.9m. These are electro -hydraulically operated. Figure 7.10(b) shows close up view of a tool on a boom fitted on the Jumbo used on the Konkan Railway. The tools were capable of drilling holes upto 4m length, but to suit the fractured condition of rock, holes of 2.5 to 3 m were driven. With this they could achieve an average pull of 2.5 m. As against conventionally drilled hole diameter of 35 mm, these tools can drill larger holes and 45mm dia. holes were drilled, facilitating use of larger 40mm dia slurry explosives (Indo tunneler K of IBP). It facilitated reducing the number holes per full face from 76 (for 32mm cartridges) to 68 thus reducing drilling and charging time and some saving in consumption of explosives also. Drilling that number of holes could be done in two hours.

Road Header

A Road Header, also known as 'boom-type road header machine' or 'header machine' is an ecavating equipment used in coal mining and now in tunnelling operartions. It consists of a boom- mounted cutting (tool) head, a loading device and a crawler travelling track to move the machine to the tunnel face. The cutting tool can be a general purpose rotating drum mounted either in line or perpendicular to the boom, so that it can excavate in forward and transverse directions. It can have 'special function heads such as jack-hammer like spikes, compression fracture micro-wheel heads like those on tunnel boring machines, a slicer head like a gigantic chain saw for dicing up rock'. It can have simple traditional jaw- like buckets of normal excavators. It was developed and patented first by Dr. Z. Ajay of Hungary in 1949, A schemaic view of the machine is in Figure 7.13 © [Sorce- wikipedia]

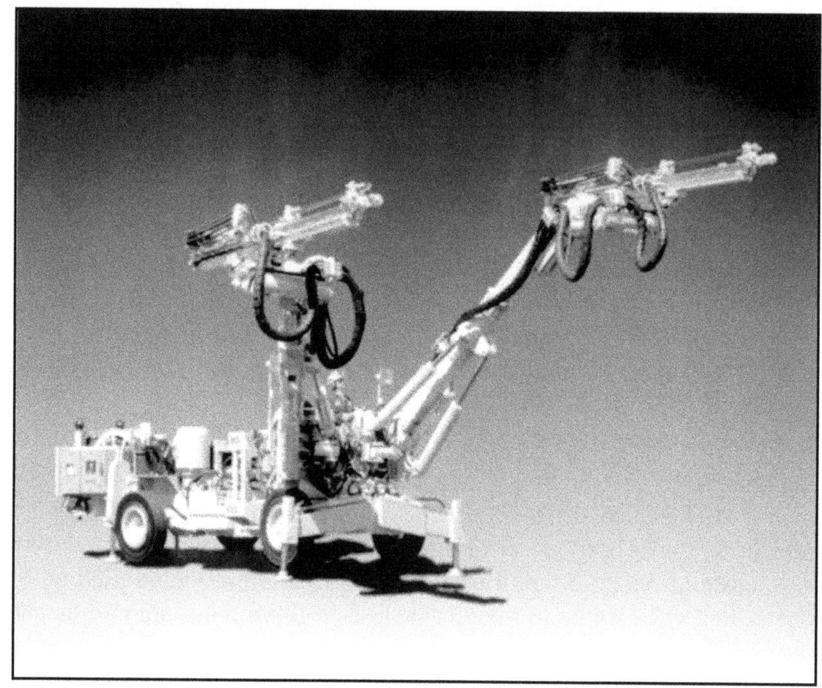

Source: Sandwik web site

Figure 7.13(a) A view of Tunnel Runner 2LF Jumbo.

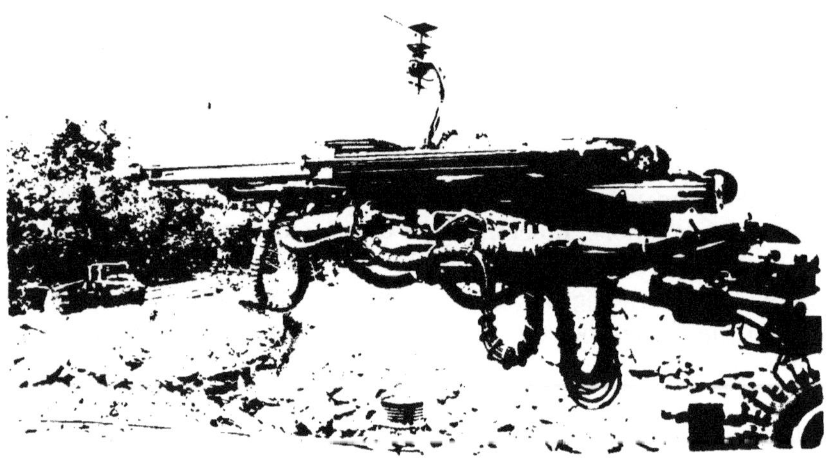

Figure 7.13(b) Tool Boom of an Elector Pneumatic Jumbo used on Konkan Railway

Source: Wikipedia

Figure 7.13(c) Road Header.

7.9.2 Mucking Operation by Road vehicles.

It has become a common practice now to do the mucking operation using road vehicles, avoiding need to lay temporary tracks and use of rail trucks and locomotives. The loading operation is another bottle neck in expediting progress of tunnelling. The loaders need high energy and use of diesel engines for operating them add to the pollution and need additional exhausting arrangements, especially in long tunnels. In order to simplify and expedite this operation, high capacity electrically operated Electro-Hydraulic 'Hoggloaders' have been developed. They come with digging blades, gathering arms, lifting arm and an inclined chain conveyor all built into a mobile unit, which can transmit the muck from face to its rear for dropping into a dump truck behind. The gathering arms simulate human arms and feed the lifting arm which in turn feeds the chain conveyor. A built-in sprinkler system automatically sprinkles water on gathered muck at the toe of conveyor, which reduces pollution. Figure 7.14 shows a schematic view of one such loader used on Konkan Railway construction (Anand, 1994)[10]. Modern such loaders have loading speed of 3 to 4 cum per minute. They are rubber tyred and are diesel powered for haulage from portal/ parking niche to the face, while working on electric power to operate the hydraulic machinery for loading.

Taking a cue from the needs on Konkan railway construction and likely future needs, indigenous equipment producers started developing and supplying easily maneuverable smaller capacity loaders, which have been used in smaller tunnels in many projects since. The tyres of the loaders are provided with protective chains to guard their tyres from damage due to sharp ended rock pieces. Dumpers of high capacity and shorter turning circle

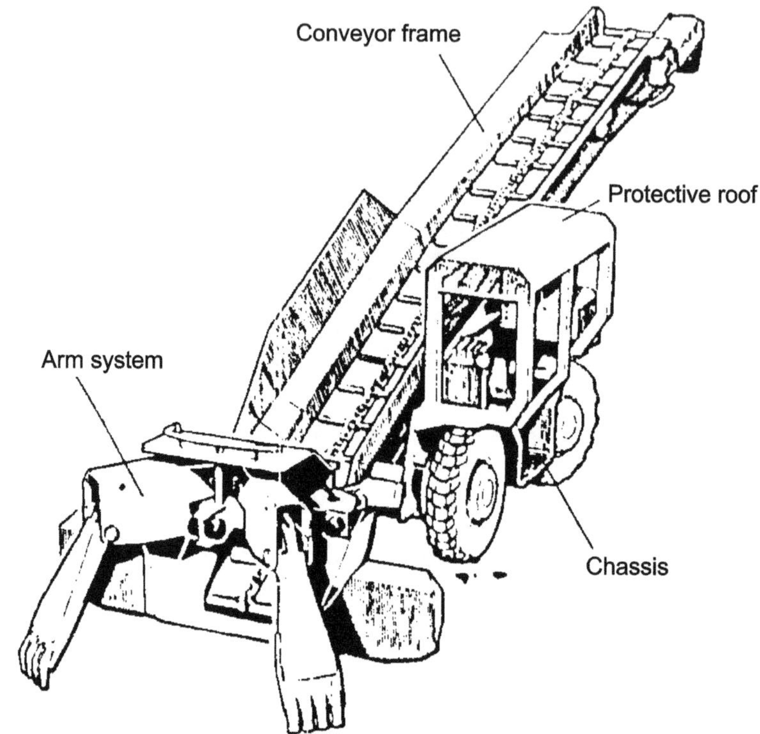

Source: Reference-10

Figure 7.14 Schematic of an electro- Hydraulic 'Hoggloader'.

for turning inside the tunnels are used for removing the muck from loader to the portal.

7.9.3 Ancillary Equipments and Arrangements

The modern drilling tools and loaders will thus be seen to have to be electrically powered at work at an assured high voltage. They will require an assured supply and requirement will be heavy also. In long tunnels, full length will have to be properly lighted for movement of men and the equipment including dumpers. The gases emitted by explosives and trucks and CO_2 from the work force have to be exhausted quickly, for which large diameter (600mm) ducts are run from the face to portal with blowers at intervals (about 250 m) which also need electric power. Water has to be supplied to the drilling tools for lubricating holes and to the loaders for dust suppression, which means provision of water supply lines with booster pumps, needing power. Seepage water from sides and water leaking from joints and fissures and that used for cooling tools and washing holes will be substantial in long tunnels, which require a high power pump and large dia (100 mm) pipe line for dewatering to portal. Hence as a first requisite, electric supply cable has to be run from portal to face and it should be extendable. In very long tunnels, intermediate booster transformers may have to be provided to ensure the required voltage at demand point.

1. *Case Study*

Various steps taken in the construction of Nathuwadi tunnel, mentioned above, are listed below, which will give an idea of steps taken for maintaining a progress of 80 to 120 m per day per face.

(i) Provision of an air duct line 900m dia (fabricated using 22 swg. GI sheets) with angle rings and rubber ring joints; Provision of 25 HP 1440 rpm blower at 250 intervals in addition to one at portal end.

(ii) 100mm dia GI water supply pipe line leading from 4 nos storage tanks of 10,000 capacity each for supply of 20,000 to 30,000litres per drilling cycle; Three bore holes had to be drilled for providing the supply of water.

(iii) Installation of a 100mm dia. Pipe line for dewatering the muddy water and a high power electric pump at the face for discharging the same.

(iv) Treating and maintaining the road bed in sufficiently good condition for the plying of tippers (7 such used) efficiently and quickly as well as reducing wear and tear on them as well as other machinery

(v) Refuges provided for use as trolley refuges later, were enlarged to accommodate jumbo/ tippers for loose checking and parking when not working. They were large enough for these to reverse as required.

(vi) Step-up transformers were installed in some of the trolley refuges for maintaining voltage.

(vii) Provision of telephone communication at face, at intervals along, portal, site office, maintenance shop.

(viii) Wireless communication facilities between working face, other face of tunnel, and connected to Chief Engineer's and Deputy chief engineer's offices

(ix) Provision of a small site workshop with sufficient spare parts for maintaining the machinery.

(x) Provision of a 250 -kVA Diesel Gen. set for meeting with power demand during shut downs and emergencies.

For the workmen, site camp was provided in a hygienic environment with board, lodge, lighting and minimum medical facilities.

7.10 RECENT DEVELOPMENTS IN DRILLING AND BLASTING

Under the Drilling and Blasting methods, discussed in this chapter there have been considerable recent developments. Present scenario is presented here, with a case study on a Railway tunnel driven through a 'competent 'rock formation near Mumbai. There are two basic methods of such tunnelling, viz.,

(i) Cyclic drilling method, and
(ii) Continuous boring machine method

The second method is in its early stages of development, but still it is competing with the former in speed and cost. The first is the old traditional method, which has been continuously improved, and is the one commonly used. Being a versatile method, it is suitable for any type of rock, soft sedimentary ones to hard igneous rocks. The case study described here illustrates present stage of development of this traditional 'drill and blast' method, using latest technologies.

Case Study - Karjat Tunnel[11]

Karjat tunnel is one of the tunnels recently done by the Central Railway for their Panvel - Karjat new BG line through the foothills of Matheron Hills done in 1999- 2004 period. It is 2.7 kms long in one straight alignment. Its minimum excavation width and height are 6.2 m and 7.7 m respectively. It comprises of two parts, rectangular section 4.5 high and arch 3.2 m above springing level. Overall cross sectional area is 43.48 sqm. for the minimum excavation line and pay line 46.362 sqm. The overburden is about 20 m average and maximum overburden is 168 m. It has two shafts for ventilation purposes, first 1212 m from Karjat end and second one 900 m from same. It has a gradient of 0.5 % for full length It has 98 holes for full face (with 60 to 63 in heading portion and 24 to 26 in benching portion). Full face or Heading + Benching method was used to suit the site conditions. Inclination of holes were kept from perpendicular to maximum 22°, as per design criteria.

Holes were drilled using a jumbo and generally depth of hole was kept at 3.6 m. Explosives used is a combination of Gelatine as main one with Aluminum Nitrate (ANFO). The types of Gelatine used were: Noble (most effective with pull 2.72 to 2.9m); Power Gel (pull 2.4 to 2.6 m); Shakti (Pull 2.2 to 2.5m); and Rajdine (Pull only 1.9 to 2.2 m). The amount of fumes emanating was highest for Noble and least for Rajdine. The diameter of both gelatin stick and ANFO stick was 25 mm. Sequence of placing them in hole was four of gelatin followed four of ANFO sticks (totaling 2.4 m) and packing the remainder of the hole with earthen sticks rammed for packing. For blasting 25 milli-seconds Delay detonators were placed in the 25mm Gelatin stick called (primer).

Sequence of work and time taken for each activity for a 'Full face' work were as follows:

Activity	**Time taken for full section**	**Brief description of activity**
Survey/ set out	1 hr	Mark centre line (vertical) on face of tunnel, bottom level and spring level (for arch) taking into account gradient etc.
Drilling	6 to 7 hrs	Mark profile on face, bring jumbo to face and drill holes (only wet drill holes) as per design pattern -
Dry shotcreting was preferred letting water in Explosive charging		Clean holes with compressed air, charge all holes, except dummy holes {Smoking / sparks strictly prohibited during process)
Arrangement for blasting	30- 40 mts.	Pull back water pipe line, air pipe, lighting arrangements etc. Connect wires from detonators in series and lead wire taken to safe distance; .
Blasting		After checking everything, exploder to blast.
Defuming	1 hr	Start and run exhaust fans for defuming.
Top scaling	3 hrs	Loose stones to be removed from crown portion exposed surface, bringing jumbo near blasted face and where not reachable men standing on debris
Mucking	8 hrs	Withdraw jumbo, bring in excavator/ loader followed by 4 to 5 dumpers and clear muck
Final scaling	1 hr	Recheck, bring in jumbo to face and remove any loose stone from and near about face
Bottom cleaning	30 mts.	Final clearing operation by excavators.

A few improvements made or found available during these tunnelling operations are worth mentioning.

(a) *Tunnel Support:* Conventional shape Bent ISB ribs were used for supporting the arch portion. The bent quadrants have a kick-up at crown for accommodating concrete slick line. At the other end (foot at springing) it has a straight piece having an off-set for rib feet. The rib feet are enclosed in concrete with provision of steel dowels. Later during benching, column legs were placed below, if required. In such cases, a straight flat piece was welded to heading rib at springing level and the column rib cap was bolted to the same.

(b) *RCC precast concrete laggings* (in concrete M20, same as of lining concrete, were over the outer flange of rib supports and space behind filled with spalls. Thus the lagging concrete was compatible with line concrete and had better integration. Alternatively, shotcreting has been done to cover space between ribs with steel mesh reinforcement for support– Dry shotcreting, by having water mixing arrangement at the nozzle. Shotcreting had to be done in two stages viz., first provide a thin layer, then embed the reinforcement mesh and hold it to rock with steel dowels installed by drilling. Provide final shotcrete layer over same. In many cases , it was found the mesh does not get fully embedded in final layer and resort had to be made for excessive shotcreting in second layer. This has been found a time consuming process not liked much by men. Robotic arms have been developed, which carries out the operation from some distance. Also, fiber reinforced shotcrete mixes have been developed, which avoids need for embedding mesh reinforcement. It is learnt these later technologies have been used on tunnels bored in same area for Mumbai- Pune Expressway.

(c) *Rock Bolting*

Fully grouted rock bolts have been provided for preventing rock falls and strengthening the exposed surface. The work requires care so as not make them redundant. On this tunnel, after drilling the holes and cleaning them, ready mix grout of correct consistency was pushed into the hole by inserting polythene grout pipe till grout fills the hole and comes out. The pipe in then withdrawn and rock bolt is pushed into the hole.

Alternatively, they have used 'cement capsules'. Required number of capsules (for filling 80% of the hole) are wetted in water till bubbles come out to indicate they are fully wetted and have been pushed into the hole. Then the rock bolt is pushed into the hole for full depth and the grout oozes out of the hole. Rock surface should be made flat around so that bolt plates butt uniformly and nuts tightened. About 10% of bolts are test checked after setting. They had used epoxy grouting also for rock bolts, following similar methodology. In such cases, after the epoxy capsules are filled in, the bolt is inserted and turned. As it is turned, it punctures the epoxy capsules and epoxy fills the hole.

7.11 PROBLEMS IN TUNNELLING

Common problems met with during tunnelling in rock are (i) finding the rock in layers or meeting with fault zones (ii) seepage flow (iii) hazards. In the first case, the loose rocks or masses are rock bolted so that they would not fall down due to vibrations of moving vehicles and trains, or if too small to be held so or too loose to remove them deliberately. In case of seepage flow, channelize the flow and provide for proper drainage to outside, if it cannot be stopped or diverted otherwise.

Hazards refer to roof or side collapse after blasting, during supporting arrangements or mucking operations. They can occur during blasting operations also for next pull after completing the previous pulls with supports, due to poor nature of rock or during benching work after completing the heading, due to excessive charge. These are common occurrences in Drill and Blast type of work in areas where rock is mixed with soil. Most common method of dealing with such hazards is to provide a shield (roof like structure) below the affected and vulnerable portion, suitably supported at ends and sides and do the work of providing permanent rib supports at closer intervals below; clearing muck and lining. Any debris falling due to accretion of the collapse would be held on the shield.

A typical arrangement used on a tunnel in the Hasan- Mangalore link project[12] done in 1960s is shown in Figure 7.15. Hasan - Mangalore rail link project covered 92 km of which 55 km passed through Shiridi Hills in Western ghats[12]. It involved construction of 50 bored and 15 cut and cover tunnels totaling 11 km overall length. The geology of the hill comprised of stratified rock with intervening patches of talus formation. Though the track on that section at that time was for a MG line, tunnels were made to BG standards. Prominent rock was gneiss with intrusions of pegmatite and quartz veins. Most of tunnelling was done by blasting. Except where there was continuous rock, rib, wall plate and post type of supports were provided at 20 cm, 100 cm and 150 cm spacings for posts and 25 cm and 30 cm spacing for ribs.

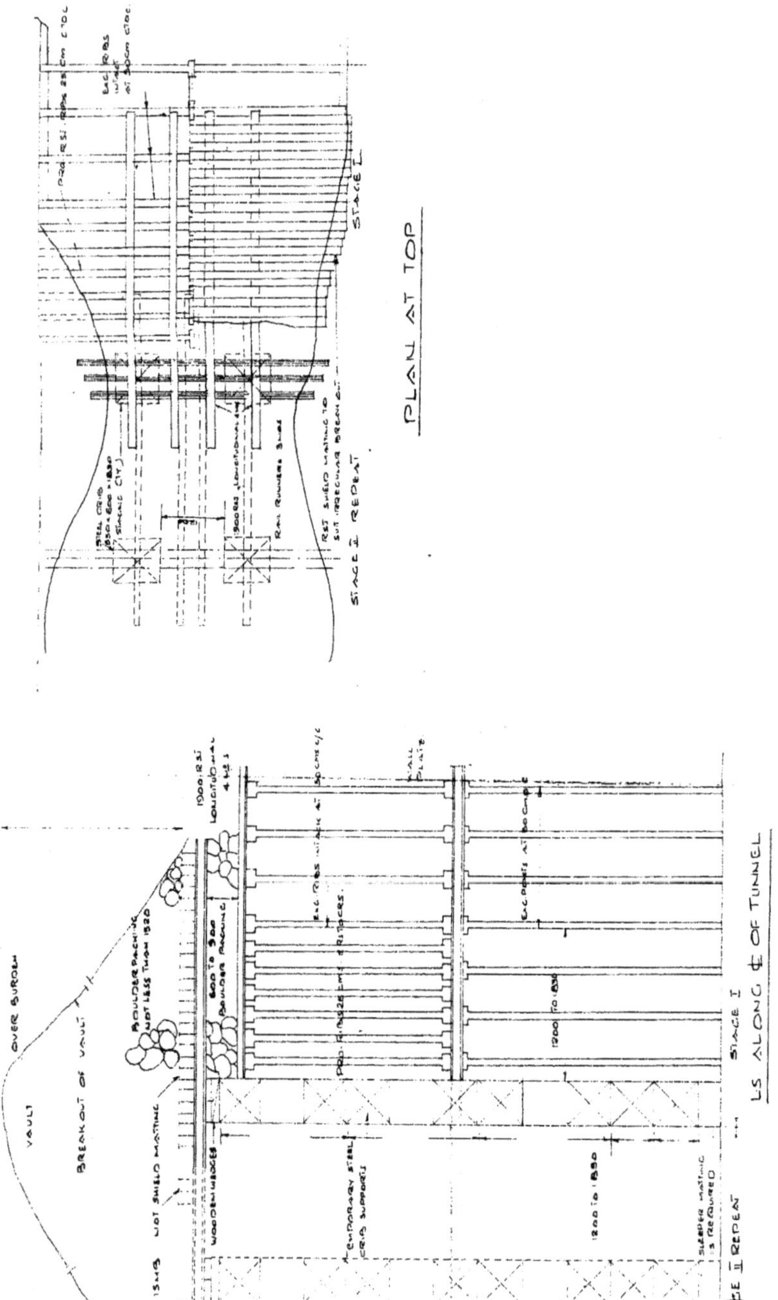

Figure 7.15 Conventional Method of Permanent shield for Hazards[12].

Hazards were met with In 6 tunnels. They were mostly dealt with by conventional method of providing a flat cover, i.e., a shield made of steel matting using RSJs touching each other transversely across over the affected length, i.e., over the length of the vault created by the collapse. After the vault becomes stable, a number of ISBM 150 mm RSJs are laid across over longitudinal rail runners at 1.2 to 1.5 m spacing, running along length of tunnel. The runners are supported in the rear over the erected stabilised ribs of the tunnel; and over temporary crib supports in front as shown in the figure 7.15. The ISBM mat is with flanges touching each other. A 1.5 m boulder packing is done over the shield. Any more debris coming down will be held over the boulder packing. With the shield protection, the debris below is removed for short lengths, (say 1.2 to 1.5 length) and permanent ribs and supports are erected and consolidated. Then next length is cleared and rib and supports erected. The process is repeated by extending shield over affected length as required, by shifting the crib supports in front.

Most cases could be dealt with in this manner, except for a hazard in Tunnel 12A, where, the debris heap was too high and this method could not be used. There, they adopted a forepoling methodology, similar to the arrangement shown in Figure 6.4 (b) to go through the heap of muck and advance the 'face', treating it as a hill formation. Sharpened scrap drill rods 2.4 to 3.6 m long were used as the poles.

There are cases where, the collapse has extended over full depth of the overburden, leading to a chimney formation. In such cases, usual practice is to stabilise the opening by driving in some piles at the periphery of hole and lay a shield (matting of ISMB at a higher level and provide a lean concrete floor (0.6 to 0.9m) thick. After it sets, remaining portion is filled up to prevent any inundation. Alternatively, in some cases, they have sunk a well of suitable size to cover the hole and have grips on sides. The remaining part of the pit around the well is filled up to natural ground level. This well can serve as a ventilation shaft later[12].

7.12 REFERENCES

1. Pequinot, C.A., (1963) "Tunnels and Tunnelling'- Hutchinson, Scientific and Technical, London
2. IS: 5878- Part II/ Section 1 (1970) Construction of Tunnels Part II- 'Underground Excavation in Rocks- Section 1- ' Drilling and blasting'- Bureau of Indian Standards, New Delhi.
3. Limaye, G.K., (1979) 'Tunnels- Notes of Lectures at the Institute of Advanced Track Technology '- Institute of Railway Civil Engineering. Pune.

4. Kulkarni, B.R., (1994) ' Construction of 6.5 km Karbude tunnel', Indian Concrete Journal. February 1994 pp 61-64, Associated Cement Companies, Mumbai.
5. SJVNL (2005), Reports and Presentation on 'Designs of tunnel Cross sections, Drainage System for Tunnel No1 and 2 of Katra- Laole section of Udhampur- Srinagar Link Project in Jammu and Kashmir', KRCL 2005.
6. Nick Barton (2008), 'Training Course on Rock Engineering for Drill-and Blast and TBM Tunnelling-----' Lecture NB-1. ' Introduction to the Q-System of Rock Mass Classification' - Indian Society for Rock Mechanics and Tunnelling (ISMTT) and Central Soil and Materials Research Station,
7. Russo,G.(2008), 'A simplified Rational Approach for the Preliminary Assessment of Excavation Behaviour in Rock Tunnelling, Tunnels et Ouvragas Souterains'pp 1-8.
8. Pokrovsky,H.M (1977) 'Driving Horizontal Workings and Tunnels', Translation by Savic, L.V., M.I.R. Publications, Moscow.
9. Parker, A.B., 'Planning and Estimating Underground construction', McGraw-Hill Publications.
10. Anand, V., (1994) 'Tunnelling Machinery for Konkan Railway Project' Indian Concrete Journal. February, 1994. pp 75-78, Associated Cement Companies, Mumbai.
11. Garg, S.K. (2014)- Tunnelling Using Conventional Drill Blast Method, and Support System. Case Study of Karjat Tunnel' National Conference on Management of P.Way works through need based Outsourcing and Design, Construction and Maintenance of Railway Tunnels'*, Jaipur 2014, Institute of Permanent way Engineers (India), New Delhi pp 275-291
12. Venkata Rao, S.N. and Seshadri, S, Paper on Hasan- Mangalore Railway Project, Shiridi Ghats - Tackling of a Hazardous tunnel 12A. (Unpublished)
13. Prakash, L., 'Construction experiences in Katra- Dharam Section of USBRL Project.', Konkan Railway Corporation.

ANNEXURE 7.1

Q Tables for Use with Histogram Logs (Source Nick Norton)[6]

1	Rock quality designation	RQD
A	Very poor	0 – 25
B	Poor	25 – 50
C	Fair	50 – 75
D	Good	75 – 90
E	Excellent	90 – 100

Note: (i) Where RQD is reported or measured as < 10 (including 0), a normal value 10 is used to evaluate Q.

(ii) RQD intervals of 5, i.e. 100, 95, 90 etc/ are sufficiently accurate

2	Joint Set number	J_n
A	Massive, no or few joints	05–10
B	One joint set	2
C	One joint set plus random joints	3
D	Two joint sets	4
E	Two joint sets plus random joints	6
F	Three joint sets	9
G	Three joint sets plus random joints	12
H	Four or more joint sets random Heavily jointed. 'Sugar cube' etc	15

J	Crushed rock, earthlike	20
Note:	(i) For intersections, use (3.0x J_n) (ii) For portals, use (2.0 x J_n)	

3	**Joint roughness number**	**Jr**
(a)	**Rock-wall contact and b) Rock-wall contact before 10cm shear**	
A	Discontinuous joints	4
B	Rough or irregular, undulating	3
C	Smooth, undulating	2
D	Slickensided, undulating	1.5
E	Rough or irregular planar	1.5
F	Smooth, planar	1.0
G	Slickensided, planar	0.5
Note:	(i) Descriptions refer to small scale features and intermediate scale features, in that order	
(c)	**No rock-wall contact when sheared**	
H	Zone containing clay minerals thick enough to prevent rock-wall contact	1.0
J	Sandy, gravelly or crushed zone thick enough to prevent rock wall contact	1.0
Note:	(i) Add 1.0 if the mean spacing of the relevant joint set is greater than 3m. (ii) Jr = 0.5 can be used for planar slickensided joints having lineations, provided the lineations are oriented for minimum strength	

4	**Joint alteration number**	**θ_r (approx.)**	**J_a**
(a)	**Rock-wall contact (no mineral fillings, only coatings)**		
A	Tightly healed hard non softening impermeable filling, i.e. quarts or epidote		0.75
B	Unaltered joint walls surface staining only	25-35°	1.0
C	Slightly altered joint walls - Non softening mineral coatings, sandy particles clay free disintegrated rock etc	25-30°	2.0

D	Silty or sandy clay coatings small clay fraction (non-softening)	20-25°	3.0
E	Softening or low friction clay mineral coatings kaolinite or mica. Also chlorite, talc, gypsum, graphite etc and small quantities of swelling clays	8–16°	4.0
(b)	**Rock-wall contact before 10cm shear (thin mineral fillings)**		
F	Sandy particles, clay-free disintegrated rock, etc	25-30°	4 .0
G	Strongly over-consolidated non-softening clay mineral fillings (continuous, but < 5mm thickness)	16–24°	6.0
H	Medium or low over-consolidation, softening, clay mineral fillings (continuous, but <5mm thickness)	12–16°	8.0
J	Swelling - clay fillings, i.e. montmorillonite (continuous, but <5mm thickness) Value of Ja depends on percent of swelling clay size particles and access to water etc.	6–12°	8–12
(c)	**No Rock-wall contact when sheared (thick mineral fillings)**		
KL M	Zones or bands of disintegrated or Crushed rock and clay (See G,H,J for description of clay condition)	6-24°	6.8 or 8.12
N	Zones or bands of silty or sandy clay small clay fraction (non-softening)	–	5.0
OP R	Thick continuous zones or bands of clay (see G,H,J for clay condition description)	6-24°	10,13, or 13–20
5	**Joint water reduction factor**	**Approx. water press. (kg/cm^2)**	**J_w**
A	Dry excavations or minor inflow i.e. <5 l/mm locally	<1	1.0

B	Medium inflow or pressure occasional outwash of joint fittings	1-2.5	0.66
C	Large inflow or pressure in competent rock with unfilled joints	2.5–10	0.5
D	Large inflow or high pressure, considerable outwash of joint fillings	2.5–10	0.33
E	Exceptionally high inflow or water pressure at blasting decaying with time	>10	0.2-0.1
F	Exceptionally high inflow or water pressure continuing without noticeable decay	>10	0.1-0.5%

Note: (i) Factors C to F are crude estimates increase J_w if drainage measures are installed.

(ii) Special problems caused by ice formation are not considered.

6	**Stress reduction factor**	**SRF**
(a)	**Weakness zones intersecting excavation, which may cause loosening of rock mass when tunnel is excavated**	
A	Multiple occurrences of weakness zones containing clay or chemically disintegrated rock very loose surrounding rock (any depth)	10
B	Single weakness zones containing clay or chemically disintegrated rock (depth of excavation ≤ 50m)	5
C	Single weakness zones containing clay or chemically disintegrated rock (depth of excavation > 50 m)	2.5
D	Multiple shear zones in competent rock (clay free) loose surrounding rock any depth	7.5
E	Single shear zones in competent rock (clay free) (depth of excavation >50m)	5.0
F	Single shear zones in competent rock (clay free) (depth of excavation >50m)	2.5
G	Loose open joints heavily jointed or sugar cube etc (any depth)	5.0

Note: (i) Reduce these values of SRF by 25-50% if the relevant shear zones only influence but do not intersect the excavation.

(b)	Competent rock, stress problems	σ_c/σ_1	σ_φ/σ_1	SRF
H	Low stress, near surface open joints	>200	< 0.01	2.5
J	Medium stress favourable stress conditions	200–10	0.01–0.03	1
K	High stress, very tight structure Usually favourable to stability, may be unfavourable for wall stability	10-5	0.3-0.4	0.5-2
L	Moderate slabbing after 1 hour in massive rock	5–3	0.5–0.65	5–50
M	Slabbing and rock burst after a few minutes in massive rock	3–2	0.65–1	50–200
N	Heavy rock burst (strain- burst) and immediate dynamic deformations in massive rock	<2	1	200-400

Note: (ii) For strongly anisotropic virgin field (if measured). When $5 \leq \sigma_1/\sigma_3 \geq 10$ reduce σ_c to 0.75 σc. When $\sigma_1/\sigma_3 > 10$ reduce σ_c to 0.5 σ_c where σ_c is unconfined compression strength, σ_1 ,σ_3 are the major and minor principal stresses. σ_φ is maximum tangential stress (estimated from elastic theory.)

(iii) Few case records available where depth of crown below surface is less than span width. Suggest SRF increase from 2.5 to 5 for such cases.

(c)	Squeezing rock : plastic flow of incompetent rock under the influence of high rock pressure	σ_φ/σ_c	SRF
O	Mild squeezing rock pressure	1-5	5-10
P	Heavy squeezing rock pressure	>5	10-20

Note: (iv) Cases of squeezing rock may occur for depth $H>350\ Q^{1/3}$ (Singh et al.1992) Rock mass compression strength can be estimated from $q = 0.7\ Y\ Q^{1/3}$ (MPa) where Y= rock density in kN/ m^3 (Singh, 1993)

(d)	**Swelling rock - chemical swelling activity depending on pressure of water**	**SRF**
R	Mild swelling rock pressure	5-10
S	Heavy swelling rock pressure	10-20
Note:	J_r and J_a classification is applied to the joint set or discontinuity that is least favourable for stability both from the point of view of orientation and shear resistance T (where T = σ_n tan (J_1/J_2)	

Source: Nick Barton[6]

ANNEXURE 7.2

Case Study-Tunnelling Through Varying Rock Conditions

A 7.2.1 Typical Case- Hills in Himalayan Region

The Udhampur- Srinagar-Baramulla-Rail Link passes through a region with young rock formation and tunnelling through them has thrown a lot of challenges to engineers. Considering the magnitude of the problems likely to be met with, very exhaustive geological and geophysical studies had been conducted and special design procedures adopted. The engineers involved in the project went through a special training course conducted by international experts in association with the Indian Society for Rock Mechanics and Tunnelling Technology in association with the Central Soil and Materials Research Station[6]. The design of the tunnel section and support arrangements for tunnels were decided based on Rock Mass Quality (Index). This annexure covers the basic approach adopted and also a few typical construction problems met with on a few tunnels on Katra- Dharam section of the project.

Original alignment of Katra- Dharam section covered a length of 71 km (Km 30 to 100.868) of the link[13]. This included 31 tunnels totaling a length of 57.33 km and 46 bridges totaling a length of 9.14 km. During progress of work in early stages, a number of difficulties were experienced, especially during tunnelling e.g., Portal 2 of Tunnel 1 and Portal 1 of Tunnel 2 through the shear and fault zones. A review was made and alignment was revised at the difficult locations, so as to minimise number of portals by combining some tunnels and relocation of some station yards and bridges. The revised alignment came to 74.072 km. The revised alignment has reduced the number of tunnels to 19 reducing number of Portals, which were most difficult to deal with. The total length of the 19 tunnels is 62.195 km. Number of bridges also could be brought down to 33 totaling 7.49 km. The terrain is so difficult that part of yard lines and platform extend into nearest tunnel or over part of adjacent bridge.

The alignment passes through areas with Dolomite overlying younger Sivalik group of rocks. Contact between the two rock zones are marked by a thrust zone called 'Reassi Thrust' (MBT), as between T1 and T2. Since tunnelling through thrust zones is time consuming and expensive, it had to be modified so that it cuts across the thrust zone in almost a perpendicular direction. The rock structure over the length has been quite varying. Also in some lengths, seepage has been a problem, e.g., Sirban Dolomite between tunnels T2 and T5 met with closely jointed / fractured, sheared dolomite with water. There were heavy over breaks and occurrence of frequent cavity formation. Km 51- 73.4 lies in Subathu formation e.g., first 400 m of T-1 consisted of red coniferous shale, silt stone, limestone, and sandstone Murree formation consisting of alternate bands of claystone, siltstone and sandstone was met with. Km 73.800 to 83.800 is along a local fault known as Sangal Kund Fault, where also some realignment was called for. In next length (upto 100.867 km) tunnels pass through a 'fair' category rock, passing through Murree formation with deep overburden/slide derived from parent Murree rocks. Overburden material itself is considered poor to very poor tunnelling medium. The length was fairly dry through, with isolated patches of sub-surface water. Beyond Km 100.867 to Km 120 the structure comprises of Salkhala group of rocks mainly consisting of slate, phillytes, quartzite, schist and granitic gneiss, which rocks are better tunnelling media, compared to the Murree rocks.

To start with, the preliminary design of D section for tunnel was reviewed and an alternative form of Elliptical section with curved invert was decided upon and a horse shoe shape for wider tunnels, as they were found structurally more efficient for the strata here. Two basic sections adopted were one narrower one with a footpath alongside the track for tunnels upto 3 km length and one with a 3 m wide service road running alongside the track for tunnels over 3 km long. The two basic sections are shown in Figure A.7.1.

The elliptical section is structurally more efficient than D section as can be judged from the following table of comparison for a single line tunnel. The arc shape of sides resists side pressure much better. The invert continuity adds to the lateral resistance of the structure. Another advantage is that the internal dimensions of the tunnel need not be changed upto 2.75 degree

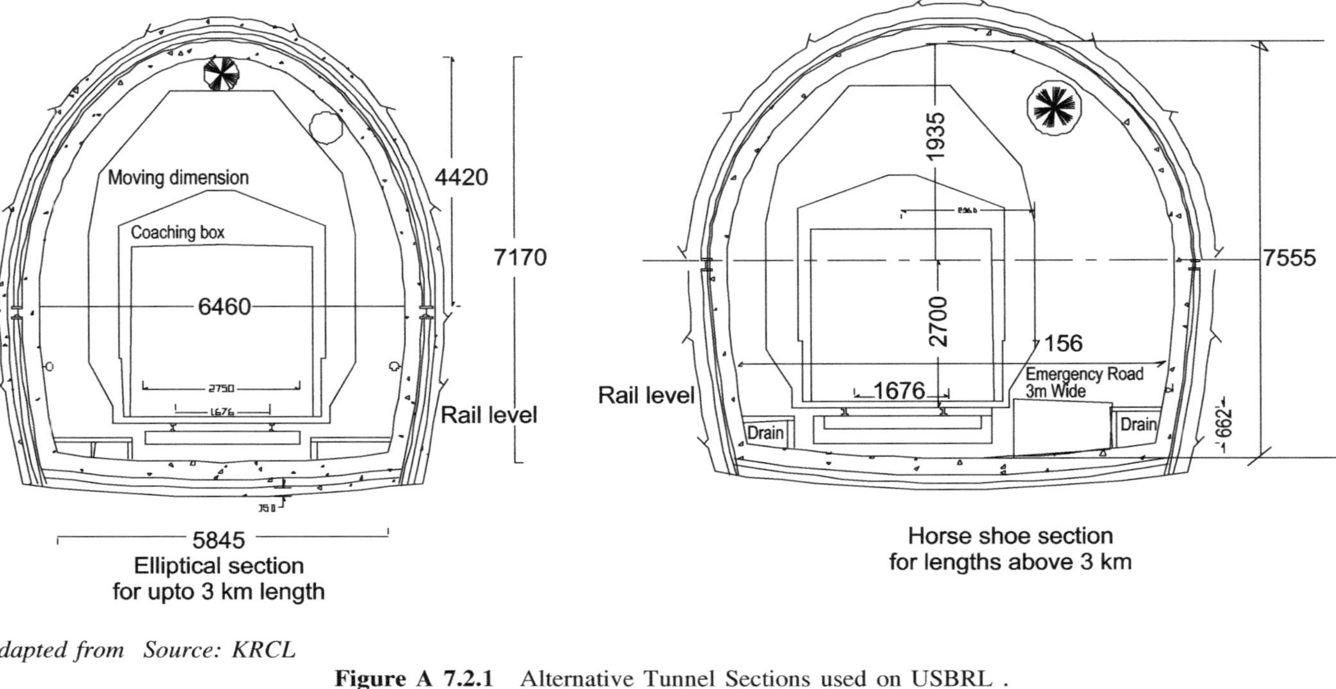

Adapted from Source: KRCL

Figure A 7.2.1 Alternative Tunnel Sections used on USBRL .

curve, as the requisite clearances required on the curves as per SOD can be achieved by shifting the centre line of track.

Comparison between D- shaped and Elliptical Tunnel in Structural performance[5]

Description	Maximum axial force	Maximum bending moment	Maximum shear force	Maximum displacement
	kN	kN-m	kN	M
	With Top pressure of 15000 kg/M and Side pressure of 5000 kg/ M			
D- shaped tunnel	412.5	137	15129	0.083
Elliptical tunnel	567	35.80	9025	0.024
	With Top pressure 15000 kg/ M			
D- shaped tunnel	412	48.10	8565	0.03
Elliptical tunnel	639	35.65	9635	0.028

Source: SIVNL / KRCL

The planners also evolved an innovative design for the tunnels passing through very poor strata, where difficulties were expected in supporting the heading arch during benching. They eliminated provision of wall beams at SDL of the arch and in lieu introducing support legs at SDL for the arch ribs, (elephant foot) as indicated in Figure 7.8. In such a case, the excavation for benching can be done in between the arch ribs without disturbing the support legs.

A 7.2.2 Construction Problems[13]

The frequent problems met with during excavation were in form of (a) deformation in tunnel cross section (b) heavy seepage of water (even reaching over 100 lit/ min) (c) frequent cavity formation and (d) squeezing of soils and low stand up time of some soils (e) difficulties in doing portals through shear/ thrust zones

1. The first type of problems were, to a large extent contained, by using the modified egg shaped design and invert treatment and use of rock bolts in conjunction with shotcreting for preliminary support. Elimination of wall beam and provision of independent supports (elephant foot) for arch ribs and providing curved section for side supports, and rock bolting and shotcreting as required, helped in reducing problems that arise during benching. Continuous monitoring with instrumentation was done using Pressure cell, Load

cell and Tape extensometer. Tunnel support systems were modified to suit and grouting, in case of water incursion, could be arranged for in time. In case of poor soils with low stand up time, umbrella pipe forepoling at roof level at fairly steep angles (50 to 150 degree) was done.

2. The general approach adopted for tunnelling in different strata was:-
   - Conventional method of excavation by heading and benching was followed. Deep probe holes (at least three times the length proposed to be excavated that day) were driven through the tunnel face in advance to find type of strata to be met with and sub soil water condition.
   - Road header was used for excavation of 'jointed rocks and soft to medium strata upto 100 UCS'. A combination of excavation by Road header with option of Drilling and blasting is followed.
   - Hard rock comprising Dolomite and Murree formation is excavated by Drill and Blast method, followed by conventional rock supports (using 150 x 150 H- beam ribs, shotcreting and rock bolting.)
   - Wherever drilling has not been possible and where the holes drilled do not stand intact, 25 and 32 mm SDAs are used.
   - Bottom closure is done with steel sections in most reaches.

Adoption of NMT and NATM methodology for tunnelling with support systems using self drilling anchor rods along with shotcreting with fibre reinforcement, especially in crown portion helps to contain tunnel movements and increases stand up time of excavated surface. Three specific cases of problems met with are mentioned here to indicate how specific cases were dealt with.

(i) Failure of Portal 2 @ Tunnel No 1.

Due to a heavy land slide at the face of this portal, the portal and tunnel for some length were damaged. The portal lies along the Reasi thrust. Fig A 7.2 shows the general view of the isolated portal land slide around this portal.

This was dealt with by clearing the debris and rebuilding the length by providing micro piles to support the structure. Side gaps near face were filled with Reinforced Earthwork.

(ii) Tunnel No2 -Tunnelling through shear zone[13]

(iii) Problems met with here were poor stand up time and water seepage. After about 250 m of tunnelling, it was found difficult to work further through the thrust zone due to extra-ordinary flowing conditions.

Source : KRCL

Figure A 7.2.2 Collapse at Portal 2 of Tunnel No1 due to land slide[13]

They were dealt with as follows. Tunnel face was temporarily closed and 4 m deep SDAs were drilled through the face and grouting done with cement grout (normal cement and super fine cement and Polyurethane chemical. Self drilling anchors 6 to 9 m long were used in lieu of normal rock bolts at the crown. Lattice girders with steel fibre reinforced shotcrete support were used for supporting arch roof. 6 m long pipe 51mm dia SDA forepoles at 300 mm crs. were driven in the crown to form an umbrella protection for advancing excavation in heading. Steel fibre Rock bolts were installed and shotcreting was done along with provision of the lattice girder supports every half a metre. Fibre concrete and strutting was done in the bottom for closure. Latter were removed during benching. The different actions taken are conceptually indicated in Figure A7.3.

(iii) Tackling cavity formation

Cavity formation has been a frequent occurrence in dolomite. They were accompanied with drainage problems also. They were dealt with by doing filling/ grouting the cavity with pumped in cement mortar/ concrete and simultaneously draining out water from the pocket. A ramp was formed with the debris filled in gunny bags till a 'cut off was made between loose muck and broken tunnel face'. Drainage pipes (upto 15 m long) were driven in through the dolomite roof to reach the cavity to drain the water collection there. Cement sand mortar in 1:3 ratio and/or lean concrete was pumped into the

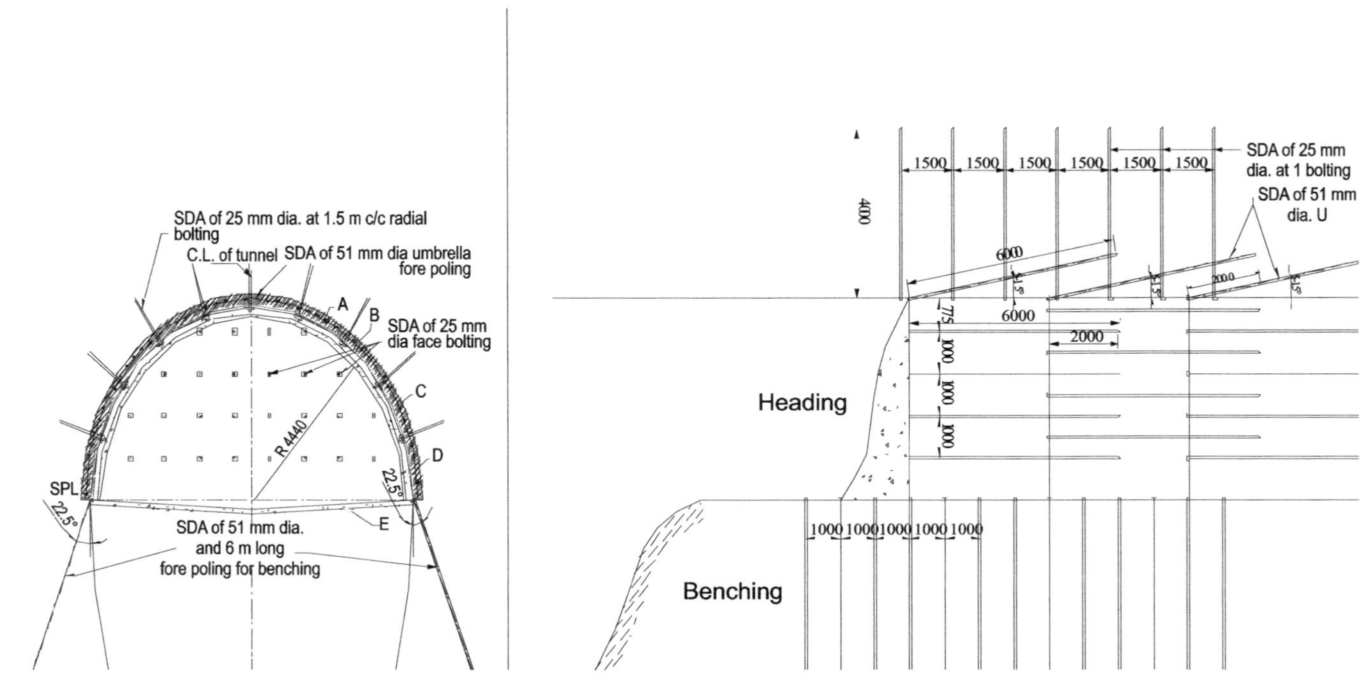

Source; KRCL

Figure A 7.3 Scheme for tackling Shear zone in Tunnel No2 Katra- Dharam section[13].

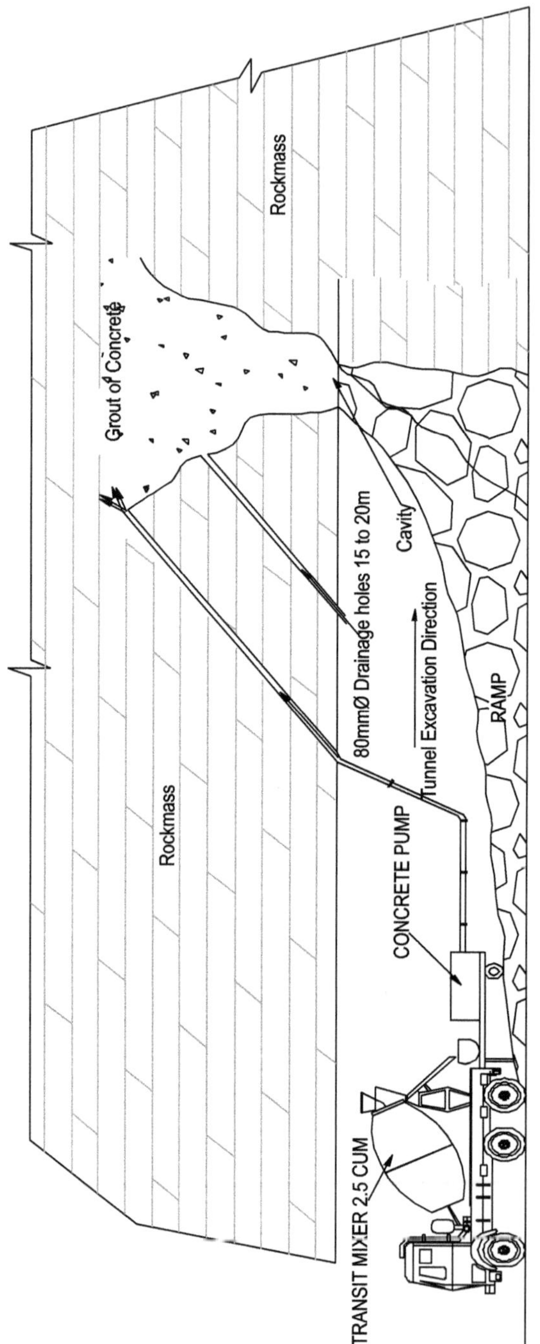

Figure A 7.4 Tackling a cavity formation in Tunnel No 5 Katra- Dharam section[13]

Source : KRCL

cavity through one or two pipes. For this purpose a transit mixer and pump were used taken close to the cavity. After filling the cavity and stabilising the area on crown/ cavity zone, work on the face was commenced by adopting multi drift excavation, with two or three openings till the cavity zone was crossed. Figure 7.4 shows the details.

Source : KRCL / IRICEN web site.

Figure 7.5 Tunnel Support failure due to swelling soil.

(iv) Swelling / Squeezing soils

In some locations in the thrust area, there were failures due to swelling pressure of soil against which the vertical supports of reverse U shape tunnel could not stand. One typical case in T1 is shown in Fig A 7.5. Such locations were avoided by changing the alignment at such locations as mentioned earlier also by going in for more efficient ellipse shaped sections. The general methodology suggested by designers for such locations has been covered in Para 7.7.5.

Similar methodologies have been to be followed for tunnelling in rocks in geotectonic region e.g., for Barat twin highway tunnels (2.8km each) near

Caracas. Each tube is providing for 3 lanes of Road shaped in a D form 12.43 m wide. Near the portals at ends and at locations where the roof of the tunnel is in weak zones, they opted to use Jumbo mechanical excavators, loader shovels and dump trucks. At other locations they used Drill and Blast method. In both cases, they followed sequential excavation method. Light supports were provided in form of projected concrete and rock anchors where rock is good. At weaker locations and locations with infiltration of water, they went in for thicker section of shotcrete, rock anchors and metal trusses as primary support. Permanent lining concrete has been of three types, varying in thickness and designed to suit the rock condition met with. [Tunnel 3/14, May 2014]

CHAPTER

8

Metro Tunnels

8.1 INTRODUCTION

In developing countries metros are generally understood to refer to underground rail systems. Metro rail systems need not necessarily be entirely underground. A major part of the metro rail in New York and Chicago has been and is being constructed on elevated structures. In London the metro rail lines are below ground in the city area and on the surface in the outskirts. BART (Bay Area Rapid Transit) in San Francisco revived the idea of using elevated structures for the metro over road medians for a considerable length (37 km elevated, 36 km in tunnel and 40 km surface, then). The cost of construction vis-a-vis aesthetics and space shortage on the chosen route now dictates choice of form of construction. The first phase of Singapore metro comprises a short surface route, a 41.4 km long elevated and a 20-km underground route. In Indian conditions the initial cost of construction would be in the ratio of 1:3:6 to 1:4:10 for the three forms (surface, elevated an underground).

Metro or subway tunnels refer to a continuous underground passage provided for movement of commuter trains within cities and suburbs. They have to be considered separately as their design is more restrictive and has to allow for a number of special features related not only to the vehicles using them, but also to city characteristics. Also, they have to be at shallow depths. The choice of construction methodology also imposes certain limitations. Alignment of the route is primarily dictated by the likely patronage pattern along the line. A subway has not only to provide space for running tracks, but also for location of stations at close intervals of about 0.80 to 1.20 km.

8.2 ROUTE SELECTION AND CONSTRUCTION METHODOLOGY[1,2]

8.2.1 Alignment

The first problem in locating a metro tunnel is route selection. While location of tunnels for roads and railways for intercity and other traffic is dictated mostly by costs (construction and operating costs), permissible gradients and curvature based on haulage consideration and topography, the route for Metro tunnels affect not only costs and construction problems, but also the traffic that will be generated and the development (commercial/residential) it will induce along the alignment. In order to cater to maximum traffic, the route should pass through the most densely populated and the busiest areas. But such a route would increase the cost of construction and aggravate construction problems. The cost of construction also depends on choice of method of construction, viz., cut-and-cover or bored tunnelling. The busier the area, the more difficult is cut-and-cover construction. For cut-and-cover, the alignment has to follow street layout. Such streets should be wide enough to allow open excavation with minimum need for dismantling buildings and without endangering building foundations. Also, there should be sufficient room on the surface to permit minimum traffic and use of machines for loading and carrying away excavated materials. Space is also needed for storage of construction materials at the site. When the alignment has to change direction, requiring curves, the street alignments may not always suit since they cannot accommodate flatter curves (radius of 250 m or over for heavy metros). Wherever stations are required, a wider space availability is a must. Hence in some areas, off-street alignments become necessary requiring bored tunnelling.

8.2.2 Utilities

Existing utilities (water supply pipes, gas pipes, sewerage lines, drains, and electricity and telephone cables) have to be diverted or supported if they cross the alignment, if the cut-and-cover method is adopted. This method will also cause disruption to traffic, calling for total or partial diversion depending on the width of the road and availability of parallel roads for diversion of traffic. Access to properties on either side will also be affected during the period of construction. Off-street alignment would avoid both these difficulties.

8.2.3 Construction Methodology and Costs

On the other hand, off-street alignment would pass under a primary artery (street) representing the dense traffic demand corridor and partly under private properties adjacent to the same. Tunnels can pass under built-up areas where curves are introduced but would have to be well clear of the foundations of buildings, i.e., below the zone of influence of such foundations. This would call for locating them at greater depths, increasing the distance of vertical travel for passengers. There is always some risk of settlements and ground subsidence affecting the buildings above. The track structure also depends on the method of construction. While cut-and-cover calls for a twin-box cross-section, the bored tunnelling method calls for two parallel tunnels.

A single circular tunnel to accommodate two tracks (necessary for any metro in order to maintain required frequency of trains) would be too large. For the bored tunnelling method, space between tunnels has to be cleared to accommodate station platforms with the concourse placed either on surface or underground. Space for platforms etc., can be provided by using cut-and-cover methodology or adopting other special methods of supporting overburden for inter connecting the tunnels and providing platforms. These aspects would also greatly influence cost. According to Morton (1982)[3], the relative advantages and disadvantages (of the above methods would be as indicated in Table 8.1. It would be seen that under street cut-and-cover alignment is likely to lead to the least overall costs and construction time, whereas bored tunnelling results in the least construction nuisance and minimum interference with traffic.

Table 8.1 Unweighted Comparison Criteria for Different Modes of Construction(Morton 1982)

Components	Criteria rating score for			
	On-street alignment		Off-street alignment	
	Bored tunnel	Cut and cover	Bored tunnel	Cut and cover
Right-of-way costs	4	4	3	2
Construction costs	2	3	2	4
Construction time	2	3	2	4
Construction nuisance	4	2	4	3
Construction interference with traffic	4	2	4	3

Note: The ratings are based on a score of 4 for best performance and a score of 1 for worst performance.

8.3 ALIGNMENT AND TRACK DESIGN

While fixing the alignment certain track geometry requirements have to be borne in mind pertaining to horizontal alignment and vertical profile. Tunnel size and floor orientation depend on the type of track bedding (ballasted or ballastless), provision for cant to cater for speed and comfort required by passengers, type of traction and location of line side signals and other equipment The permissible minimum radius of a horizontal curve depends on length of coaches, wheel arrangements and speed to be provided for. Maintenance problems, such as wear on rail, and noise nuisance due to screeching wheels on sharp curves have also to be considered. Generally the minimum radius of curvature is limited to 250 m but desirable practice requires 300 m or over. The sharpest curvatures in some of the metros are 120 m in BART, 200 m in Singapore and 200 m in Kolkata. The maximum operating speed in the metro rail can go up to 80 / 100 kmph and permissible maximum permitted cant and cant deficiency (maximum permitted unbalanced super elevation) are specified to suit local circumstances. A few examples are given in Table 8.2.

All curves have to be transitioned. The transition length would be such that the super elevation is run out at a maximum rate of 1 in 333.33. A minimum length of 30 m of straight is generally provided between two reverse curves. In stations, the radius of curvature is limited to 1000 m as otherwise, due to the inside and outside throw of the vehicles on curves, a wider gap will occur between the platform edge and the side of the vehicle at the doors, endangering the passengers boarding/alighting. This gap should be limited to 100 mm. Any change of curvature or introduction of curves at station approaches should be at least 25 m away.

Table 8.2 Design Criteria for Passenger Comfort on Horizontal Curves System

System	Maximum operating speed (kmph)	Maximum permitted speed (kmph)	Cant deficiency mm	Maximum permitted lateral acceleration at max. speed
San Francisco (BART)	128	150	68	0.04g
Toronto	88	100	62	0.04g
Washington	120	100	112	0.08g
Singapore		150 (Ballastless) 125 (Ballasted)		0.06g
Kolkata	80	150 (Ballastless)	75	0.06g
Delhi	80	150 (ballastless)	75	0.06g

8.4 VERTICAL PROFILE[3]

The gradient permissible is dependent on the tractive capacity of vehicles. A minimum drainage gradient of 0.3% is to be provided. At stations, for the purpose of stability of stopped vehicles, this should also be treated as maximum gradient albeit the desirable maximum is 0.2%. There are systems with up to 5% maximum gradient in-between stations on some heavy metros (Boston and Philadelphia). Washington has a 4% grade in its tunnel, Singapore 3% and Kolkata 3%. The gradient should be compensated for curvature (0.04% per degree of curvature). A vertical curve is always introduced when the gradients change. Parabolic vertical curves are preferred. Lengths will be (G1 - G2) x 100, where (G1 - G2) is the algebraic difference between the grades of connected lengths; this is limited to a minimum length of 80 m. The constant profile of any grade is also limited to a minimum length of 30 m (in Toronto it is 150 m). The minimum radius of vertical curve adopted in metros ranges from 1500 m to 3000 m as against 3000 to 4000 m on main line railways. At 80 kmph, which is the normal maximum speed of metros with close station spacing, this would give a vertical deceleration of 0.033 g for 1500 m to 0.017 g for 3000 m. Hence this range is satisfactory. It is also advantageous to provide a vertical grade that ascends to the station and descends out of the station thus providing a camel-back type of track surface at the station (but of uniform grade or level over the station length plus 25 m at either end). This not only helps reduce vertical transportation distance for passengers, but also aids in quick deceleration and acceleration of the vehicle, resulting in some saving of energy. Typical profile for such treatment at stations is given in sketch below.

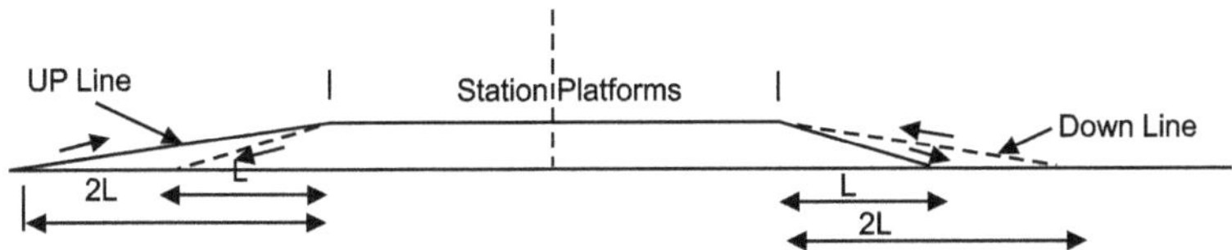

A Typical Grading Up and Down at Subway Stations

8.5 COVER/OVER BURDEN

A minimum cover of 2.0 to 2.5 m above the roof of a subway structure is required when cut-and-cover method is used. This allows passage of most utilities above the roof. Any deeper utilities would need to be diverted. If the bored tunnel method is used, a minimum cover of one tunnel diameter and more depending on the nature of the ground is provided above the tunnel crown. In rock and hard strata one diameter depth will do. In alluvial soil a cover of twice the tunnel diameter depth is required. If the tunnel has to pass under buildings, the cover should be below the zone of influence of building foundations. On Chennai Metro a minimum cover of 9 m has been adopted for bored tunnels, which goes upto 20 m or more where one line passes below another and in some areas with line taken down across built up structures or water bodies.

8.6 INVESTIGATIONS[4]

8.6.1 Geotechnical

Once the possible alternative alignments are chosen, geotechnical investigations are carried out by drilling bores, taking out undisturbed samples and testing them. Boreholes are drilled at intervals of 90 to 150 m, or closer if any sharp change in strata is noticed. Depth should extend a few metres below the proposed lowest level of the tunnel. In addition, SPT and permeability tests are carried out. Chemical examination of the soils is also done to ascertain whether any gases are likely to be encountered and if the soil is likely to adversely react with the materials of the structure. The type of soil would also determine the type of construction to be adopted and even depth of construction in some cases. During boring, records are maintained of the type of subsoil encountered and depth of groundwater tables. Wherever groundwater is encountered, observation pipes, or a piezometer are used to take long-term readings so as to establish seasonal fluctuations in water level. This helps in planning the type and quantum of pumping requirement (or freezing arrangements) during construction. It further aids in design of the structure as well as the waterproofing arrangements to be made.

8.6.2 Utilities

On-street alignment can cause disruption to various utilities which are normally below the road surface. These may be sewers, surface drains, water and gas pipelines, and electric supply and telephone cables. They invariably would run in the longitudinal direction and at some places across the road alignment. The underground structure has to be clear of all these and may even have to be graded down to clear transverse crossings.

The possibility of diverting other utilities which might come right over the alignment on a permanent or temporary basis would have to be studied in detail, in the case of cut-and-cover construction. Wherever this diversion possibility does not exist, the bored tunnelling method will have to be adopted. Some utilities, such as water pipes, G.I. sewers, telephone and electric cables and ducts can be supported temporarily from deck beams. But for high risk utilities such as gas mains, H.T. electric cables, masonry sewers running over alignment, diversion would be preferable. The cost of diversion of utilities would form a major portion of the cost of cut-and-cover construction.

During the investigation stage, a careful inventory of all such utilities as described above has to be made in consultation with the concerned maintenance agencies and methodology for their diversion/ relocation/ supporting planned in consultation with maintenance agencies..

8.6.3 Road Traffic and Diversions

In on-street alignments, for the cut-and-cover method to be adopted, diversion of traffic in part or in full is necessary. During this time period the side supports can be installed and the trench excavated to sufficient depth to enable people to work below. Full diversion of traffic away from the width of construction is mandatory. In addition, about a lane width on either side would be required for working of machinery and plying trucks to take the excavated earth away. Once this is done, temporary decking plates can be laid and traffic largely restored but at slow speed. On completion of the underground structure, during filling and restoration of the road surface, traffic diversion will again be necessary. A study of volume and direction of traffic as well as availability of parallel roads for diversion is necessary to devise a traffic management scheme. The amount of disruption to access to properties on either side and alternative arrangements of regulating traffic and their impact on local business have to be studied simultaneously.

8.6.4 Building Survey

The choice of on-street alignment means it will be clear of adjoining properties for most of its length as it would follow an artery of sufficient width. But at station locations the width of excavation would be more (about 20 m) in order to accommodate these. While choosing the station location, this requirement should be kept in mind. Alternatively, some buildings may have to be dismantled. On the other hand, at some locations in the running sections and at stations the load pressure distribution from the foundation of some buildings (when reckoned on a 45-degree dispersion line below base) may infringe upon the excavation profile. The excavation can also cause settlement of surrounding ground, accompanied by some heaving in the excavated trench, especially in cohesive and wet soils. Such settlement can cause cracks in the adjacent buildings. During cut-and-cover operations, in some cases work can be done with suitable underpinning of lighter buildings. In such cases, while assessing easements the value, that the owner places on loss of future foundation capacity and redevelopment has to be considered.

In off-street alignment, in order to avoid the high cost of acquisition of buildings if cut-and-cover is adopted, bored tunnelling is resorted to. As already highlighted, the tunnel has to be placed well clear of the building foundations and their zone of influence. Details of foundation and height / loading details of all buildings coming over the tunnel, and for a width subtended by a 45-degree line drawn from the outer lines of the tunnel opening, are collected during the investigation stage. Presence of any open wells or bores within the zone of influence of the tunnel boring should be noted, so that they are plugged in advance of the tunnel bore. Any chemical or grout used for strengthening the weak soil, in advance of TBM may, otherwise, escape through such holes with resultant problems of pollution, settlement etc.. Buildings coming within zone of influence of excavation should be carefully surveyed for pre-existing cracks and inventory made of same. Such information should preferably be authenticated or vetted by the owners so as to avoid any undue claims for damages later.

8.6.3 Construction

Table 8.3 presents the different methods of construction that can be used for underground metro rail structures, using either cut-and-cover or bored tunnelling methods. The different methods applicable for various types of soil are summarised in Table 8.4..

8.7 CUT-AND-COVER CONSTRUCTION

8.7.1 Method- Description and Types

Cut-and-cover construction basically consists of, cofferdaming, i.e., putting up two vertical retaining walls to prevent the ground caving into the trench dug for construction of the RCC box through which the rail tracks or road (carriageway will pass). In order to reduce the requirement of working areas, utilities relocation and number of diaphragm walls or piled side supports to be provided, the box profile should be for twin track. These retaining walls, diaphragm or row of piles would be about 10 m apart between stations and 20 m to 22 m apart at stations. In station areas, a row of intermediate supports

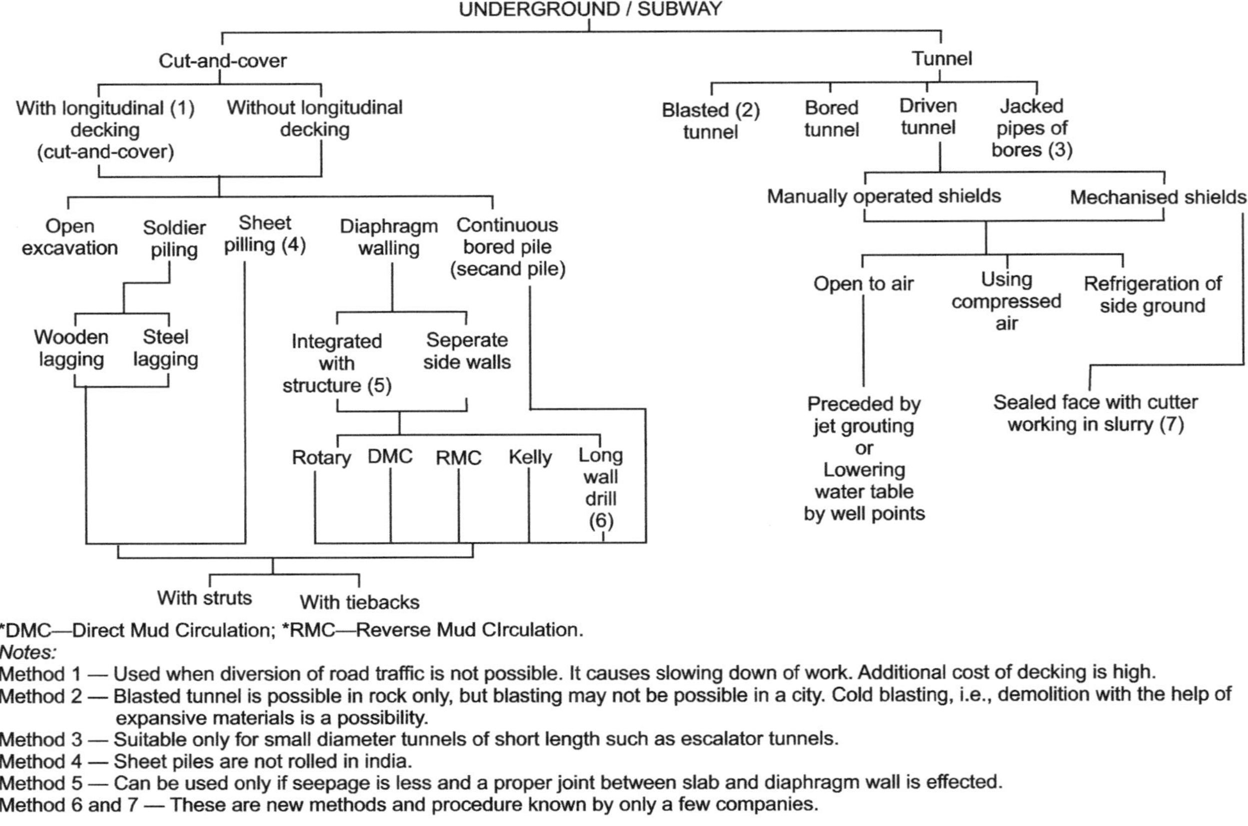

*DMC—Direct Mud Circulation; *RMC—Reverse Mud CIrculation.

Notes:

Method 1 — Used when diversion of road traffic is not possible. It causes slowing down of work. Additional cost of decking is high.

Method 2 — Blasted tunnel is possible in rock only, but blasting may not be possible in a city. Cold blasting, i.e., demolition with the help of expansive materials is a possibility.

Method 3 — Suitable only for small diameter tunnels of short length such as escalator tunnels.

Method 4 — Sheet piles are not rolled in india.

Method 5 — Can be used only if seepage is less and a proper joint between slab and diaphragm wall is effected.

Method 6 and 7 — These are new methods and procedure known by only a few companies.

Table 8.3 Methods of Underground Metro Construction

Table 8.4 Suitability of Construction Methods for Different Soils

Locations	Type of soil	Methodology for construction	Limitations of method, if any	Advantages	Disadvantages
Between Stations	Soft clay, silty, clay, clayey, sandy clay	i) Cut-and- cover with diaphragm wall or sheet piles	i) D/Wall to be taken sufficiently below the excavation to prevent heaving	Box design can be more economical. The box could be integrated with D/wall to reduce cost	Disturbance to the surface cannot be avoided in cut-and-cover
		(ii) Tunnelling	(ii) Generally not possible without compressed air		
	Solid or sparsely fissured hard rock	Blasting with unlined or partially lined tunnels	Blasting in busy city areas likely to be hazardous	Cost likely to be competitive. Progress likely to be good as work can be allotted to an independent contractor	Because of proximity of buildings many precautions have to be taken, particularly for old buildings.
Stations	Fissured rock, soft rock, friable rock, conglomerate, gravel etc.	i) Blasting with fully lined tunnels	i) Normal blasting not advisable.		(i) Same as above
		ii) Cut-and-cover diaphragm walls	ii) Excavation may be expensive		(ii) Disturbance to surface roads and high social costs
		(i) Cut-and cover with soldier piles	(i) Heaving possible	(i) Cost would be competitive	(i) High social costs. Likely damage to surface properties
	Sands, silty sands, fine kankar, loose conglomerate etc.	ii) Grouting and shields tunneling	Not to be used with loose conglomerates	(ii) No disturbance to surface or utilities	(ii) Technology of a high degree required
		(i) Shield tunnelling	Silt percentage important. Expert opinion required	Cost likely to be competitive. No disturbance to surface	Delays due to compressed air
		(ii) Cut-and cover with temporary decking			Disturbance to surface

Source: RITES Report ,1993

(temporary or forming part of station structure) may have to be provided to reduce length of struts. The walls are designed to take full earth pressure behind them when braced at different levels. The stations are easy to build if not located at cross-roads or at intersections, due to complications involved in supports for decking. Though decking can be used in exceptional cases even at cross-roads the progress of work would be slow and the cost high.

The cut-and-cover method can be adopted in all types of soils except rocky strata. To use this method of construction effectively, a complete relocation or temporary support of utilities is often essential. The two tracks allow location of emergency turnouts and points and crossings without construction of special structures, as required in the case of bored tunnelling. The main advantage of the cut-and-cover method of construction is ease in setting out and future extension of side platform, if required subsequently.

Excavations for diaphragm walls or rows of piles (retaining walls) are normally 12 to 14 metres deep (Rudrakshi)[4]. The two walls are held back by anchors and / or braced by steel struts butting against each wall over the waling piece. Three to five layers of such struts have to be used depending on depth and nature of soil behind the walls as well as the surcharge due to other surface structures such as buildings. In better types of subgrade soil/ ground anchors have been used with success at some places[4,5]. The basic sections of cut-and-cover in a station and a running section are shown in Figures. 8.1 and 8.2 respectively.

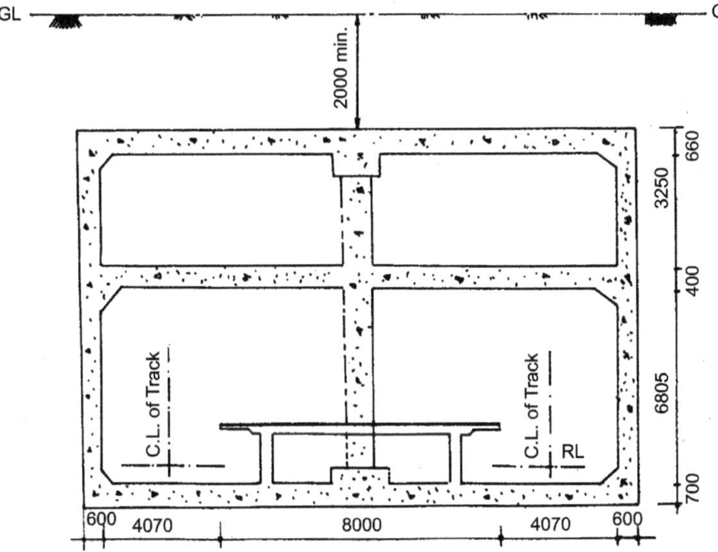

Figure 8.1 Typical Section through a Station Box in Kolkata Metro

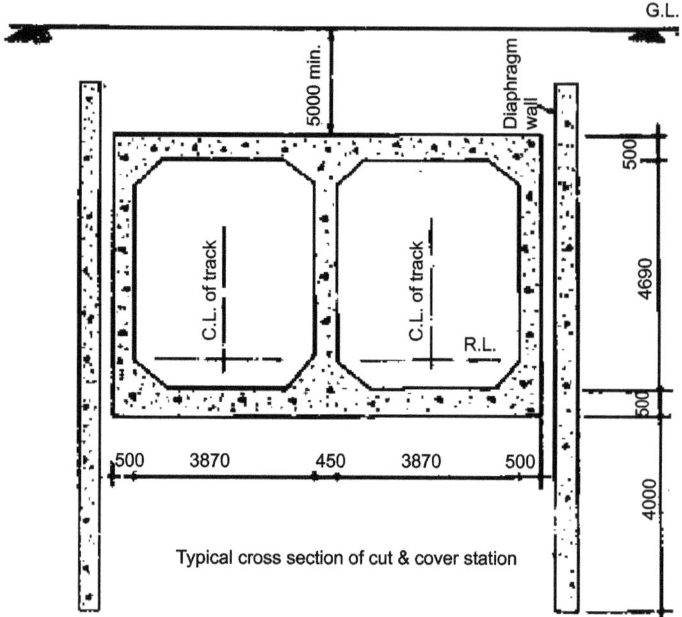

Figure 8.2 Typical Box section in between Stations in Kolkata Metro

Four types of construction are normally adopted for providing the retaining walls: (a) diaphragm walls, (b) steel sheet piles, (c) steel H-posts with steel or timber lagging and (d) Bored piles adjacent / linked to each other.

8.8.2 H-Piles and Lagging or Sheet Piling[3]

H-piles are driven for depths of about 1 m more than the depth required for casting the box. After the two rows of piles are driven for a length, excavation is done so that the top layer of struts can be fixed. Excavation is then continued and timber lagging inserted between piles and pushed down. As excavation proceeds, struts are added at lower levels at suitable intervals and timber lagging added. The section showing arrangements made using H-piles is shown in Figure 8.3. H-piles with timber/steel lagging are generally suitable for sandy soil, shallow cuts and in areas where soil is good, having a high angle of repose.

For deeper cuts, sheet piling methods are used. They are suitable for cohesive soils. A typical cross-section of the structure using sheet pile wales is given in Figure 8.4. Interconnected sheet piles are driven on outer peripheries of the section for depths extending about 1 to 3 metre below base level of the structure depending upon soil conditions and likely heaving. The piles are supported using waling pieces/runners and struts at designed levels (similar to struts for H-piles). The piles are withdrawn in both cases after the structure is completed and as filling above the structure proceeds. But in many cases difficulty is experienced in withdrawing all the piles and some have to be left and cut off below road crust level. This adds to cost and time of construction. The piling method was tried in Kolkata and given up in favour of diaphragm walls, which are more popular now.

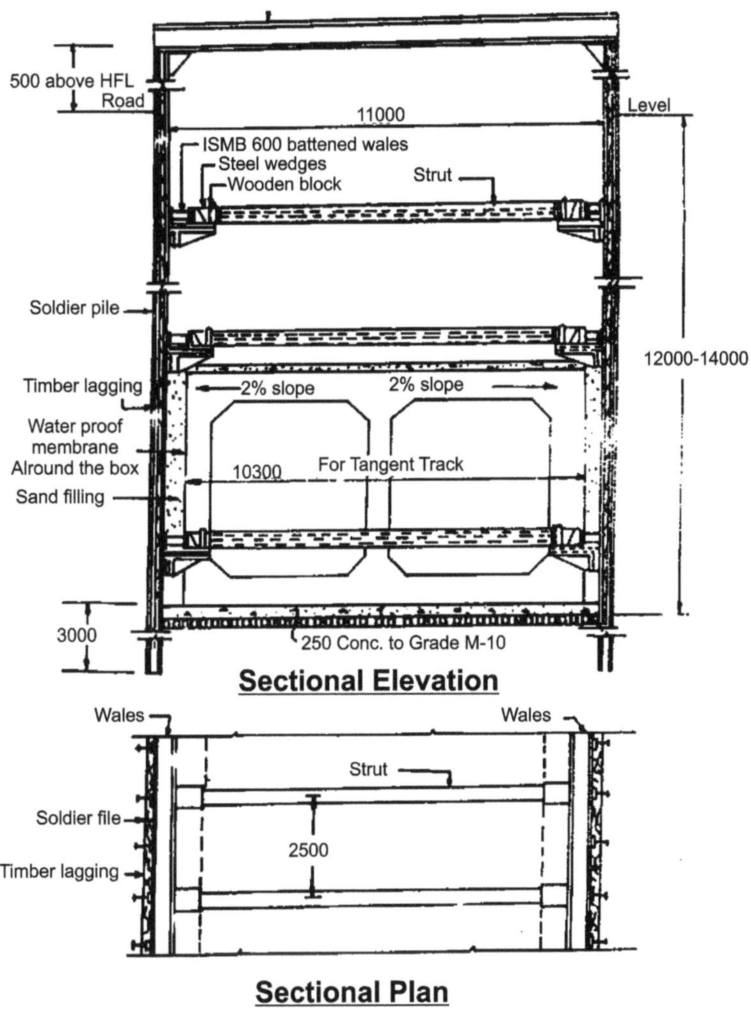

Figure 8.3 Cut-and-Cover Construction Method using Soldier Piles and Timber Lagging[3].

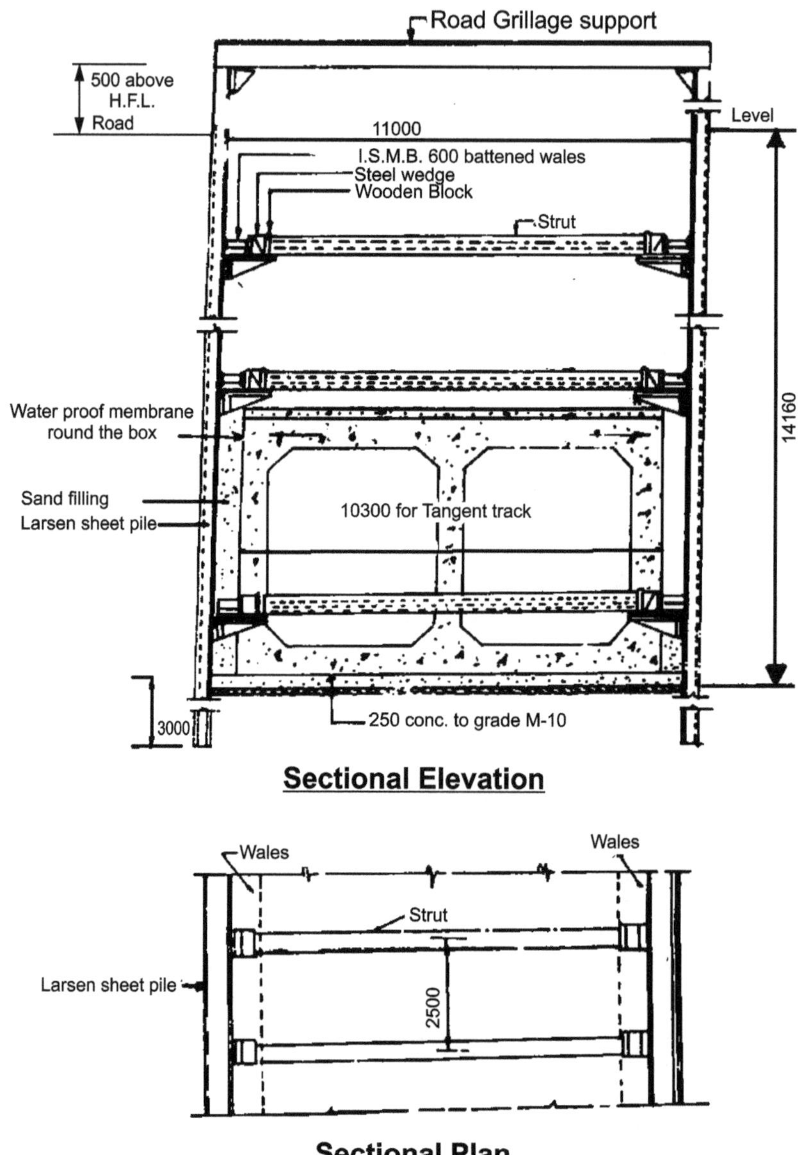

Figure 8.4 Cut-and-Cover Construction Method Using Sheet Piles[3].

8.8.3 Diaphragm Walls[4]

Diaphragm walls have to be designed as continuous slabs spanning between horizontal beams supported by struts. Reinforcement is provided in the walls to take care of the bending moments as calculated. Construction of diaphragm walls is done using the bentonite slurry circulation method, by digging / boring a narrow trench up to the required depth of 12 to 14 metres. These trenches are then filled with tremie concrete after the prefabricated reinforcement cages are lowered into them by cranes. Precast concrete diaphragm wall panels can also be lowered into the trenches using cranes and adjacent segments made watertight using special techniques such as rubber/ PVC water stops. Figure 8.5 depicts a vertical section showing prestressed tiebacks for supporting diaphragm walls. The panels are laid alternatively (with space between edges filled with grout or made watertight by match casting).

There are two methods in diversion of traffic; sequencing the excavation; and in the construction of the boxes viz., from bottom or top down. In the former, traffic is diverted on either side of proposed excavation; diaphragm walls taken down full; preparing the base; construction of the box; refilling the pit above and around the boxes; restoring the carriageway and restoring the traffic. This method is feasible in case the ROW of road wide enough and or the excavation width is limited as for the box between stations. This was the method followed for Kolkata Metro.(Rudhrakshi)[4] Figure 8.6 (a) shows the sequence of operation adopted and Figure 8.6 (b) shows provision of diaphragm walls and box in between.

The other method is the preferred one where the existing ROW of road is not wide enough for diversion of traffic to sides and maintaining same over a longer period and also where the width of box is large and in more than one layer, as in case of underground stations. In this case, the diaphragm wall construction has to be completed in two stages and providing the roof slab and restoring the road surface initially. Excavation below roof slab follows and box is formed in two stages from top as shown in Figure 8.7, which is self explanatory. This method is being followed for construction of station boxes in Chennai Metro.

8.7.4 Bored Piles

This method employs an earth auger that can bore holes through all types of soils with or without the use of bentonite slurry. After completion of drilling or boring hole, reinforcement cages or joists is lowered into the holes and concrete poured through tremies. A series of piles are so constructed adjacent to one another as to form a wall which gives support to the earth during excavation. They also have to be laterally supported by runners and struts as excavation proceeds downwards. This method of construction

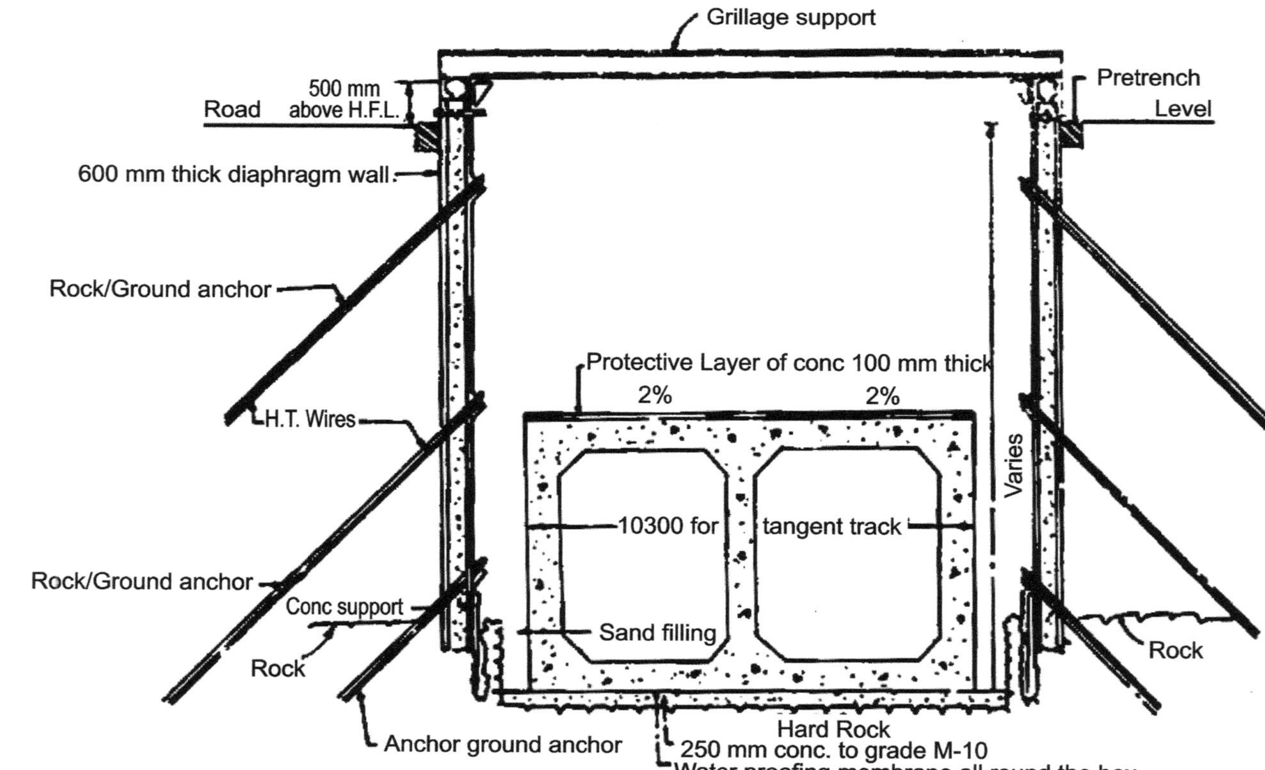

Figure 8.5 Typical Section Showing Prestressed Tiebacks for Diaphragm Wail.

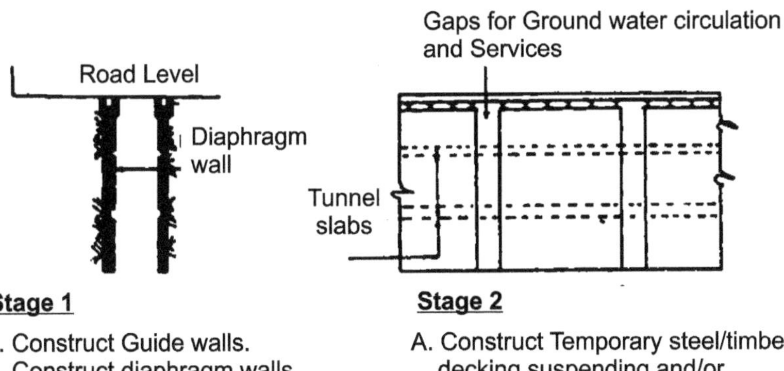

Stage 1

A. Construct Guide walls.
B. Construct diaphragm walls upto road level and upto top level of roof slab in alternate days.

Stage 2

A. Construct Temporary steel/timber decking suspending and/or diverting services as required.

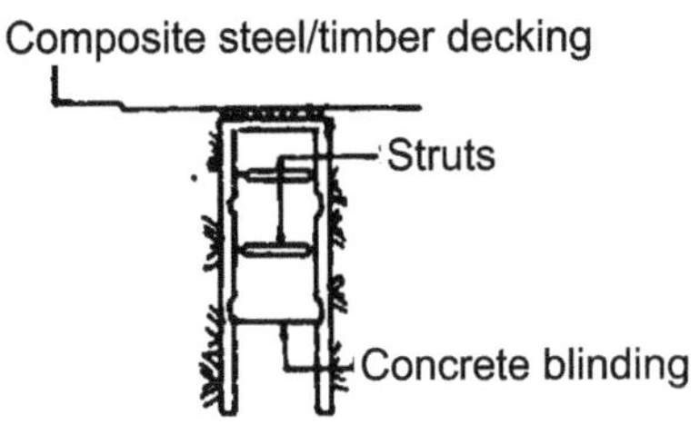

Stage 3

A. Excavate to underside of base slab strutting the walls as excavation proceeds.
B. Pour concrete blinding.

Stage 4

A. Construct base slab.
B. Construct roof slab.

Stage 5

A. Dismantle temporary decking in stages
B. Reconstruct road surface

Figure 8.6 (a) Sequence of construction of cut and cover Box by normal Process4.

results in a large number of joints, waterproofing of which becomes extremely difficult, especially where the water table is high. The main advantage of this method is that it calls for no special equipment.

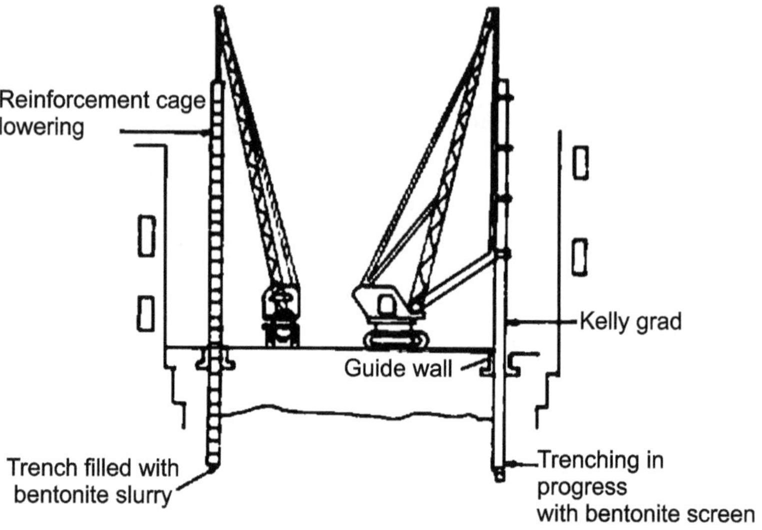

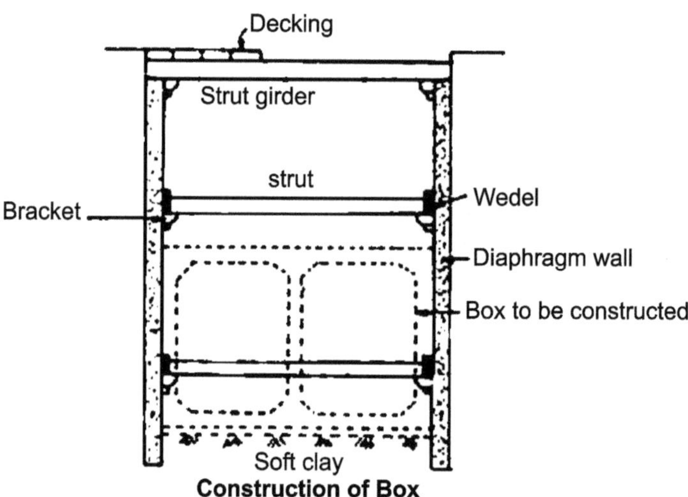

Figure 8.6 (b) Tunnelling by Cut-and-Cover- Diaphragm wall Construction for Kolkata Metro

8.7.5 Ground Movement Caused due to Excavation[5]

The primary requirement of the side supporting system, viz. H-piles with lagging or sheet piles, diaphragm wall etc., is to prevent large movement or collapse of the sides of the excavation. Excavation removes a mass of soil and water and produces a reduction in total stress along the sides and bottom of the cut. This causes the soil at the sides to move inward towards the excavation and the soil at the bottom of the cut to move upwards. The upward movement at the bottom is accompanied by an inward movement of the soil below the excavation level. Such movements depend on depth of excavation and soil conditions, rigidity, method and sequence of installing the support system and the time-period the excavation is left open. The maximum ground settlement on the sides is also likely to be about 3% of the excavation depth. Sufficient precautions have to be taken to minimise these movements. There have been cases where the soil intrusion from sides, especially in sandy strata, have caused heavy settlement on sides, resulting in caving in and heavy settlement and breaking of utilities running parallel and/or settlement of buildings on sides.

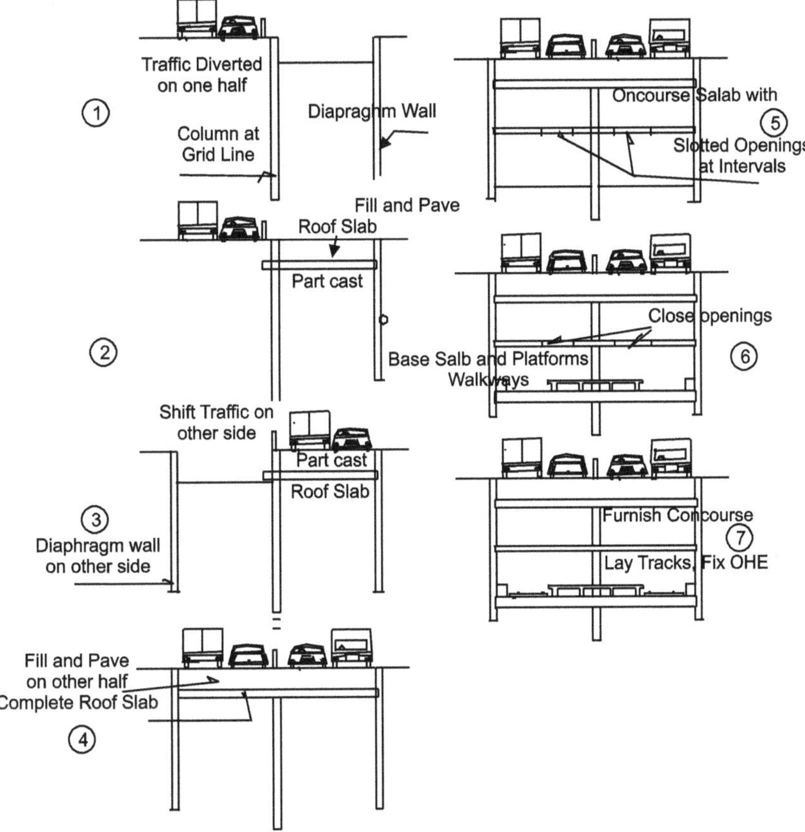

Figure 8.7 Cut and Cover Top down Method.

8.8 CONSTRUCTION JOINTS AND WATERPROOFING6

Waterproofing does not become a serious factor in over-ground (hill) tunnels since the seepage water can be drained into side drains by providing suitable gradient towards the adits. Some additional control measures are called for in the case of heavy infiltration. But in underground tunnels disposal of drained water has to be done by using heavy pumping arrangements to lift the water to the surface for siphoning off. Since seepage water can cause high humidity and hence discomfort to passengers, special ventilation arrangements are necessary. It can also have a deleterious effect.

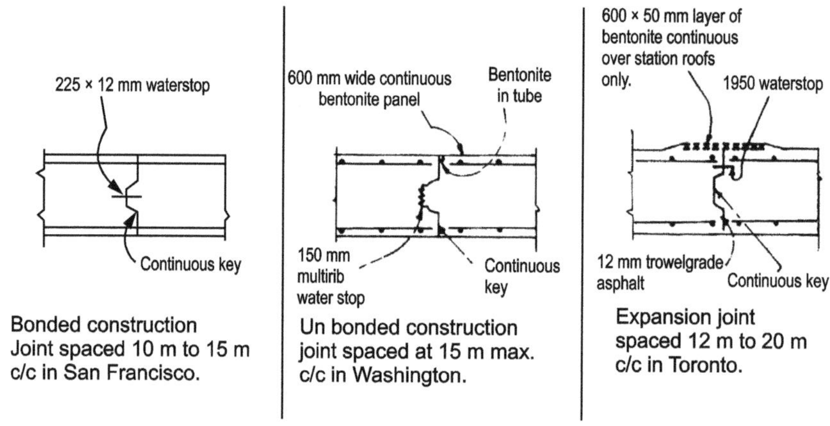

Figure 8.8 Some Typical Joints Used in Subway Construction[3].

on concrete and reinforcement as well as affect track and signal components. Hence in underground tunnels special care is taken to waterproof the structure and to provide efficient leak stops at construction joints.

Use of dense concrete of grade M40 (concrete with 28-day cube strength o140 MPa) in the surface and lining, adopting a minimum thickness of 60 cm for the walls, roof and floor or lining, careful laying and vibration provide the desired waterproofing. On the roof slab, especially in station areas, some waterproof membrane such as butyl sheet or brush-on asphalt crating is used. This is essential as any water infiltration would damage electric fittings, cause current leakages and also leave ugly patches on the finish.

Expansion and construction joints are provided at intervals to take care of shrinkage stresses and thermal variation in concrete. They are generally spaced 10 to 15 metres apart All such joints are carefully match cast or bonded and provided with non-metallic water stops. In the Washington metro a bentonite-filled tube was embedded in the concrete on the exterior face. In Toronto the joint space provided was 6 mm and plastic water stops inserted. A few typical joints are shown in Figure 8.8. Such joints notwithstanding, some seepage and leakage water will still flow in due to rain falling on open portals, street flooding due to abnormal rain, or broken water or fire-fighting pipes. (Street floods may go above the raised kerb provided at portals.) Such water is drained through side drains or invert drains under the track, collected in sumps at intervals. Submersible pumps of the order of 2000 litres /minute discharge capacity are installed to pump this water to surface drains above ground.

8.9 SHIELD TUNNELLING AND USE OF TUNNEL BORING MACHINES

The construction of tunnel boxes using cut and cover method requires opening part of the road and keeping it unavailable for road traffic. The road traffic will have to be diverted on other parallel road or if the ROW of existing road can allow part width to be used for construction and traffic diverted on other portion. For this to be possible, the ROW of the road should be at least 25 to 30 m. In very busy areas, this may not be possible due to traffic density. In such cases, especially in soft soils tunnelling is done adopting circular cross section and use of shield. Annexure 6.1 gives details of TBM and its working. In case of very soft soils, stabilisation of soil is done while progressing shield is done by use of bentonite slurry. Figure 6.9 shows the working of the method using bentonite slurry conceptually. Para 8.12 covers a case study on Chennai Metro Rail construction.

8.10 DESIGN OF STRUCTURE

The design of a subway structure has to incorporate the following: Dead loads comprising weight of structure, weight of backfill above, horizontal forces due to lateral earth pressure and water below subsoil water level.
Live loads comprise of vehicles passing over tracks, subway equipment, and other general occupancy including crowd load in times of emergency and surcharge due to road vehicles above ground and Seismic forces in earthquake prone areas.

The structure has to withstand uplift pressure on the bottom slab (inverts) and the entire structure weight has to counter uplift when there is no load on the track, with an adequate factor of safety. All loads may not act at one time. Hence the design has to be checked for these combinations.

Full vertical load + full horizontal load

Full vertical load + full horizontal load on one side and half horizontal load on the other side

Full vertical load + half horizontal load on either side

Full vertical load + full horizontal load (balanced) + horizontal seismic force.

While considering vehicle loading, impact forces are generally taken as 30% of the static load. The horizontal forces due to braking or traction are applied in a longitudinal direction at the centre of gravity of the vehicle or at 1.80 m above rail level. They are usually taken as 25% of static vehicle load. Lateral forces are applied in the form of an equivalent couple acting downward on one rail and upward on the other, assumed to be equal to 10% of static vehicle load. Example of a typical design for one of the combination of loads is given in Annexure 4.1.

Lining for tubular construction may be flexible or rigid. Most commonly used ones are (flexible) segmental ones made of steel, cast iron or concrete. (Morton)[3] The flexible lining is then to serve as temporary and permanent ground support. Design should take into consideration the following factors:

(i) It should be able to provide immediate resistance against the external loads i.e., earth pressure.

(ii) Since it is to serve as permanent one also, it should be able to provide resistance to long term developments in form of earth and water pressures, 'without detrimental deformation or leakage'. Future developments include influence of deep excavation or subsequent tunnelling alongside it.

(iii) The lining should form a watertight structure against any subsoil water present, which means the manufacture of segments should permit caulking the joints to make them water tight. Segments should form a watertight membrane. In extreme cases, it should be possible to provide an inner liner with waterproofing membrane inserted between, after the preliminary lining has adjusted and there is no further movement within.

(iv) The lining section should have sufficient capacity to resist the axial stresses induced by the propelling jacks.

8.11 SUBAQUEOUS TUNNELS[6]

This is a form of tunnelling done under water, not by using a tunnelling machine but by precasting full section units in part lengths, floating, laying on the sea or river/creek bed and joining them together. This method avoids taking the tunnel very deep and overcomes the difficulty of keeping seepage water out while tunnelling. This form of construction is, used when the road or rail has to cross deep channels or creeks, rivers in estuaries and sea-bed across bays and connecting closely spaced islands. For example, this form is used for extending railway tracks across a number of islands in Japan. It has also been used extensively in the BART metro in San Francisco and in Detroit. It has proven to be a better alternative to high-level bridges across bays where a wide clear waterway is required for flexible ship routing. The first such construction was adopted in 1909 for the Michigan Central Railroad in Detroit. Since then almost 100 such tunnels have been built (Vos, 1992) 'Also known as 'immersed tunnels' or 'submerged tunnels', they are very flexible in cross-section. They only require a simple dry lock for casting units, towing equipment and a dredged trench for laying the unit along the alignment. Units are floated and laid on a properly prepared foundation in the trench. In the bed of very deep channels they can be laid on a levelled and prepared base on the bed and after proper connection to one another and covered with sand. A typical method used for construction of an immersed tunnel for a road is shown in Figure 8.9. Wherever the soil is too soft and silty and no firm base is available, two rows of closely spaced piles are driven along the alignment and the tunnel units laid on them. This scheme was adopted for the Rotterdam metro line. Another recently developed method of founding the submerged tunnel is to lay the structure over a temporary foundation made up of concrete tiles spaced at intervals over the prepared bed. After the units are geometrically positioned to fine tolerances and jointing, the space below tunnel 'bottom and above bed (in between tiles) is filled tight by jetting a mix of sand and water from the sides.

In the floatation method the tunnel sections are made up in dry docks or on shore. If of RCC, the ends are closed so that the unit can float. It is taken to the site where the foundation bed would have already been prepared. Once the unit is aligned over the position where it is to be laid, water is let into the interior and the unit slowly lowered. It is properly matched with the end of the previous unit, alignment checked and corrected to close tolerances. The joints are then caulked or provided with leak stoppers and finished. Alternatively (American method), the tunnel walls, base and roof are made up of a cellular (two-walled) section with ends closed. The skin plate joints are properly welded or riveted. The assembly is then floated to the site, ballasted with water or concrete filled to the extent required inside the shell and lowered into position. After it is correctly positioned and joined to the

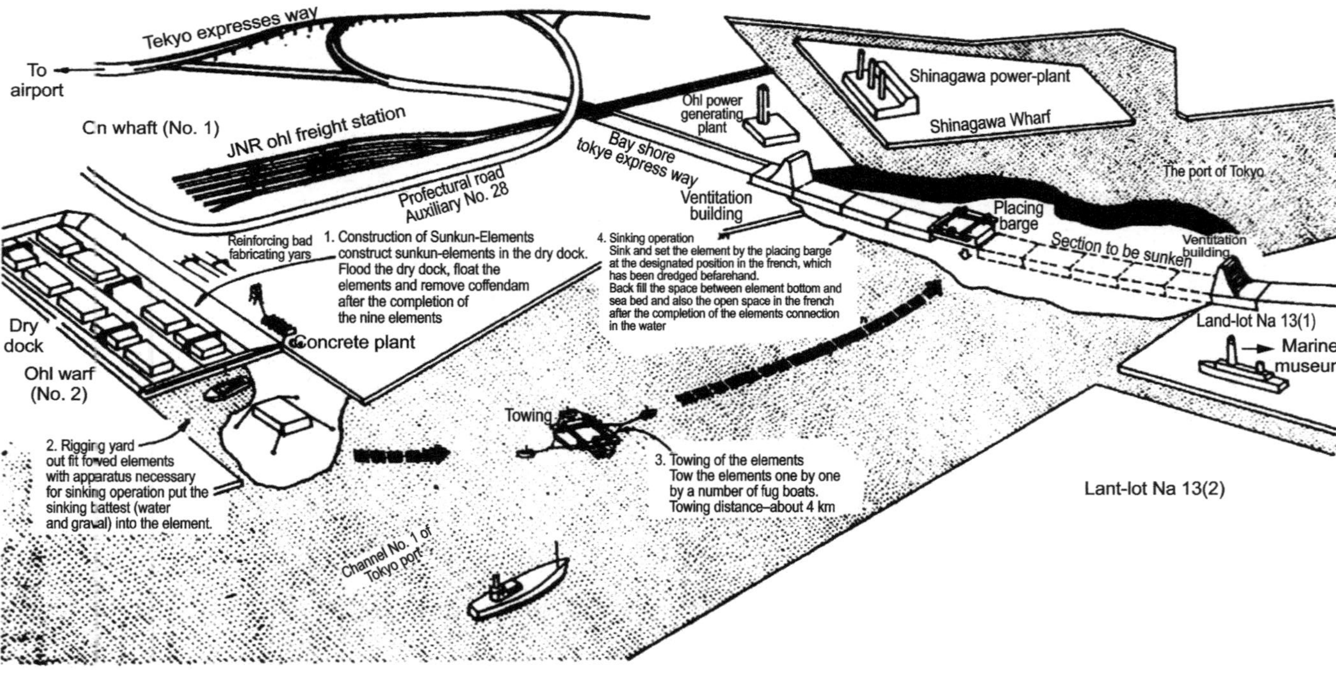

Figure 8.9 Schematic Diagram of Subaqueous Tunnel Construction.

already laid length, the remaining concrete is poured inside the shell and the structure completed.

Waterproofing submerged tunnels is a very important and the most difficult aspect of this type of construction. The steel lining used in the American method provides the best waterproofing protection. In other types of tunnels, an additional waterproofing layer in the form of coatings on the outside or additional lining inside is provided. But such procedures have been superseded, especially after technology has been improved for providing denser and better controlled concrete for structures. Only joints provided for construction, expansion and rotation are now attended to. The fresh stripped surfaces at the ends are fitted so as to give a satisfactory construction joint. The joints between independently immersed units are provided with good leak proof double water stops. In joints providing for expansion or rotation, special measures are taken while pouring concrete to avoid honeycombing or air entrainment in the vicinity of water stops at the joints. In addition, the possible leaks in-between concrete and water stops are injected with leak proof compounds or resins.

8.12 CASE STUDY- Practical Problems in Bored Subways through Soft Soil[7]

8.12.1 Choice of Methodology

Amongst the early Metros, London, started providing some sub-surface lines using 'cut and cover' methodology on wide roads and small circular underground sections (called tubes) using shield tunnelling technologies 3.5 m dia) at deeper levels. New York and Chicago metros started using elevated steel structures provided at edges or median of roads to start with. But with growing road traffic and then prevalent lower labour costs, they replaced many elevated sections on busy areas by providing shallow box like structures using 'cut and cover' methodology. The mole like circular tunnels were provided as 'deep tubes' also in Moscow, Paris etc They generally met with cohesive / clayey soils which could stand on their own longer for providing the linings. With the advent of TBMs (Tunnel Boring Machines) and use of cast iron or precast concrete segmental linings, engineers have found boring circular tunnels through sandy/ non-cohesive soils also possible. But use of 'cut and cover' technology for building more economical box type shallower tunnels have become more and more difficult with increased road traffic which cannot be easily diverted. Hence, starting with the BART (Bay Area Transit) in San

Francisco in early 1970s, they found use of a mix of 'elevated / aerial tracks' over wide roads, carried by a single row of slender columns provided over medians preferable. In busy areas, deeper 'twin circular tunnels' generally aligned under sidewalks or edges of roads using TBM are adopted. Stations in such cases are being built using 'cut and cover' methodology, using stage construction as schematically shown in Figure 8.7. With the dense growth of cities, especially in older parts of the city, it is difficult to find road alignment to follow using either methodology. Some lengths of tunnels have to be taken across built up areas under the existing buildings. In such cases, the depth of location of tunnels have to be increased and alignment deviated to avoid important and heritage structures. In many cities, especially coastal ones, one more difficulty met with is that the soil at that level would consist of fine sand and silt with high sub-soil water level. While going through such soils, care has to be taken to strengthen/ solidify such soils in the zone of influence of tunnel opening by chemical grouting/ freezing etc. ahead of the TBM. Precautions have to be taken to ensure that such grout does not either disturb the stability around nor does it escape through any pre existing holes like bore holes, wells etc.. Thus providing metro lines in such cities has to be well planned, carefully implemented and monitored. The case study given in the following paras gives salient features of such planning, problems met with during implementation; how they were tackled with and how work was guided and monitored using instrumentation in Chennai. The various issues discussed above are applicable to metro rail lines in cities in developing countries and coastal cities elsewhere. The project consists of two lines, totaling about 47 km, about half of each being underground construction using twin circular tunnels. One of them is a North-South alignment closer to coast line and the other East-West. Major problems arose specially in the old city portion of the North-South line, which lies closer to coast line.

Sharpest curve on line is with 150 m radius and in depots 100m. Steepest gradient adopted is 4%. Underground stations are provided a gentle grade up for retardation and down grade at exit for acceleration. The station roof slab in UG stations is kept 2.5 to 4 m below road / ground level to accommodate utilities like water, sewage pipes etc.. Minimum depth to top of tunnel tube is 9m and maximum 30.5 where it crosses River Cooum. Chennai Central underground station is at two levels, one below the other for the two lines.

All the underground stations are being done using cut and cover methodology and sections in between are being bored through using TBMs, which have been imported from China or Germany, depending the contract agency doing the work. Though in the preliminary study stage, much longer lengths in both the cases were proposed for viaduct construction, considering that the ROW of the existing roads on the alignment at a number of locations was hardly 20m and so would become bottlenecks for road traffic if viaduct structures were used on them, the costlier underground option was chosen.

The soil in the underground construction lengths is quite varied, consisting of Achaean rocks underlying younger alluvium of varying depths. Detailed geotechnical investigations were conducted at the investigation stage to decide on lengths suitable for TBM work as well as to design the TBM requirements. Investigations included Survey of utilities coming in the way; enumeration of bore holes/ wells etc., within the zone of influence of the tunnel and structural condition of buildings likely to be affected. Before starting tunneling work, utilities coming in the way were diverted and all bores/ wells within the area of influence plugged. These and continued soil investigations in the underground lengths helped in arranging for choice of grouting type and pressure, and stabilisation of soil ahead of the TBM cutter head. Earth Pressure Balance type of TBMs are used. Parameters for TBM such as face and grout pressure and volume of grout was continuously monitored.

8.12.2 Profile

Two circular tunnels, each of 5.8 m internal diameter, one for the Up and the other for the Down line are provided at 14.05 m (centre to centre) apart and are given sufficient cover, locating them 9m clear below the road surface. They, thus, run almost below the edge/ walkway of the road above. The tunnel section provides for space to provide a 600 mm wide walkway on one (inner) side for use of passengers in case of emergencies or breakdowns. Cross passages are provided between the two tunnels at 250 m intervals for passengers to access the other tunnel. These can work as refuges also for maintenance staff. Each TBM starts from one station and bores through upto next station. Typical layout and section of the tunnels is given in Figure 8.10(a). Figure 8.10(b) show details of one tube.

8.12.3 Major Problems

The TBM used is one suitable for working in densely populated urban environment. In some lengths, (Figure 8.11) the tunnel cuts across built up areas (two to three storeyed buildings having shallow foundations) and work had to be planned so that minimum risk/ settlement/ damage to buildings above is caused. In such areas the tunnel is taken 12 to 16 m below ground, giving additional cushion. The EPBM is capable of tackling mixed face of

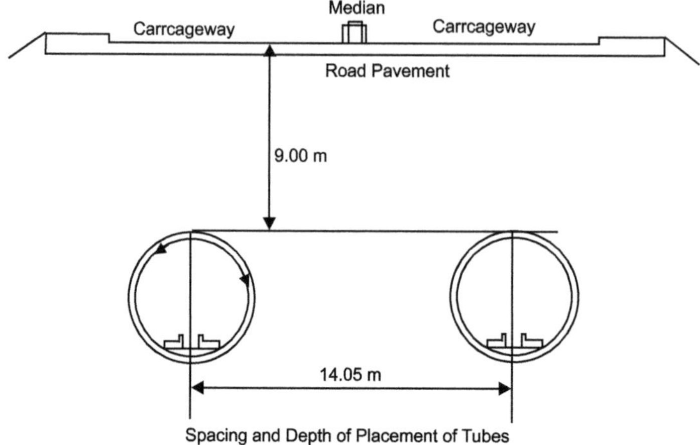

Source: CMRL

Figure 8.10(a) Typical Section of layout of Circular Tunnels in Chennai Metro Adapted from CMRLDrg

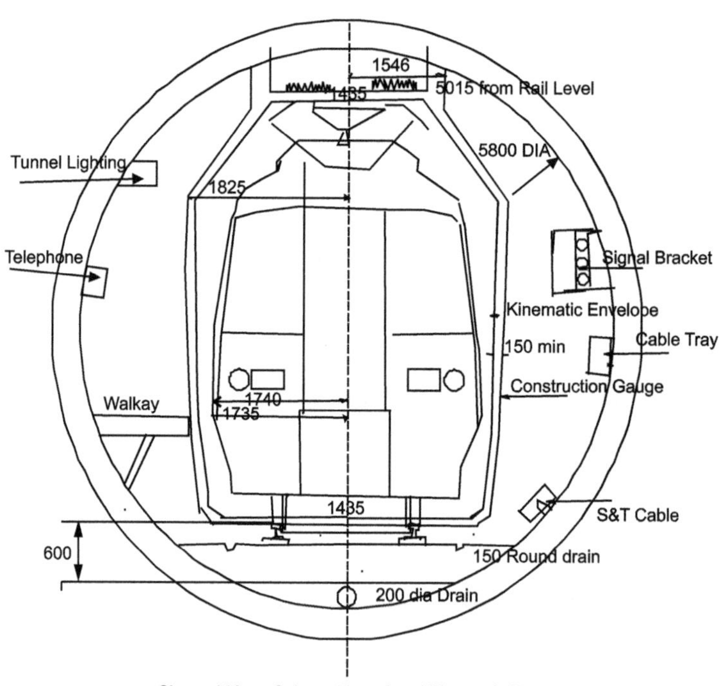

Figure 8.10(b) Clearance Diagram and details of one Tunnel[7]

soil, sand of different consistencies, clay etc.. Buildings in the area of influence have been surveyed in advance and where required monitoring devices fixed. A field team visits them twice a day during TBM operations for monitoring them and takes precautionary and remedial measures. In case of heritage buildings additional care is taken and in one case, a church coming too close to the original alignment was avoided by deviating the alignment as shown in Figure 8.12. In spite of such caution, there were two other cases of old buildings, in which some cracks appeared. They were attended to with technical advice from structural experts from IITM.

Courtesy: CMRL

Figure 8.11 Typical Alignment through Older area in Chennai

The tunnel lining is made up of precast concrete rings comprising 6 segments. Rings are 1.2 to 1.5 m long. There is a wedge ring, provided in a staggered manner as shown in Figure 8.13. Lining segments are 275 mm thick, reinforced suitably and interconnected with bolts at joints.

8.12.4 Station Tunnel

The tunnel tubes are so spaced that at stations, an island platform can be fitted in between. A typical section at a station is shown in Figure 8.14. It is implemented by using Top- down Cut and Cover method as illustrated in Figure 8.7.

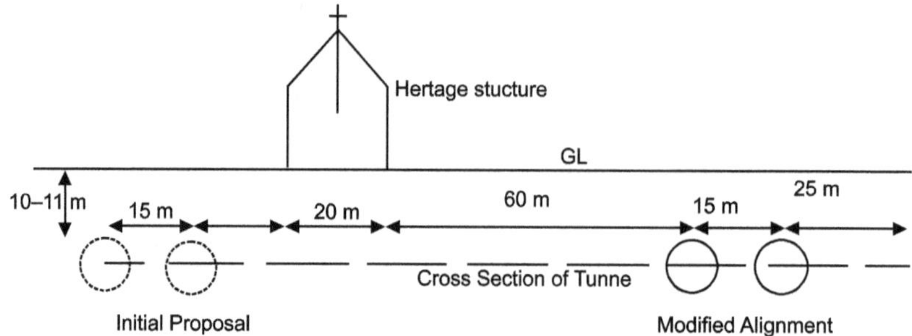

Source: CMRL

Figure 8.12 Deviation of Alignment at a Heritage Building Location

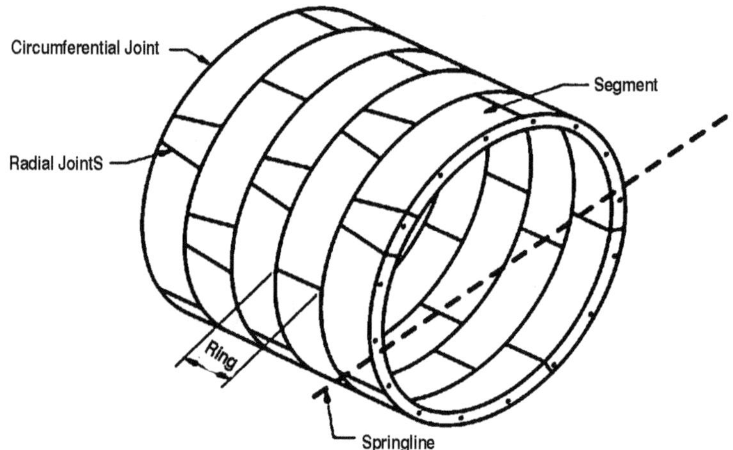

Courtesy: CMRL

Figure 8.13 Arrangement of Precast Segments of a Lining Ring

8.12.5 Instrumentation

Instrumentation is being done extensively for monitoring during construction and for later observations where required. Details of instrumentation done normally on sub way tunnels are discussed in detail in Para 11.5.Figure 8.15 shows views of different instruments used by CMRL.

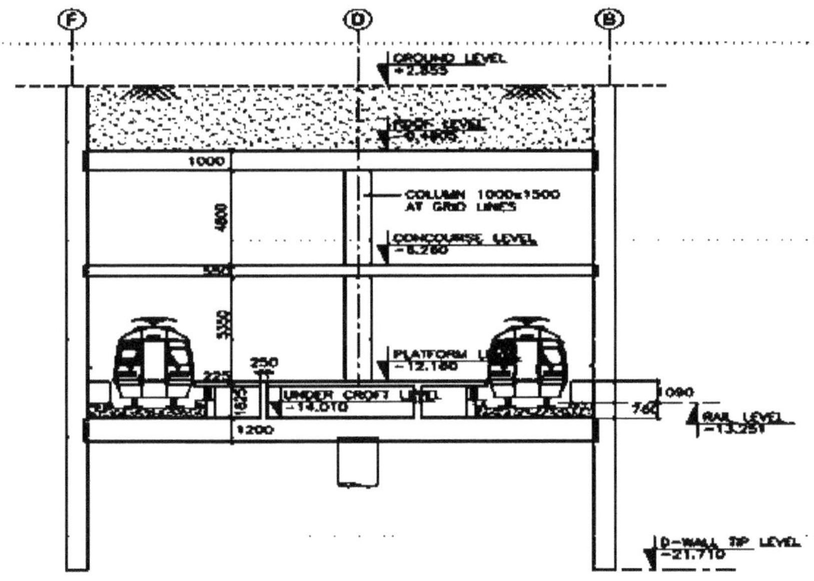

Courtesy: CMRL

Figure 8.14 Typical section of an Underground Station.

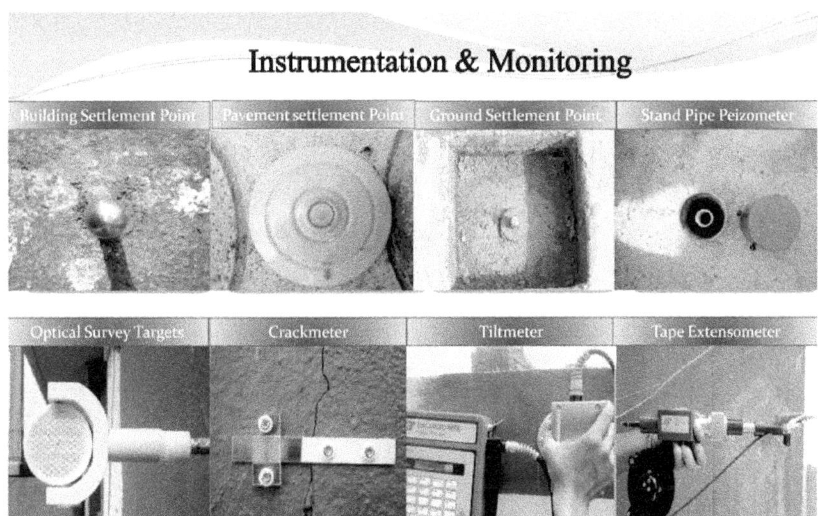

All figures in Case Study; Courtesy CMRL

Figure 8.15 Instruments used for monitoring different aspects on CMRL.

8.13 REFERENCES

1. Parker, A.D., "Planning and Estimating Underground construction', McGraw Hill Publications, New York.
2. Pequinot, C.A., (1963) "Tunnels and Tunnelling'- Hutchinson, Scientific and Technical, London
3. Morton, D.J., 1982, 'Subway Tunnels', In: 'Tunnel Engineering Handbook', Bickel, John 0. and T.R. Kuesel, (Eds.), Van Nostrand Reinhold Company" New York, pp. 417-444.
4. Rudrakshi, M., (1985), 'Construction Method- Cut and Cover '- Paper 26, Proceedings of International seminar on Metro railway- Problems and Prospects, Metro Railway, Kolkata
5. Phadke, G.N., (1985) 'Constraints and Problems in Construction of Metro Railway, Calcutta with special Emphasis on Civil Engineering Works'- Paper 1, Proceedings of International seminar on Metro railway- Problems and Prospects, Metro Railway, Kolkata
6. Vos, Charles (1993) 'Submerged Tunnels Examples of Maxima Structures', Seminar Papers of International Association of Bridge and Structural Engineers Conference, New Delhi.
7. Ramanathan, R, (2014) 'Chennai Metrorail Project- An Overview', Tech Times, April, 2014, Chennai, pp 44-48

CHAPTER

9

Lining

9.1 GENERAL REQUIREMENTS

9.1.1 Purpose and Types of Lining

Lining is provided inside the tunnel section with masonry or concrete and /or by shotcreting to provide support to the cavity against thrust from surrounding ground. The internal dimensions and shapes of tunnel after lining should be in conformity with the minimum section required for the tunnel. Conveyance tunnels, such as water and sewage tunnels, have to be necessarily lined in order to give a smooth surface for the flow of water, to prevent any leakage/seepage and also to serve any structural purpose that may be needed for withstanding inside/outside pressures. Traffic tunnel bored through hard rock which can stand on its own need not be lined unless it is intended to have a pleasing and smooth surface inside for the purpose of aesthetics and cleanliness.

According US Tunnel Inspection Manual[1], they can be broadly grouped under 7 types[1], viz., Unlined rock; Rock reinforcement system; Shotcrete; Ribbed system; Segmental linings; Placed concrete; and Slurry walls.

(i) Unlined rock: In hard rock, the exposed rock surface is left with no further treatment since structurally it is self sustaining. This was the earlier practice mostly on Railway tunnels.

(ii) Rock reinforcement system: is same as the unlined rock, but some additional strengthening of rock is done in case some structural defects exist in same and there may be some rock weak/ loose pieces. They may be strengthened to act in an integrated manner by fixing some rock anchors or bolts and/ or shotcrete with mesh and steel dowels or thin layer of concrete.

(iii) Shotcrete: In case exposed surface has low stand up time, as a first step shotcrete is provided with steel mesh reinforcement. Alternatively fibre reinforced concrete may be used. It is done in layers for required thickness. It is used as a primary lining.

(iv) Ribbed system: This is adopted in Drill and Blast system. As excavation proceeds, ribs with or without laggings behind are provided. Laggings may be of timber, steel sheets/ plates of precast concrete slabs. The space behind the lagging upto exposed rock/ ground surface is packed with broken debris/ stones. Later reinforced or fibre concrete is poured between ribs or in a layer embedding them as permanent lining.

(v) Segmental lining: is used in conjunction with tunnelling with shield or tunnel boring machine for soft ground. This is done within the TBM/ shield under its protection. Segments are bolted together transversely and longitudinally. They can be of steel; cast iron or precast (match-cast) concrete blocks.

(vi) Placed concrete: These are final linings provided after cavity takes final position and properly designed taking into account measured movements/ strains/ forces. They are reinforced. A water-proof membrane shall be inserted between this and primary lining mentioned above.

(vii) Slurry walls: are used in case of cut and cover construction. It consists of excavating a narrow trench using bentonite slurry at the edges of the profile. It is done in short panels. After trench is made, prefabricated rebar cage is lowered and trench filled with tremie concrete. These walls may themselves form side wall of tunnel or provide protection for the tunnel box to be cast within.

Lining of tunnels on Indian Railways used to be mainly decided from the point of view of utility. Many tunnels bored through hard rock have been left unlined except at the trench portion where a proper drain profile is provided. For example tunnels through hard basalt in Bore Ghats have such exposed rough interior. But wherever the rock is not of sufficient strength/ hardness, in soft rock and also in soft soil, lining is provided based on structural strength considerations.

9.1.2 General Requirements of Lining

Though the lining should be designed to withstand pressures that will be exerted by the rock after tunnelling, the practice thus far in India had been guided by past experience. One rule of thumb for the thickness of lining in rock was 1 mm / 1.2 mm (1 inch per foot) of diameter of the circle of the tunnel and twice as much in soft rocks. In earlier days lining was done with stone or brick masonry, according to local availability of the material. In such cases the lining was built (particularly in the roof portion) for a minimum thickness and the space in-between the lining and the exposed rock surface was packed with loose rubble. In the case of rocky strata the surface may be gunited with cement mortar also. Shotcreting has become the norm presently in such cases also. It can be done with fibre concrete for providing better support in case of NATM.

Wherever possible, suitable weep holes should be left for seepage water to flow into drains, instead of blocking it completely and inducing hydrostatic pressure unless the lining is designed to withstand the same, as in subaqueous tunnels.

The lining should be so designed as to prevent collapse of the strata above and around the tunnel. Reinforced concrete has to be used in soft strata. If a box section is used reinforced concrete is a must. The practice has been to use concrete lining by pumping concrete in place between telescoping shutters which can be moved forward. Alter-natively, precast concrete or cast-iron segmental lining is used. These are most suited for circular sections and/or soft strata.

Present practice is to go in for two stages of lining. Primary lining, soon after excavation is done, is provided to help the exposed face to resist and reduce convergence of the surface and conserve the integrity of the structure. As mentioned above this will be in from of shotcreting etc., which may be about 10 to 20 cm thick of fibre concrete or with mesh reinforcement to strengthen same. Suitable instrumentation is done to observe the movements and forces developed around, which aids in proper design of the permanent (secondary) lining. This lining will be of preferably reinforced concrete or fibre concrete. In hard rock, the steel rib supports embedded in concrete should serve the purpose of reinforcement.

9.1.3 Linings to suit Different Rock/ Soils

Latest trend is to provide lining on a more scientific basis, taking into consideration the characteristics of rock/ soil it passes through and subsoil condition. In the large scale tunnelling works being done by the Indian

Railways in the Himalayan Region, they are adopting three alternative designs to suit Good Rock, Fair rock and Very poor rock/ soils. Primary lining is made of Shotcrete of different thicknesses with rock bolts and with or without steel ribs, designed to suit the Rock Mass Rating of soil. This will be flexible to extent required. The Secondary lining is of concrete of different mixes and thickness to suit. Reinforcement must be provided for poor rocks and lower quality soils. The different types are shown in Figure 9.1.

In areas subjected to seepage and presence of subsoil water, a waterproof membrane of suitable thickness should be provided between the Primary and Secondary linings.

9.2 TIMING PLACEMENT OF LINING

The correct time for providing lining is governed by the nature and condition of the soil bored through. Whenever, due to tectonic forces, rock is likely to dilate or the soil is likely to heave, it is advisable to allow sufficient time for such dilations to reduce to reasonable limits, in order to ensure that concrete lining will not fail by cracking due to heavy and unequal external forces. In some cases (especially in horseshoe-type sections) the practice is to provide the lining, leaving some space for the rock to breathe and filling that space with loose packing of rock spoil. If this is not done, the breathing forces will induce undue pressure on the lining and cause cracks in the same.

In cases where the rock may not dilate much but may deteriorate in structure causing spalling, a thin layer of concrete lining provided in-between the steel ribs which may have been used for supporting the full surface of the rock and for transferring the forces to steel supports would be necessary as primary lining. The supports can later be embedded in lining also as a permanent measure. From the point of view of safety, lining should be done wherever essential, as soon as possible after excavation. On the other hand, for rock of better type earlier practice was to leave it exposed and in some cases lining was proposed from aesthetic considerations, the same being done at any time convenient in the construction programme. With electric traction, lining was not considered an essential requirement from operation and maintenance points of view. However, with diesel traction, the exposed surfaces get covered with soot, making the tunnel unhygienic and darker for maintenance men to work. Modern thinking is to line all tunnels and even periodically clean and white wash and make tunnels brighter for maintenance staff.

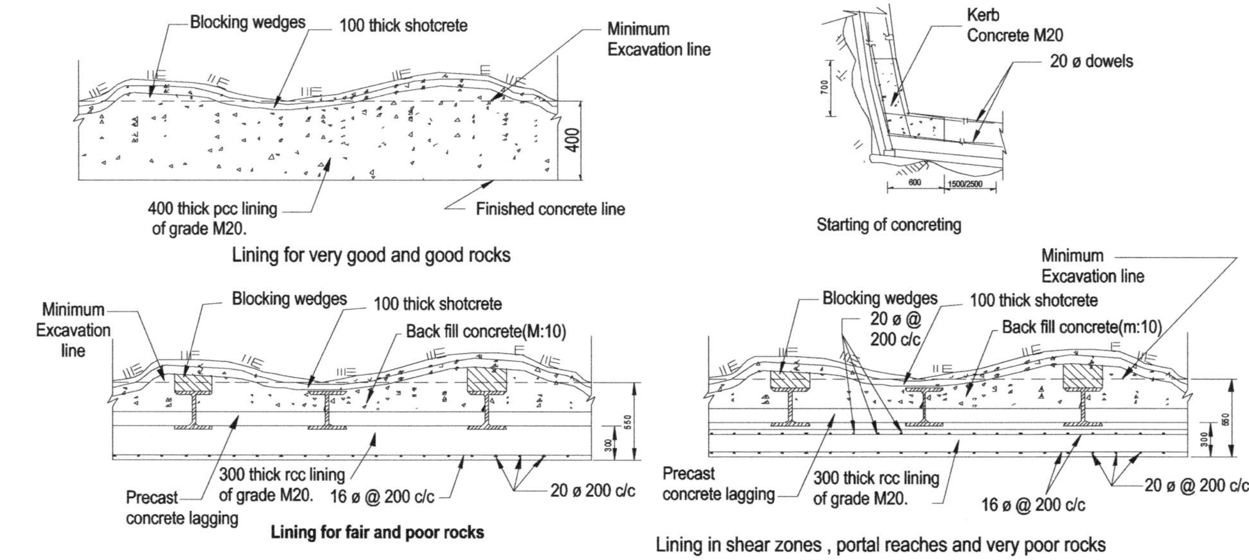

Source SVJNL/ KRCL

Figure 9.1 Types of Lining to suit Different qualities of Rocks/ Soils.

9.3 SEQUENCE OF LINING OPERATION[2]

The alternative sequences of operations generally adopted for in-situ lining are:

(i) Placing concrete to form the kerbs first, followed by side walls and arch and finally the invert
(ii) Placing concrete to form the invert first, followed by sides and arch; or
(iii) Placing concrete for the invert, side walls and arch all at one time.

Sequence (i) is suited for horseshoe, Ω-shaped and other flat-bottomed and wide tunnels. The kerb shall be built up to a section of sufficient width to serve as a base for the erection of forms for the sides. The latter shall be properly anchored and stabilised to withstand the loads of concrete lining and formwork that will come from above. Once the sides are cast and have attained sufficient strength, the arch portion can be cast. The arch portion concrete lining will generally have to be filled with concrete from the exposed face only and properly packed. Subsequently, the invert will be cast and the track or road surfacing to be provided laid over the same.

Sequence (ii) is suitable when the bottom of the tunnel is narrow or when the section is circular. In this, invert concreting is done first so that a regular base for erection of the formwork for the sides and arches is obtained before further work is taken up. This work involves removal of track and other service lines laid for the purpose of movement of form as well as lining material. In tunnels through weak strata where the tunnel floor tends to wear out fast or likely to heave, this sequence is required to be adopted. Where large horizontal thrusts are encountered, invert concrete should be placed in advance of lining the sides so that it will serve as strutting between the sides also.

Sequence (iii) is adopted for smaller circular and oval conveyance tunnels and is not relevant for large traffic tunnels, except in the case of underground railways, sub-aqueous tunnels and tunnels through soft strata where boring is done with shield tunnelling machines. Here the lining closely follows the boring, either with use of precast segments or a telescopic formwork.

9.4 FORMWORK[2]

9.4.1 Types of Formwork

It is preferable to use collapsible steel forms made up in standard size/unit lengths in the interest of speed and economy. These forms can be used for multiple operations and easily removed and added on for proceeding ahead. The types of formwork generally used are rib and plate, rib and laggings and travelling shutters, with or without telescopic arrangements. In transitions, junctions and sharp bends, timber formwork is necessarily used due to the need for flexibility and change of sections. The ribs in the first two types of formwork mentioned are made up either of channels or T-sections placed at intervals of about I m. This spacing may be varied according to the thickness of the concrete lining. The ribs will be spanned by steel plates suitably stiffened or timber lagging firmly fixed from the bottom upwards as the concrete rises. In a tunnel in which the solid rock is to be lined and concrete placed mechanically, forms can be removed even 16 to 20 hours after placing the last batch of concrete. The usual precautions of oiling and greasing the formwork must be taken before final assembly and concreting, to facilitate easy removal. A travelling type of formwork is generally used in the full-bore method of tunnelling or when the invert is done first and the sides and arch are done together, as in the horseshoe-type tunnel. Such forms are otherwise known as travelling jumbos. Such jumbos are used for lining tunnels even during service by blocking traffic for short intervals of time. There are two types of tunnelling formworks: non-telescopic and telescopic.

9.4.2 Non-telescopic Formwork

The entire formwork is preassembled and mounted on a travelling frame (jumbo) fixed with wheels running on a track and screw jacks for collapsing the formwork when required. The sections are hinged to permit collapsing and the jacks are required for bracing and aligning. The forms are 2 to 4 m long and can be struck and reassembled quickly depending on the requirement of construction, matching concreting equipment.

Figure 9.2 shows the typical formwork used in the Indian Railways for concreting the arch and roof (Padmanabhan, 1965)[4]. The travelling steel shutter running on narrow gauge rails is provided with turn buckles and jacks to adjust and ease the shuttering. This provides concrete free from blemish on the surface and ensures high standards.

9.4.3 Telescopic Formwork

This type of formwork is so designed that the back unit can be collapsed and moved forward through the front unit without disturbing it. The side plates are so hinged to the arch plate that they can be collapsed. The formwork should be self-supporting while travelling in a collapsed condition also. Another formwork (slip formwork) or monolithic formwork is used for circular or near-circular section tunnels and lining of the full section is done in one operation.

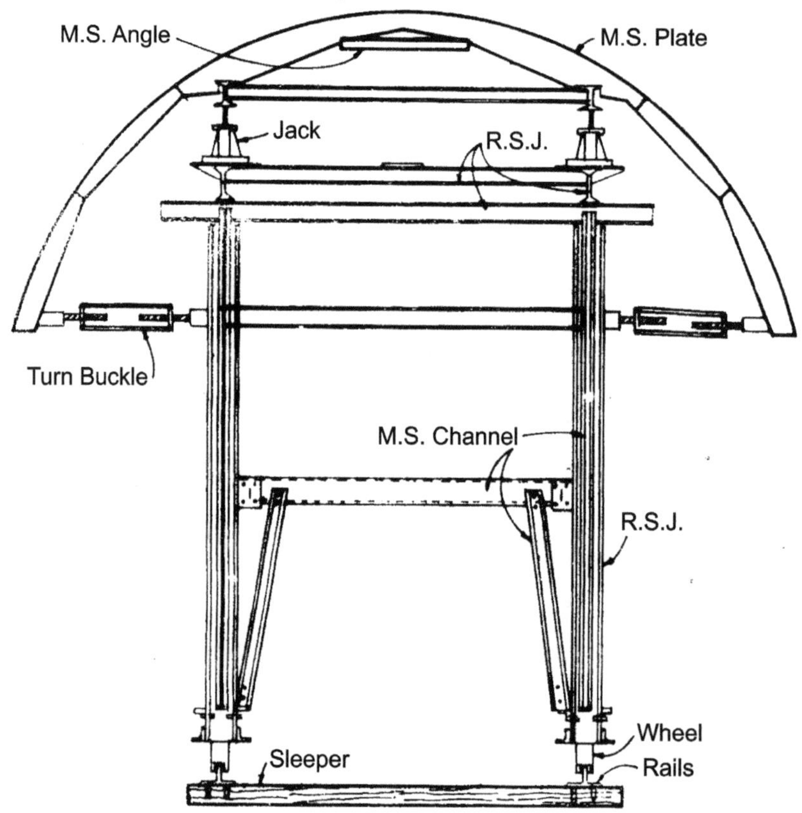

Figure 9.2 Travelling Steel Shutter for Arch Concrete4.

All formworks should have inspection windows about 0.5m x 0.5m in size and not more than 3 m apart. Horizontal and vertical spacings of such windows depend on the thickness of lining, method of placement of concrete and workability. These windows are used to place and vibrate the concrete and have to be so located that dense, compact concrete is ensured. The shutters to these should be strong and be capable of being easily fitted and removed. The joints should be leakproof to prevent cement slurry flowing out, causing honeycombing or leaving projections in the finished concrete surface.

At either end of the formwork, except when non-stop concreting is adopted, bulkhead shuttering has to be used; timber is generally more convenient for this purpose. Neat construction joints are also mandatory.

9.5 CONCRETING[2,5]

The mix of concrete for lining depends on the location and design requirements. Generally a concrete mix of grade more than M 20 or more is used for the invert and sides and M20 for roof. Of late RCC Box and circular linings are designed with M40 mix. Use of M30 concrete with fibre reinforcement is also the present trend. The slump of concrete should not be less than 100 mm so that the concrete can flow easily or be pumped in, avoids segregation and also properly fills the space. However, wherever concrete is placed directly, as in inverts and kerbs, the slump may be reduced to 50 mm. The maximum size of aggregate shall not exceed 40 mm. The minimum cement content in the concrete may vary from 350 to 450 kg/m^3 when natural aggregates are used and there is no hindrance of supports or reinforcements. As far as possible, concrete should be placed through a pneumatic placer. For tunnels of short length and jobs requiring comparatively large volumes of concrete, the latter should be mixed in a 'batching plant located at a suitable site outside the tunnel and the mixed concrete conveyed as quickly as possible to the site of placement by means of a short belt conveyor and poured into the hopper of the concrete pumps or placer situated close to the working site. In the case of tunnels of longer length, it is advantageous to batch and mix the aggregate in dry condition outside the tunnel and then convey it inside to locations of placement by tipping type wagons or dumpers. The dry mixed aggregates are then remixed at the site adding the required quantity of cement and water (and any plastciiser) to obtain the specified slump and water cement ratio. The mixer at the site of placement has to be so located that the dry mixed aggregates can be readily dumped into the upper part, and the mixed concrete can then flow into the hopper of the placer or pump. Suitable retarders may be used where ready-mixed concrete is transported in transit cars.

Following structural design specifications have been suggested for the second (permanent) lining by ITA[5]:

(i) Thickness cast-in- place lining can have a lower limit of 25-30, to avoid difficulties of placing, compaction and possible honey-combing-
-20 cm if it is unreinforced
-25 cm if reinforced

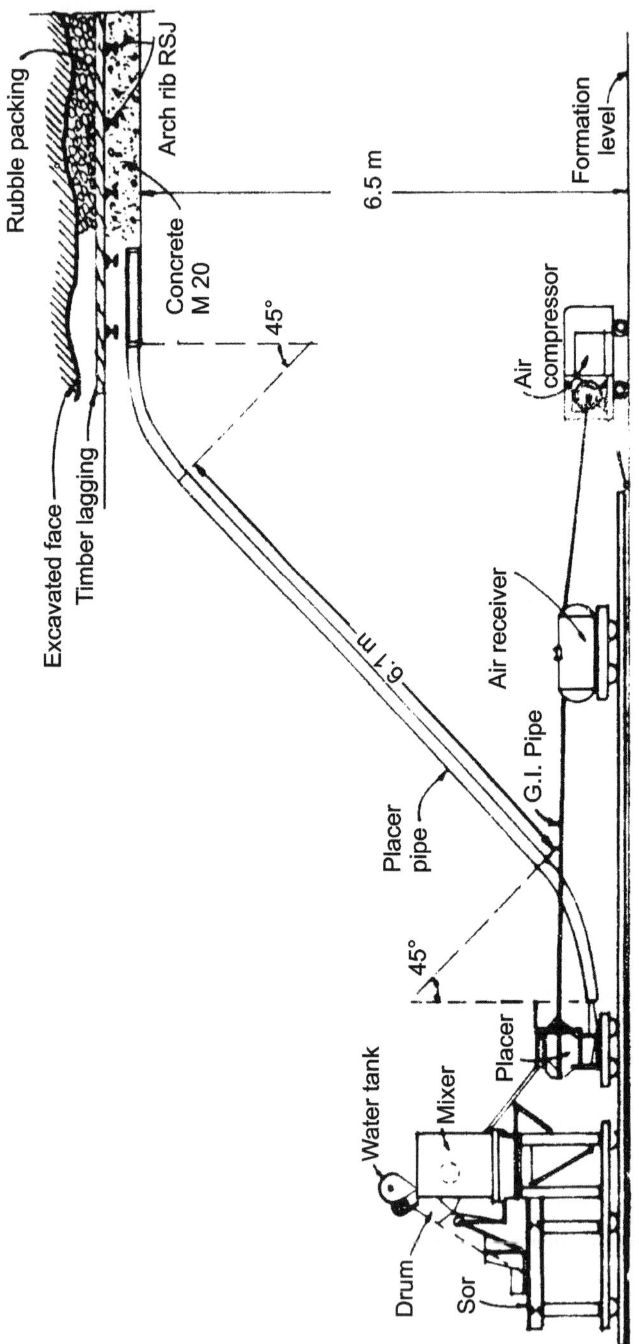

Figure 9.3 Layout of Lining Equipment for Arch Concreting[3].

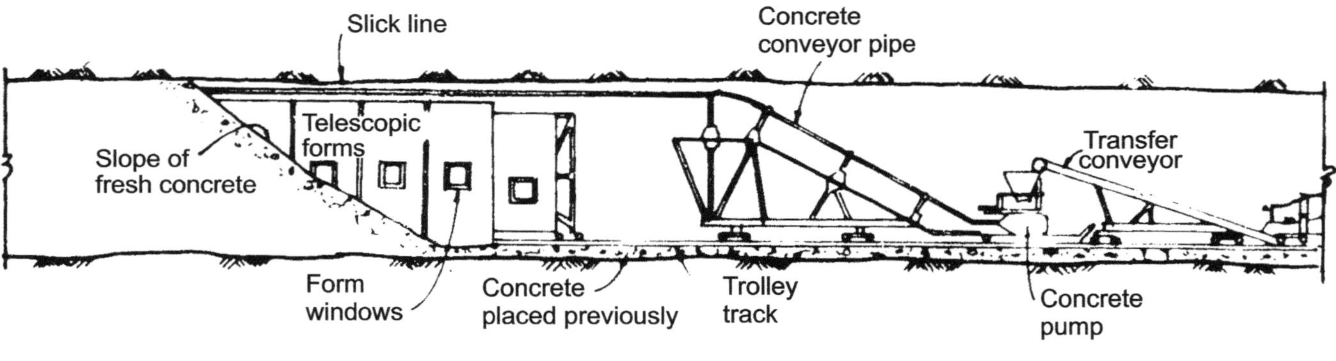

Figure 9.4 Concrete in-situ Lining with Telescoping Formwork[3].

- 30 cm for watertight concrete

(ii) Reinforcement is desirable for crack control, even if not required from stress point of view. From point of view of placement such reinforcement may have following quantum of closely steel mesh in both directions.
-at least 1.5 cm^3 / m of steel on outer surface
- at least 3.0 cm^3 / m of steel on inner surface

(iii) Minimum cover recommended for reinforcement are:
30 mm at the outer surface if a waterproof membrane is provided
50mm-60 mm -at the outer surface if it is directly in contact with ground and ground water
40 mm- 50 mm - at the inner surface
50 m - for tunnel inverts, where water is aggressive.

For segmental lining specifications (i) to (iii) are not applicable, if it is the outer primary lining. Special attention should be paid to avoiding damage during transport and damage and design detailing should take care of this

(iv) Sealing against water (waterproofing sheets) may be necessary in following cases.
-when there is threat from aggressive water to damage steel/ concrete
- when sub-soil water is more than 15 m above crown
- when there is possibility of freezing of ingressing water along tunnel section close to portals
- for protecting inner installations of tunnel

For achieving water tightness of concrete, special specifications for the mixture; avoiding shrinkage stresses and temperature gradients during setting and quality control are more important than crack control.

They have also suggested that long term durability of shotcrete concrete used in Primary lining should be preserved if it is to be considered to provide stability to tunnel. This requires absence of aggressive water, limitation of concrete additives (liquid accelerators) and avoiding 'shotcrete shadows' behind steel arches and reinforcements.

A typical layout of compressor, placer, pipes etc. for laying arch concrete is shown in Figure 9.3. Concreting the arch and sides together is shown in Figure 9.4. Concreting should preferably be done using a tremie for sides and invert. Precautions normally adopted in tremie concreting should be followed in the case of concrete being placed by pumping through pipes. The open end of the conveyor delivery pipe must be buried at least a few cm inside the freshly placed concrete to avoid segregation. Care should be taken while placing through side doors in forms so that no hollow pockets remain. In the case of monolithic forms, as the concrete will fill the invert first, there may be a tendency for the form to float and hence the formwork should be rigidly strutted down from the sides and the roof of the tunnel as well.

During concreting with the pump or placer, if placement is interrupted for a period of more than one hour, a batch of mortar sufficient to cover the area by a 15 mm layer should be pumped to cover the cold joint before the next batch of concrete is pumped in.

9.6 COMPACTION

As far as possible flexible shaft immersion-type vibrators with a vibrating needle of 50 mm dia and 3000 vibrations/minute frequency should be used for vibration of the concrete. In addition, the concrete should be vibrated by external form vibrators of minimum 0.5 KVA capacities. The spacing of form vibrators depends on the size of vibrators, make of formwork, thickness of concrete etc. Any hollow space should be grouted under pressure not exceeding 5 kg/sq. cm. through holes drilled at intervals. Where grout intake indicates a gap of more than 10 mm, very fine sand or rock dust should be added to the grout to fill the gap. In certain locations the addition of bentonite may also be helpful to hold cement in suspension.

9.7 CURING

Curing of concrete should be done by spraying water at short intervals to maintain a wet surface. The minimum period will be the same as for the curing of concrete in other locations.

9.8 SEGMENTAL LINING[6]

A faster method of lining circular tunnels is to use segmental lining. They can be of Cast Iron or of Precast RCC segments. They are quicker to install and it is easier to ensure quality in them. The major advantages of use of segmental lining are:

- They provide a flexible lining, which adjusts with the movement of soil around and distributes forces better. They can take the loads soon after erection, without need for waiting for concrete to age and gain strength.
- Being produced in factory conditions, there is better quality assurance.

- The dimensions are uniform and surface finish better.
- Installation time is reduced and the opening can be supported sooner after excavation is complete.
- No in-situ curing is called for.

Main disadvantages are:

- It calls for establishment of a separate casting yard and good transport arrangements for transport to site. The segments have to be match-cast and properly marked for matching during erection.
- For small jobs and isolated tunnels, they will be less economical
- They need special equipment, highly skilled staff for erection at site, and in casting yard.
- After erection, the gap left behind them has to be grouted to ensure proper contact with the excavated surface.

They are flexible, with hinged joints and hence can deform to suit the ground movements and help convert the forces into thrust mainly.

For sub-way bored tunnels, segmental linings are used invariably. They are preferable for circular tunnels driven using shields.

Typical details of the segmental lining and details of a circumferential joint of RCC segments used at Hanover tunnel are shown in Figure 9.4(a) and (b).

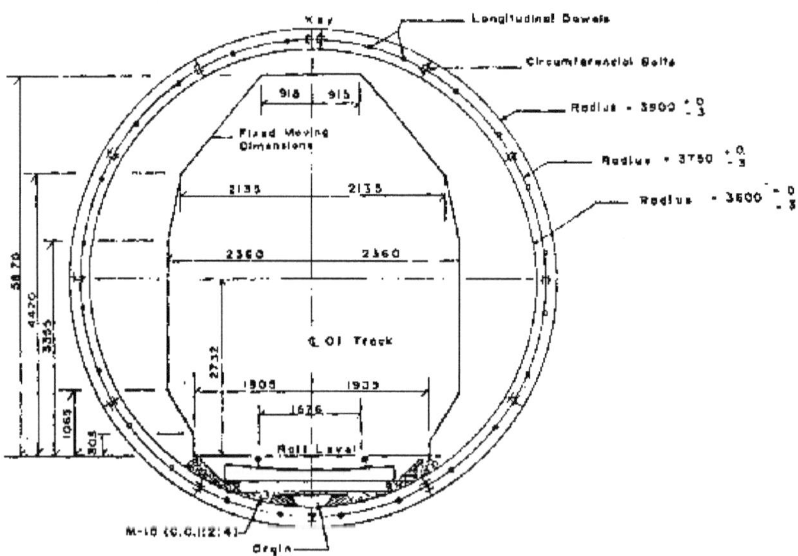

Source: ICJ, February, 1994

Figure 9.4(a) Details of Tunnel Ring used on Hanovar Tunnel.

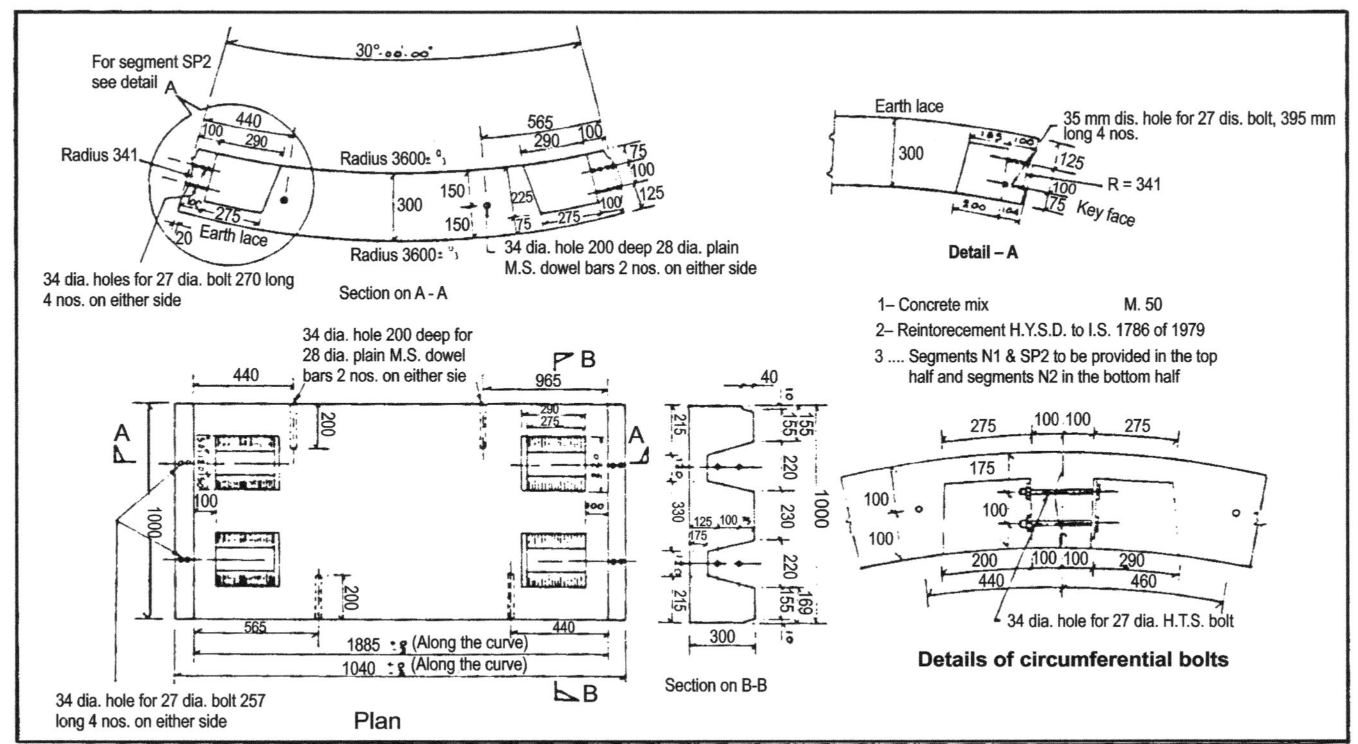

Source: ICJ, February 1994

Figure 9.4(b) Details of a Segment and a Circumferential joint[6].

9.9 REFERENCES

1. FHWA, 'Highway and Rail Transit Tunnel Inspection Manual'- Chapter 2 'Tunnel Construction and Systems', FHWA, Department of Transport, Washington.
2. IS 5878: Part V , (1971). 'Construction of Tunnels Part V - 'Concrete Lining'- Bureau of Indian Standards, New Delhi.
3. Parker, A.D., 'Planning and Estimating Underground construction', McGraw Hill Publications, New York.
4. Padmanabhan, V.C.A., (1965), 'Notes on D.B.K. Railway Project', Unpublished Indian Railways Report, 143 P.
5. Heinz Duddock, (1988) 'Guidelines for Design of Tunnels' Feature Report of ITA Working Group on General Approaches to Design of Tunnels, Tunnelling and Underground Space Technology, Vol3, No 3, PP 237- 249
6. Raju, C., Narayanan, G., and Kurien, A.P., 'Soft soil tunnelling using shield for Honnavar tunnels'. Indian Concrete Journal, February 1994, Associated Cement Companies, Mumbai

CHAPTER

10

Ventilation, Lighting and Drainage

10.1 VENTILATION

Explosive charges and plying of vehicles or locomotive for moving drilling equipment or for mucking leave fumes and unhealthy gases inside the tunnel. An efficient ventilation system to drive away these fumes and gases from the tunnel should be provided for all tunnels during construction[1]. The effect on humans of the carbon monoxide emitted by petrol-ignited engines and, to a small extent, by diesel engines is given in Table 10.1.

The acceptable limit for CO is 0.04% according to US practice. The ventilation arrangement should be such that a minimum fresh air of 2.8 M^3 per minute per person and 2.1 m^3/min for each BHP of diesel equipment working in the tunnel is provided for. Such ventilation can be done by means of ducts through which either fresh air is blown towards the face of the shaft or the stale air exhausted from it, leaving the fresh air to flow through the other face or faces along the length of the tunnel. Whichever system is adopted, it must fulfill two major requirements during construction:

(i) Fumes from blasting must be cleared quickly from the working face so that work can be resumed after blasting without delay.

(ii) Accumulation of dangerous concentrations of fumes anywhere along the length of the tunnel is prevented.

During construction, fans delivering 600 to 850 m^3 of free air/min to blow over the face after blasts help to reduce the delay due to waiting for fumes to clear. For small works it is difficult to provide for such powerful equipment and generally more time is spent waiting for the gases to escape by using smaller boosting fans. Such action would result in carrying out only one blasting cycle per day.

Table 10.1 Effect of Carbon Monoxide

CO %	Exposure time (min)	Effect
0.01	Several exposures	Ultimate poisoning of the system
0.04	240	Headache and discomfort
0.12	60	Palpitation
0.20	30	Unconsciousness
>0.50	up to 30	Death

In addition, in certain types of ground like shale and in some cases of rocks (with presence of coal seams) some harmful gases may escape from cavities, like methane which is a very dangerous gas, since it is easily combustible and harmful to breathe also. The ventilation of traffic tunnels has to keep the tunnel free from such obnoxious and harmful gases and smoke and gas arising out of blasting to such an extent as to keep their levels inside the tunnel at any time within safe limits.

Methane[2] is a colourless, odourless gas lighter in density than air (Specific Gravity being 0.553) and tends to remain in cavities of roof. IS 4756:1978 lays down guidelines in dealing with this gas. In such cases, the executive has to take action in consultation with the Director General of Mines Safety. Methane is highly explosive in the concentration of 9% with 4.5% and 9% being Lower and Higher explosive limits. Some immediate precautionary steps have to be taken on its detection. When its concentration exceeds 1.25% at any location, it is advisable that supply of electricity is cut off, diesel engines to be stopped and battery terminals disconnected. No form of spark (from welding, cutting etc..) or naked fire will be permitted nearby. The gas should be cleared by blowing fresh air and preventing any recirculation of exhausted gas. Tunnel instruments to detect presence of gas and its laboratory analysis should be available in suspicious locations. Methanometers are available now in market for detecting presence of Methane and hand held Multi- gas Detectors for detecting presence of Methane, carbon monoxide, Carbon di-oxide and Oxygen. Hanging of two Flame Safety lamps near the face of excavation can also help in detection of CH_4 , since its yellow flame would turn bluish in presence of methane.

If concentration increases, the flame size also will increase. It is desirable to have services of an experienced mining engineer while tunnelling in such locations. Whenever there is any coal present, presence of CH4 is a possibility.

Actions to be taken in case of presence of methane are:

> 5% of LEL- increase ventilation

> 10% of LEL - suspend hot work such as welding, gas cutting etc.,

> 20% of LEL - Stop all works, cut power supply, withdraw all men and wait till concentration by ventilation falls below this limit.

(LEL- Lower Explosive limit)

Vent of ventilation system in case of presence of methane should not be more than 5 m from face of excavation, installed near crown. Exhaust type ventilation system for tunnel, capable of sucking the gas and pushing into the exhaust system may also be fixed at about 20 to 25 m from face of excavation. Smoking or use of mobile phones should be prohibited.

After construction and commissioning of tunnel, during service with traffic the carbon monoxide content in the tunnel should be within 100 to 250 ppm range. To achieve this in highway tunnels the ventilation system should generally be capable of producing a minimum fresh air supply of 0.85 m^3/tlm (tlm = tunnel lane metre).

In the case of long and curved tunnels special blowers are required during service for quick clearing of the fumes emitted by the vehicles and motive power. This is particularly important in highway tunnels and also in railway tunnels when steam or diesel traction is used. Ventilation is achieved artificially by blowing air through ducts from one end of the main body of the tunnel and exhausting it by suction through exhaust fans at the other end in short tunnels. In long tunnels ventilation shafts are provided to act as natural vents or for assisting in driving the foul air through.

10.2 VEHICLE EMISSIONS

10.2.1 Exhaust Emissions

The exhaust emissions from spark-ignited and compression-ignited vehicles differ. The major constituents of the exhaust from these two types of vehicles and their relative percentage by volume are listed in Table 10.2. During service more care is called for as the public are exposed to the effects of these gases. Their effects are therefore discussed in more details here.

10.2.2 Carbon Monoxide

Carbon monoxide (CO) is an odourless toxic gas. It has a very high affinity to blood haemoglobin (300 times that of oxygen) to form carboxyhaemoglobin (COHb). This adversely affects the capacity of the blood to transport oxygen to body tissues;

Table 10.2 Typical Composition of Emission from Road Vehicles[1]

Exhaust gas component	Spark-ignited engine (%)	Compression-ignited engine (%)
Carbon monoxide	3.000	0.100
Carbon dioxide	13.200	0.000
Oxides of nitrogen	0.060	0.040
Sulphur dioxide	0.006	0.020
Aldehyde	0.004	0.002
Formaldehyde	0.001	0.001

Source; Bickel and Kuesel, 1982

instead the carboxyhaemoglobin is absorbed by the tissues, which at 63% concentration can be fatal. Even at 10% the first toxic effects become evident. The process is reversible.

10.2.3 Carbon Dioxide

This too is toxic but only at levels which are well above those found normally in vehicular tunnels, as can be seen from Table 10.3 which shows the dilution effects.[1,4]

10.2.4 Oxides of Nitrogen

Of the several oxides of nitrogen, only two, being toxic, are of concern here, namely nitric oxide (NO) and nitrogen dioxide (NO_2). Their effects are similar except that NO_2 is five times more toxic than NO. Nitric oxide is a colour-less, odourless gas formed during high temperature combustion, its quantity increasing with flame temperature.

It is soluble in water and has great affinity to blood haemoglobin, which produces shortage of blood oxygen, as does carbon monoxide, but with less severe effect.

Nitrogen dioxide is a pungent reddish-orange-brown gas which is almost insoluble. Since 95% of inhaled NO_2 remains in the body, concentrations of 100 to 150 ppm are dangerous for exposures of 30 to 60 min. It also

Table 10.3 Dilution of Engine Exhaust Gases*

Gas	Spark-ignited		Compression -ignited		Threshold limit value-time weighted average** TLV-TWA (ppm)
	At exhaust (ppm)	Level after dilution (ppm)	At exhaust (ppm)	Level after dilution (ppm)	
Carbon monoxide	30000	200.00	1000	6.70	50
Carbon dioxide	132000	880.00	90000	600.00	5400
Nitrogen dioxide & nitric oxide	600	4.00	400	2.70	5
Sulphur dioxide	60	0.40	200	1.30	25
Aldehyde	40	0,27	20	0.13	NA
Formaldehyde	7	0,02	11	0.07	2

- Diluted to maintain .200 ppm of CO using a dilution ratio of 150 to 1,

** Time weighted concentrations to which workers may be exposed 8 hours per day without adverse effects

combines with water in the lung to form nitrous and nitric acid. The latter can destroy the alveoli, thereby reducing the ability of lungs to transport oxygen. It can also combine with hydrocarbons present in exhaust gases and sunlight to form smog, which reduces visibility (as smog absorbs light).

10.2.5 Sulphur Dioxide

Sulphur dioxide (SO_2) is a non-flammable, non-explosive, colourless gas. It oxidises in the atmosphere and with moisture forms sulfuric acid (H_2SO_4), and thereafter reacting with other pollutants forms toxic sulphates. When SO_2 is present at over 3 ppm it has a pungent odour. Since its proportion is low in vehicle exhaust, it does not present a serious problem.

10.2.6 Hydrocarbons

Though these may form a small proportion of vehicle emissions, hydrocarbons are the most complex. They include methane, ether, propane, ethylene, acetylene, pentane and hexane. Most are inflammable at high concentrations. But they rarely reach the threshold limit levels in tunnels. The main problem is that they aid in formation of photochemical smog, which affects visibility.

10.2.7 Aldehydes

Aldehydes are organic compounds present in exhaust gas; they are irritants to either the skin or the mucous membranes or both. The levels at which they act as irritants provide sufficient warning to preclude health hazard.

10.2.8 Particulates

Particulates are caused by incomplete burning of hydrocarbon fuels (petroleum products). They remain in suspension for a long enough period to be respired, thus affecting the lungs. These are more significant in diesel engine exhausts and constitute one of the major problems of diesel engine trains in tunnels.

10.3 VEHICLE EMISSION RATES

Design of a ventilation system requires a good knowledge of vehicle emission rates. The earliest work done in this respect was in 1920 by the US Bureau of Mines. Further tests were done by the Colorado Department of Highways in the US in 1990 relating to the Straight Creek tunnel. But this was at elevations of 330 to 1650 m. For elevations about 2135 m the Institute for Highway Construction of the Swiss Institute of Technology prepared a report. The results can be adapted for other elevations, based on the knowledge that the consumption of fuel increases at higher elevations due to reduced air pressure, producing more smoke and gas.

The content of carbon monoxide emitted gives an idea of the emission of other gases; if the level of carbon monoxide is kept at or below 200 ppm (equivalent to a 150 to 1 dilution ratio), all other constituents will stay well within threshold limits. Hence the role of emission of carbon monoxide is mainly considered for tunnel ventilation design.

Emission rates vary also with vehicular speed. The rates of emission are taken as 1.5 x 10^{-3} m^3/kN/km (0.88 ft^3/ton/mile) for spark-ignited engines and 0.046 x 10^{-3} m^3/kN/km (0.026 ft^3/ton/mile) for diesel engine vehicles. When traffic is stopped and an engine is idling, up to 10% carbon monoxide is produced in the exhaust gas. But since an idling engine consumes less fuel, the total amount of carbon monoxide will be less. For a normal ignition-powered engine consuming one litre per hour of gasoline, the rate of emission will be about 0.3 m^3/car/hour (10.6 cft./car/hour).

The emission of a compression-ignited engine is more complex. To reduce the amount of smoke emitted in tunnels, such engines have to be operated with 20% to 40% excess air. But this means a reduction in power and efficiency. Hence a certain amount of smoke in exhaust has to be accepted. In determining this amount (commensurate with safety and comfort), a stopping distance of 90 m for a speed of 60 kmph can be taken as standard. To give good visibility and to keep smog low, a smoke level of 2 to 4 Dmg/m^3 is considered satisfactory, depending on the lighting level in the tunnel. Regarding emission of other gases, in view of the complexity of determining exhaust emissions of diesel engines, the same procedure as adopted for ignition engines is adopted.

10.4 METHODS OF VENTILATION

10.4.1 Types of Ventilation[1]

Ventilation is required for all vehicular tunnels. Natural ventilation cannot be relied on for tunnels longer than 150 m. For long tunnels, and for tunnels with anticipated heavy traffic, mechanical system of ventilation using fans must be considered. There are 3 types of ventilation: longitudinal, semi-transverse and full transverse[1,3].

10.4.2 Longitudinal Ventilation

This is a system whereby air is introduced or removed from a tunnel at a limited number of points along it, creating a longitudinal air flow. Typical ways of doing this are sketched in Figure 10.1. It is most effective in a unidirectional traffic system. In the simplest form, it can be done with no shafts. The addition of shafts induces air flow due to the stack effect. In this system the air velocity is uniform. The contaminant level is lowest at the entrance and highest at the exit.

In a longitudinal system provided with two shafts at the centre, one for exhaust and the other for supply, the contamination level can be kept low in the second half of the tunnel also. The system is affected by external adverse atmospheric conditions. In bidirectional traffic tunnels, using this system the contamination level is highest near the shafts. The system may be one of natural ventilation or forced air using booster or exhaust fans. The length that can be managed depends on traffic volume. For example, the 1.8 km tong East River (USA) road tunnel has a volume of 100 vph (vehicles per hour) while the 0.45 km long Lake Washington (USA) tunnel has a peak volume of 5280 vph.

Ventilation can also be obtained in this system by providing booster fans fixed in the ceiling. This eliminates the need for shafts but requires a larger (taller) section for tunnel profile. The most economical one is the system without booster fans but with supply or exhaust at a limited number of locations. The existing tunnels using this system vary in length from 0.25 to 3.00 km (e.g., Tende in France/Italy is using axial boosters).

10.4.3 Semi-Transverse Ventilation

In a semi-transverse ventilation system air is passed through a duct with supply flues that release it into the tunnel at about the exhaust pipe level. A pressure differential is maintained to eliminate piston effect and to counteract the effects of atmospheric winds. This system seeks to produce a uniform level of carbon monoxide throughout the length of the tunnel. In unidirectional tunnels the maximum contamination level occurs at the exit portal and in bidirectional tunnels mid-length where a zone of zero fresh air is created. It may be a supply only, or an exhaust only or a combined system (Figure 10.2). These systems are suitable for tunnel lengths up to 900 m. Beyond this length the air velocities at portals become excessive.

10.4.4 Full-Transverse Ventilation

For large tunnels, full-transverse ventilation is the most suitable. In this, a full exhaust duct and a full supply duct are provided. The system achieves a uniform distribution of supply air and uniform exhaust of contaminated air. A uniform pressure of air throughout the length of tunnel and more or less uniform level of contamination result. Nevertheless adequate pressure differential has to be maintained between air in the tunnel and air in the duct ways. Here the best level for admission of air is at vehicular exhaust emission level. Figure 10.3 shows a typical arrangement.

This system was first developed in 1924 for the Holland Tunnel. Figure 10.4 shows its longitudinal profile. The lengths of existing tunnels using this sys-tem vary from 0.36 to 14.4 km. It has been adopted mostly in subaqueous tunnels and subway metro tunnels. In main line railway tunnels artificial ventilation is used only in long tunnels since the vehicles provide necessary suction and pressure effect in smaller lengths. Urban tunnels call for artificial ventilation due to frequent services and long lengths.

A special artificial end-to-end ventilation arrangement was used in the New Cascade tunnel (USA) by providing a portal gate type door closing the higher (eastern) end and installing two number 850 HP electrical fans capable of delivering about 14,866 m^3 (525,000 cu ft) of air per minute. The door is kept closed with one fan blowing when the east-bound train climbs the 1.57% gradient. When the train is within 975 m of the eastern portal, the door opens automatically. After the train clears the tunnel, the door closes

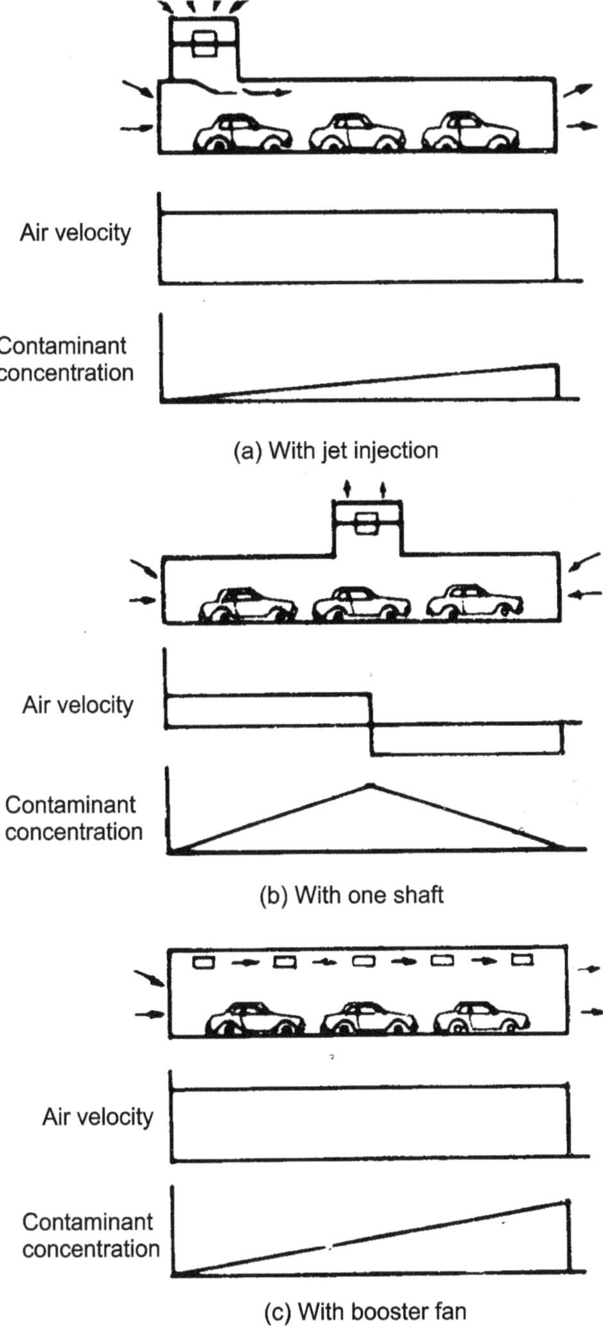

Figure 10.1 Longitudinal Ventilation System[1]

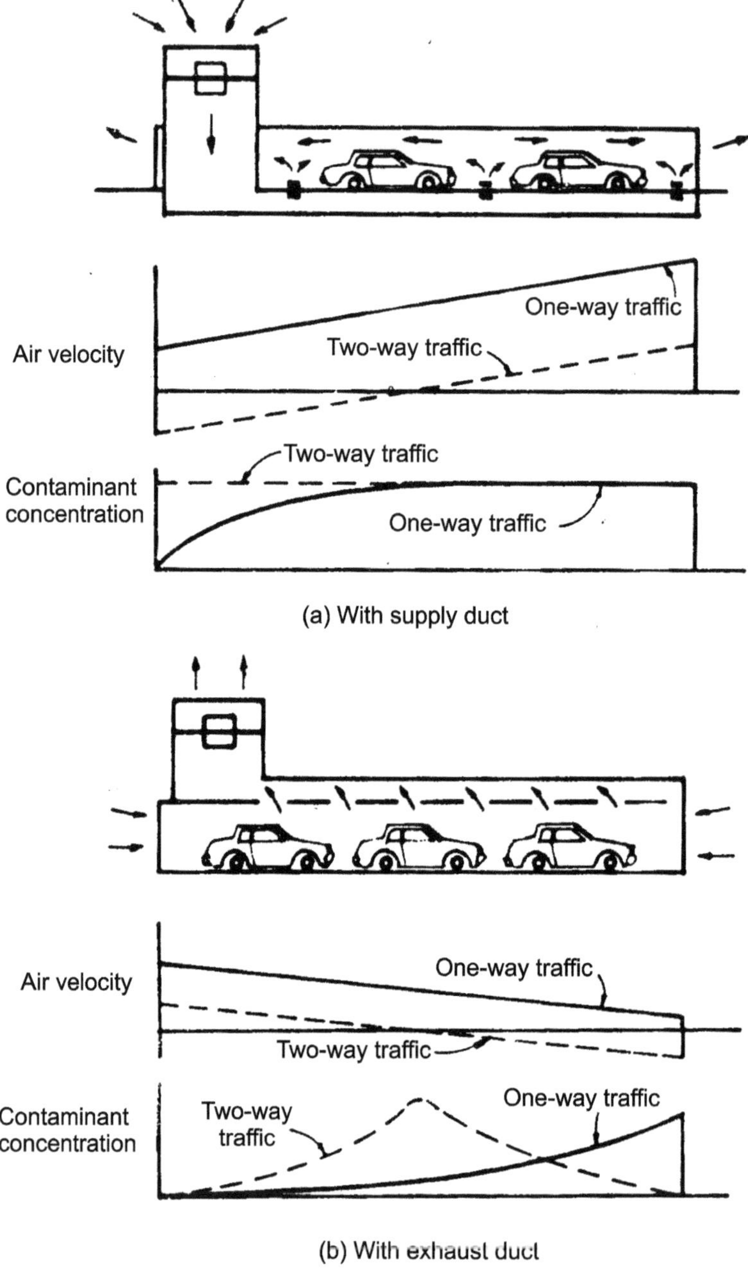

Figure 10.2 Semi-transverse Ventilation Systems[1].

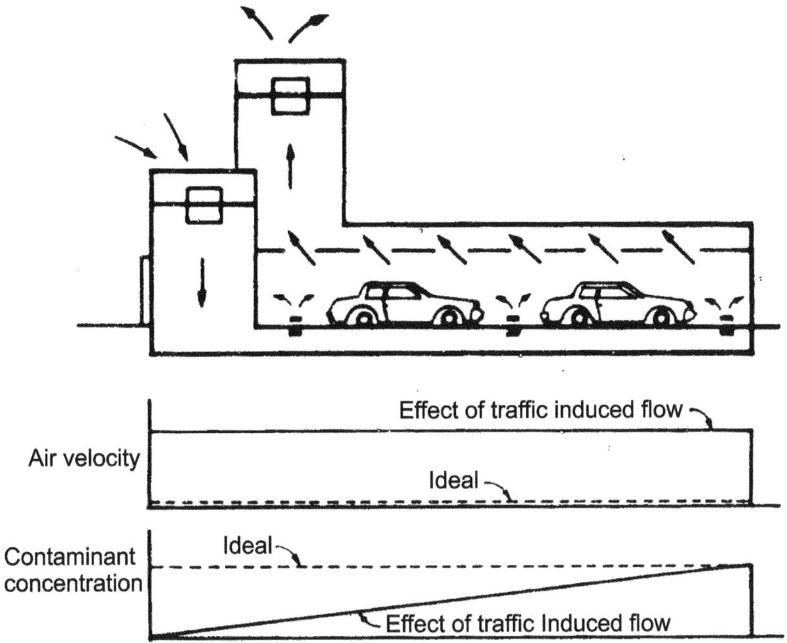

Figure 10.3 Transverse Ventilation Systems[1]

and both fans work to clear the tunnel of the fumes. West-bound trains being lighter and descending downgrade by coasting need no such operation.

10.4.5 Design of Ventilation System

Design of ventilation system for long tunnels by itself is a complicated one and a number of Tunnel codes or Guidelines have been issued by UIC and standards bodies. A few typical ones are listed below for information[4] :

- UIC 779-9 (2003) for Railways
- UIC 624-V (2001) - Exhaust emission tests for diesel traction engines
- US EPA document- Control of Emissions of Air Pollution from Locomotive Engines and Marine Compression -Ignition (2008)
- ‘‘Good engineering standards"- Technical Standards in tunnels (rail/ road) - a comparison to other country guidelines and actual experience like:
  - PIARC, 2003, C5 -Road Tunnels
  - ASTRA 13001, 2008, Luftugder Strassentunnel
  - RABT, 2006, *Richtlinean fur die Ausstatung und dien Betrieb von SraBentunneln*

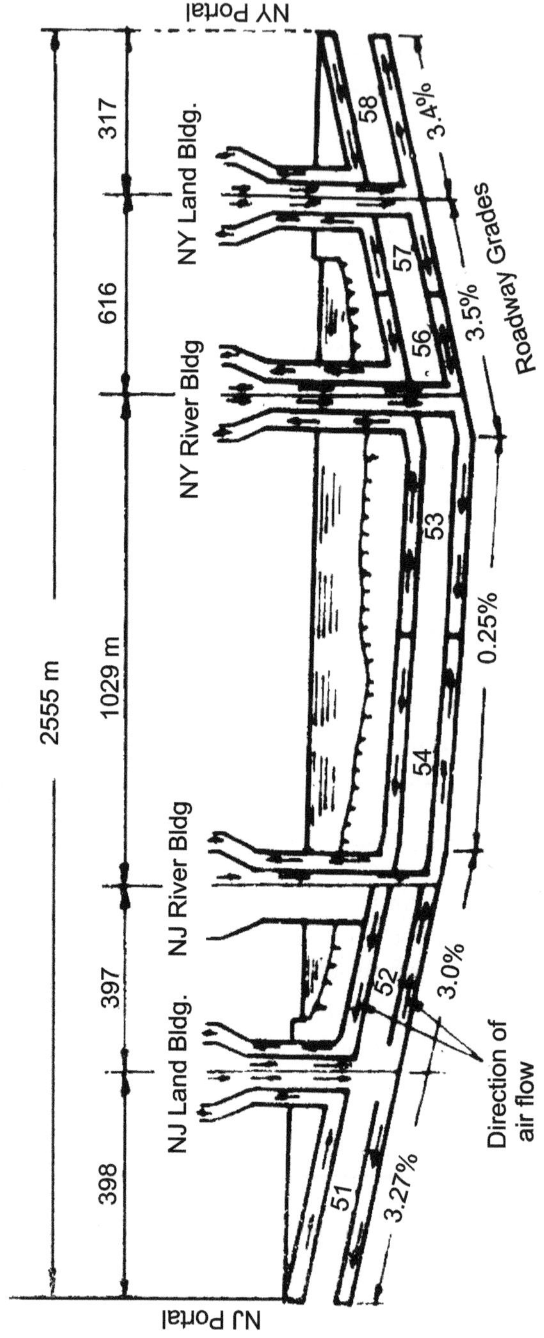

Figure 10.4 Longitudinal Section of Holland Sub- Aqueous Tunnel[3] (showing Ventilation details)

- RVS 09.02.31, 2008, Tunnelasrutung - Beluftung
- and some actually constructed tunnels.

Reference 4 gives a case study of design of ventilation system for Pir Panjal Tunnels, longest tunnel on the Jammu and Kashmir rail line.

10.5 CASE STUDIES

10.5.1 General

Three case studies, one on ventilation of a road tunnel; on the ventilation of a metro rail (subway) tunnel and the third on a main line rail tunnel are presented in this section. The Holland (road tunnel) is situated on the highway connecting New York with New Jersey (USA). It was opened to traffic in 1927 and was designated a National Historic Civil and Mechanical Engineering Landmark in the USA in 1984. The ventilation system has undergone renovation necessitated recently by need to replace the ceiling which has deteriorated over 60 years of use (Lesser et al., 1987)[3]. The subway rail tunnel is the metro tunnel in Kolkata. This metro system was in 1989 adjudged the best in the world in terms of maintenance. The main line railway tunnel Karbude tunnel 6.52 km long, considered longest on India, was built on Konkan Railway in early 1990s.

10.5.2 Holland Tunnel [Lesser et al., 1987][3]

Once described as the eighth wonder of the world, it is approximately 2550 m long. At the time of construction, it was the longest underwater tunnel, made possible due to the provision of an unprecedented ventilation system. The ventilation arrangement is a fully transverse system divided into a number of sections along the tunnel length. Each section is served by multiple fans, each of which can operate at three or four speeds. Outside air is supplied through a chamber below the roadway deck through openings on the kerb side spaced 4.5 m apart. Foul air is exhausted through ports provided in the ceiling at 4.5 m intervals. There are four intermediate ventilation shafts (Figure 10.4). The ports in the ceiling vary in size so as to balance air quantities, i.e., ports nearer the shafts are smaller. Figure 10.5 shows a cross-section of one of the two tunnels, one provided for each direction of traffic. Each tube accommodates a two-lane roadway and an inspection gangway. In emergencies and for maintenance operations two-way traffic is permitted in one tube for short periods.

For emergencies, say, a fire in a section of the tunnel, the ventilation system has been configured in such a way that smoke movement would be forced in a specific direction. Given one-way traffic movement, this poses no problem for car occupants since all vehicles downstream (traffic flow) would have exited. The exhaust would be downstream of the ventilation section over the affected length. The air supply upstream of that section would provide minimal air in the section containing fire and all other sections downstream. Fire-fighting vehicles would approach from downstream and personnel would be assured sufficient oxygen.

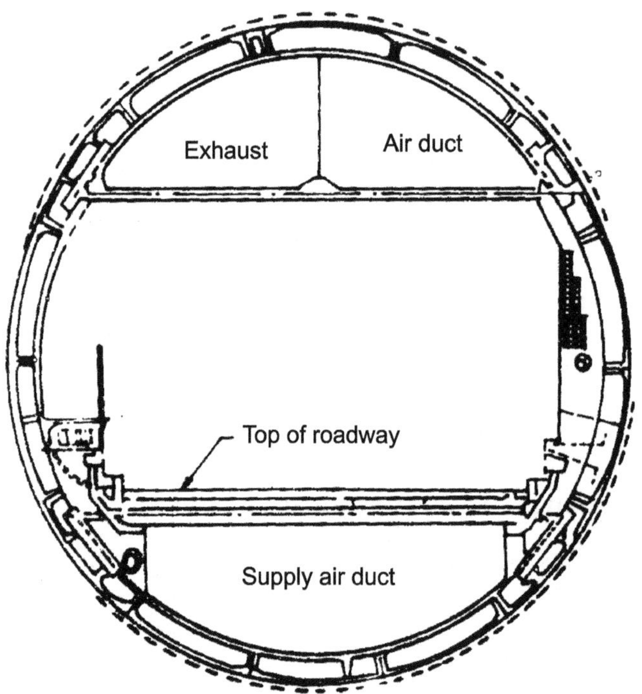

Figure 10.5 Cross-section of Holland Tunnel[3].

But for two-way traffic during maintenance and reconstruction of the ceiling, this type of system was not found satisfactory. For the major reconstruction activity it was decided to divert traffic through a neighbouring tunnel (Lincoln tunnel). Tests of four configurations were conducted to decide the best strategy. For this, an artificial fire was created in section S5 of the south tunnel (along 300 m from the nearest downstream ventilation building).

Test 1: Air supply maximum in sections S1 and S2; almost maximum in section S4 with no exhaust, minimum in S5, S6, S7 and S8; with maximum exhaust in S5, and near maximum exhaust in S6, S7 and S8. This resulted in a smoke-free tube upstream and downstream up to 120-150 m, after which smoke dropped downwards affecting visibility.

Test 2: Air supply minimum in S1 to S4 and S6 to S8 and at S5 kept near minimum. Exhaust also kept nil in S1 to S4 and minimum in S5 to S8. Smoke moved downstream at 75 m/min but also moved upstream.

Test 3: Maximum air given in all sections. No exhaust provided in SI to S4, with maximum exhaust in S6 to S8 and near maximum in S5. This resulted in insufficient air for diluting the smoke-laden air, which quickly dropped to the roadway level causing a visibility problem.

Test 4: A moderate quantity of air was supplied to S1 to S4 with no exhaust and in S5 a moderate supply with exhaust twice the amount of air provided. In other sections a minimum air supply and exhaust were maintained. This created a velocity of 150 m/sec in S5 and smoke began to drop down beyond 135 m. Still adequate visibility was maintained and fire-fighting vehicles could reach the site of the fire and quench it. This was the final strategy selected for adoption during emergencies and ceiling replacement.

It is a semi-transverse system to be used for the interim period only and after the ceiling has been replaced, the full transverse system will be restored.

10.5.3 Kolkata Metro Tunnel[5]

In metros in developed countries, a fully air-conditioned system is adopted for vehicles and stations. This requires a system with a large ventilation capacity for the tunnel and stations. Taking into account economy in construction and maintenance, manufacture cost of vehicles and need for keeping power requirement as low as possible, it was decided to provide only that uniform temperature (and humidity) level and fresh air supply within the vehicles and stations as would be obtained under shade outside on a normal day. Shade temperature in Kolkata on a normal day is as follows:

	Summer	Monsoon	Winter
Dry bulb (temperature), °C	36	33	17
Relative humidity, %	55	70	57

The system has to absorb heat of different kinds as detailed below:

(i) Traction heat load
(ii) Passenger heat load (sensible heat and latent heat)
(iii) Auxiliary heat load from fans, lights, working machines
(iv) Heat transfer from structure to soil and vice versa

(v) Solar gain in shallow metro.

Only the first three were computed with 10% to 20% more provided to take care of the others. The heat load on a time basis per passenger was taken at 130 kC/h/passenger. The total heat load for all worked out to 1.31×10^6 kC/h.

Air circulation inside the subway was designed to remove thermal load caused by the equipment and also to provide the biological needs of the commuter in terms of fresh air and maintenance of temperature within the subway at ± 6°C of that obtaining outside. The installed system supplies filtered air passed through cooling (chilled) coils and provides 16 to 20 air changes per hour. There are two types of tunnels in the Kolkata metro. In the rectangular twin-box sections a maximum air quantity of 80 m^3/sec and in the circular and independent boxes 50 m^3/sec is maintained. Station areas have 220 cum /sec. of intake and 210 cum/ sec. exhaust air.

Longitudinal ventilation system with air supply at stations with partial exhaust and full exhaust through ventilation shafts at midpoints has been provided. The average station-to-station distance is one km. A schematic diagram of the ventilation system is shown in Figure 10.6.

The equipment used for each section comprises:

Intake			
	Four 30 m^3/s station fans	–	100 mm SP
	Two 50 m^3/s tunnel fans	–	50 mm SP
Exhaust			
	Two 80^3/s midpoint exhaust fans	–	50 mm SP
	Two 25 m^3/s under platform exhaust fans	–	100 mm SP
	(all are centrifugal pumps)		
	(SP - Static Pressure)		

A ventilation mezzanine is provided at each station. Just below this adequate area was reserved between tracks for locating the refrigeration plant and its control panels. At each end of the station platform an exhaust fan room is provided under the platform.

Fresh (ambient) air is taken and cleared at the air washery located at the entry of the ventilation mezzanine before being admitted into the moisture trap.

The clean, drier air is then passed through cooling coils to lower its temperature to design level before being admitted into the ducting system. The ducting system is an extensive network over the platform and mezzanine area using 56 grills fixed in the false ceiling for letting air into the subway.

The working hours of the system are distributed as follows:

(i) Chiller package during summer and monsoon only

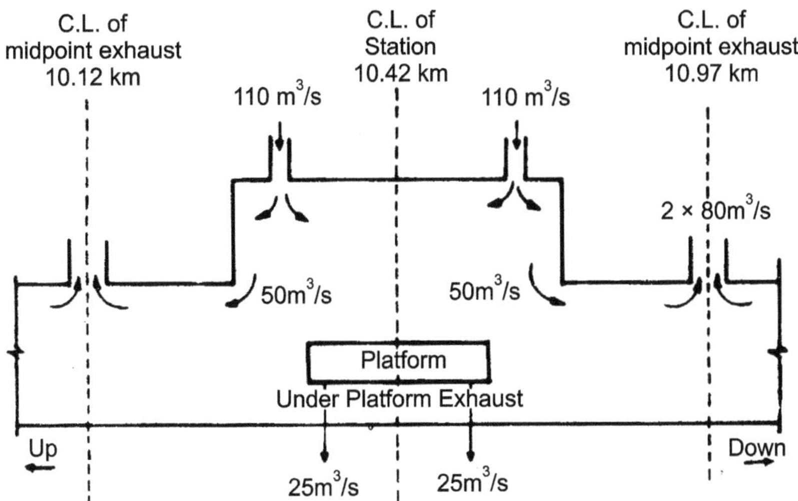

Figure 10.6 Air Schematic Diagram for a Typical Metro Station.

	(151h March to 15th Nov.)	–0700 to 1000 hrs
(ii)	Full ventilation	–0050 to 2300 hrs
(iii)	Half ventilation (except under platform)	–2300 to 0500 hrs

Each station is provided with 2 air intake shafts 7m x 4 m raised 3 m above ground level to take in ambient air. Two under-platform exhaust shafts 2.5m x 2 m rising to a height of 6 m are provided adjacent to the intake shafts. Midpoint exhaust arrangements comprise underground fan rooms located adjacent to each track and above track with necessary approach to surface. Two fans 80 m^3/s (50 mm SP) are provided for exhausting air on either side of the tunnel direct to the atmosphere. Each fan has an independent shaft 4.5 m × 4.5 m rising to 6 m above ground level. No hot air is exhausted near any residential building.

10.5.4 Konkan Railway Tunnels[6]

Based on an Expert Group study done for these tunnels, KRCL adopted the following policy for ventilation of tunnels:

(i) Unlined Tunnels upto about 2 km length and lined and shotcreted tunnels upto 3 km can be allowed without any ventilation arrangements like shafts for forced ventilation.

(ii) In all other cases, either shafts to reduce lengths of tunnel segments or forced ventilation arrangement is to be provided.

In accordance with this guideline, they provided shafts/ adits in 5 out of 9 long tunnels. They provided 3 shafts in their longest Karbude Tunnel. Segment lengths were less than 2 km but this was adopted to provide better passenger comfort. Forced ventilation facilities system in form of three centrifugal fans of 150 kW capacity each was provided near the middle shaft, along with mechanical dampers, and air curtains were provided to control direction of flow inside the tunnel. Figure 10.7(a) shows the functioning of the system conceptually at different conditions of traffic.

For three other long tunnels (4.425 km, 3.39 km and 2.96 km respectively) the proposals provided for forced ventilation with jet fans, as shown in Figure 10.7 (b). These jet fans are suspended from introdas of roof. Though jet fans deliver relatively smaller quantity of air, it is provided at high velocity, the momentum created by which, is transmitted to the larger air present in the tunnel and necessary flow velocity can be maintained.

Similar arrangements of fixing powerful jet fans at a number of locations have been made in the 11 km long Pir Panjal Railway tunnel, which has Adit and a Shaft added in between the end portals for air flow. Brief details are given in Annexure 6.2, a case study on Pir Panjal Tunnel.

10.6 LIGHTING[7]

Lighting requirements may be considered temporary during construction of the tunnel and permanent for the tunnel in service.

For efficient working (during construction) inside the tunnels, good lighting is needed and hence artificial lighting must be resorted to. Provision of lighting which can be allowed is a light intensity 258 lumens/m^2 in the working area. For example, for a tunnel size of single lane width or single track, it is required to provide light intensity of the order of 1000 cp.

Tunnel lighting during construction is a complicated process as there is need to introduce lighting arrangements at the time of drilling and charging of the holes, and then withdraw it prior to blasting; and repeat the process.

In India, electric lighting with incandescent bulbs is generally arranged with a portable diesel generator kept outside and cables running through the length of the tunnel.

Most short-length railway tunnels, especially those in remote areas where power supply is difficult to obtain, are not lighted permanently. However, railway tunnels should be provided with adequate lighting for workmen to do patrolling and to carry out maintenance work and also to serve during emergencies. Incandescent lamps at suitable intervals can be provided on side walls. Since traffic is guided and controlled, the engine headlights serve the drivers' purpose.

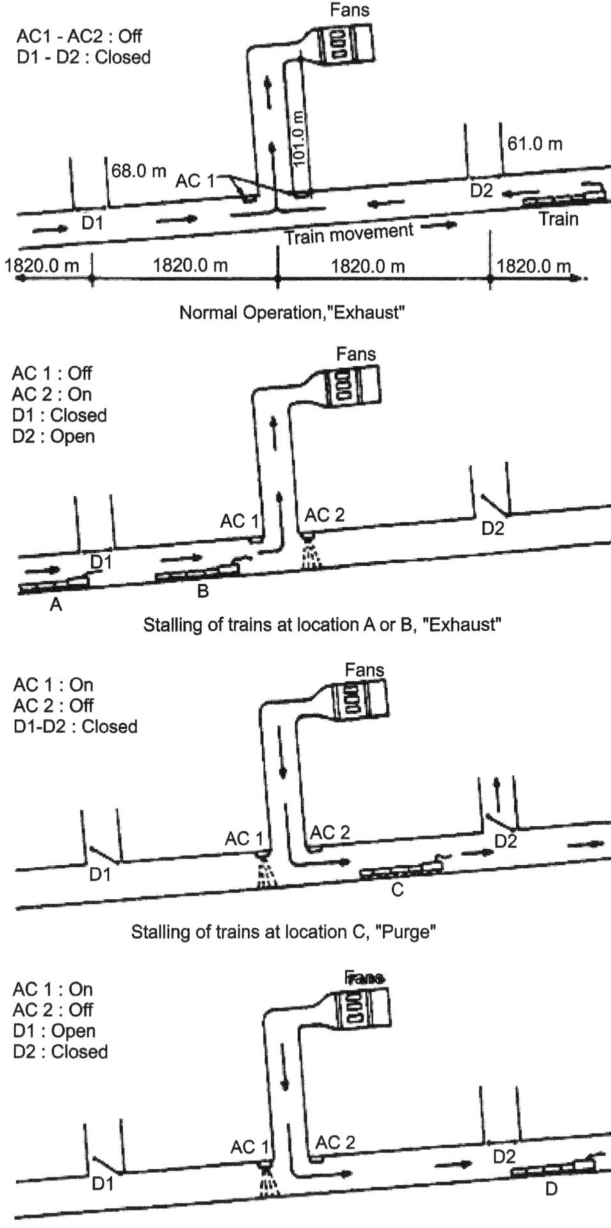

Source : Indian Concrete Journal, February 1994.

Figure 10.7(a) Working of Ventilation Arrangement with Train at Different Positions

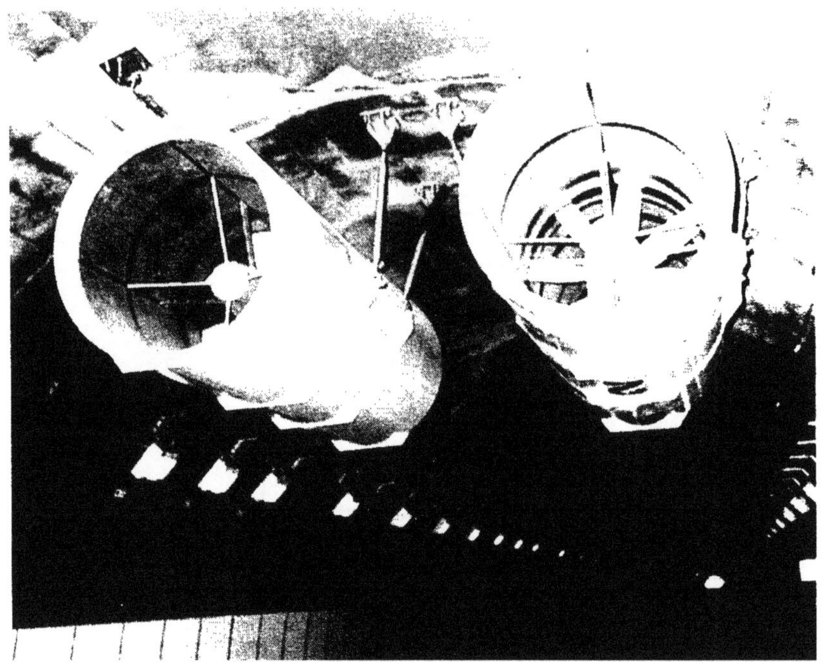

Source: Indian Concrete Journal, February 1914

Figure 10.7(b) Jet Fans fixed at Tunnel ceiling

For highway tunnels, adequate lighting has to be provided as a permanent arrangement also. This lighting should be commensurate with the importance, type and level of traffic. Vehicular traffic should be able to pass through the tunnel without having to use the high beam of headlights. Further, tunnel lighting should not throw glare on the drivers. In a curved alignment, lighting should be sufficiently diffused so as not to suddenly impinge on the viewer. It is usual to adopt sodium vapour or mercury vapour lamps for this purpose.

The design of lighting for a long tunnel is quite complex. The lighting should be economical and suited to the degree of adaptability of the human eye under the varying conditions of light and speed of vehicle as one travels from the open highway through the tunnel to the other end. The problem is most severe during the day when the driver has to adapt from a bright illuminated open road to the much lower level of illumination of the tunnel. For the purpose of design of lighting, a long tunnel is divided into five zones: (a) approach zone, (b) threshold zone, (c) transition zone, (d) interior zone and (e) exit zone (Gallagher, 1985)[1]. The length of zones (a) to (c) and (e) and the roadway luminance required for each zone depends on the design speed and carriageway width.

The approach zone is the area outside the portal for a length corresponding to the safe stopping distance for the design speed. The lighting level may be about 2000 cd/m^2 in daylight. The threshold zone is the area of most serious design consideration and needs most elaborate lighting and control systems. The lighting level in the threshold zone may be 1/15 of the approach zone. The length of the threshold zone is usually the safe stopping distance at the design speed. The lighting level should be reduced gradually from the threshold to the transition zone. The transition zone has a lighting of about 20% that of the threshold zone and should have a length approximately equal to the safe stopping distance for the design speed. The interior zone represents the length of the tunnel from the transition zone to the exit zone. The recommended level of luminance for the interior zone is 5 cd/m^2 for Portland cement concrete surfaces. The exit zone is the final zone before transition to daylight. The length of the exit zone is nearly equal to the safe stopping distance.

Design for night lighting is relatively simple. The lighting at night should be uniform throughout the tunnel. The lighting intensity inside the tunnel should be about half to one-third the lighting on the approach road out-side the tunnel. Relative intensities of lighting required in different zones as adopted in St. Gotthard tunnel are indicated in Figure 10.8.

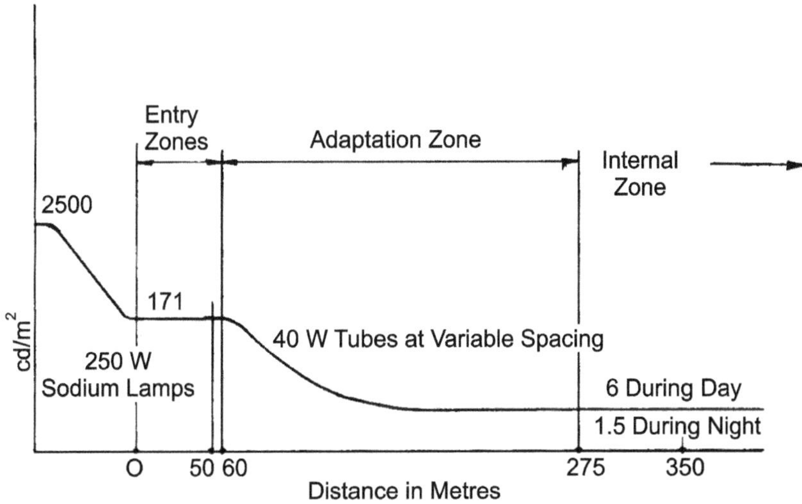

Figure 10.8 Typical Lighting Requirements in Different Zones of Tunnel.

10.7 DRAINAGE

It is very rare that a tunnel is bored through for its entire length without coming across seepage. Should the gradient fall towards the tunnel, wherever water coming from the approaches is encountered, suitable side drains and intercepting drains are to be provided to lead it away before it can enter the tunnels,. Wherever necessary, suitable pumping arrangements have to be provided.

Inside the tunnel a suitable drainage arrangement should be provided so that the water collected can be drained without flowing over the surface of the road or track. Normally, in single-line railway tunnels with ballasted tracks, properly lined side drains are provided for water collection and drainage. The lining should be sufficiently wear resistant to withstand the high velocities likely to be attained in steeply falling gradients. Modern practice is to go in for ballastless (paved) tracks, in which case the floor section is shaped to include such drains running in the middle of track or on either side, to suit the design of pavement. On highway tunnels, with provision of camber on carriageway, such drains will be paved and run along the sides.

In short tunnels the water will drain out to the open exit and naturally can be led away easily. In long tunnels collection may be considerable and suitable intercepting sumps and pumping or cross-drainage works by way of drainage tunnels or bores may have to be provided.

In the case of double-line tracks drainage is generally provided in the middle between the two tracks, which may be covered, and suitable manholes provided at 30 to 50 m intervals. Alternatively, open-jointed covered channel drains (with perforated slabs on side for water) may be provided between the tracks. In addition, side drains are provided and connected at intervals to the central drain. It should be remembered that good maintenance of the track or road surface depends on the drainage system provided. Improper drainage can also cause seepage of water below and cause settlement of track or pavement due to disturbed moisture balance in the soil below (particularly cohesive loose soils, shale etc.). In subaqueous tunnels across rivers, creeks etc., since the gradient will fall towards the middle, water tends to collect in the central length of the tunnel. In such cases it is necessary to provide separate drainage tunnels leading away from the central portion to the shafts at the end, through which the collected water is pumped out. Similarly, in long tunnels in hills also the shafts provided for the purpose of ventilation can be used for collection and disposal of drainage water.

The general shape of the side drains provided in railway tunnels in India can be seen from the tunnel section shown in Figure 3.4(a) to (c) and A.4.2.1.

10.8 NICHES AND REFUGES

In long tunnels, some provision has to be made at intervals for the maintenance staff to take shelter when trains pass over the tracks. These are provided in form of recesses cut into sides of the tunnels for maintenance staff to take rest and also store their tools and equipment. On Indian Railways, provision of such shelters known as 'trolley shelters'/ 'safety refuges' at not more than 100m intervals is mandatory. In curved lengths of tunnels it is desirable to provide them at about 50 m intervals. Even in highway tunnels, such recesses will be necessary and some planners advise such provision at not more than 50 m intervals. Simple man refuges may be 2 to 2.25m high, 2 to 2.5 m wide and 1m deep. Refuges where some equipment or maintenance trolleys will have to be sheltered, should be at least 3m wide, 4 m deep and 2.5 m high.

10.9 REFERENCES

1. Lesser,N (1987)-' Tunnel Ventilation System', in : 'Tunnel Engineering Handbook', Ed. Bickel, John O. and T.R. Kuesel, (Eds.), Van Nostrand Reinhold Company" New York,
2. Pawan Kumar Singh, 'Experience on Tunneling- Problems and Solutions' - presented in Course No 13026, IRICEN Journal of Civil Engineering, Indian Railways Institute of Civil Engineering, Pune pp 36-44
3. Lesser, N., Horowitz, F., and King, K., 1987, 'Transverse Ventilation System of the Holland Tunnel Evaluated and Operated in Semi-Transverse Mode', Transportation Research Record 1150, Transportation Research Board, Washington, D.C.
4. Hitesh Khanna, (2014), 'Tunnel Ventilation and Fire Safety in Tunnels', Proceedings of National Conference on National Technical Seminar on 'Management of P.Way Works through Need Based Outsourcing & Design, Construction and Maintenance of Railway tunnels' Jaipur 2014. (Volume II)
5. Narayanan, G.A., 1988, 'Air-Conditioning and Ventilation of Calcutta Metro', International Seminar on Transit Technology of Tomorrow, Bombay
6. Chopra, I.D., (1994), 'Ventilation of long tunnels on Konkan Railway', Indian Concrete Journal, February, 1994, pp113-116.
7. Gallagher, V.P., 1985, 'Tunnel Lighting Design Procedure, Report No. FHWA-IP-85-9', US Department of Transportation, Washington,

CHAPTER

11

Instrumentation in Tunnelling

11.1 NECESSITY FOR INSTRUMENTATION

11.1.1 The observation of physical and structural properties and behaviour of the soil and the tunnel itself are integral parts of design and construction of underground structures. This becomes particularly important in case of tunnels which, necessarily have to pass through varying type of soils and rocks and in many cases close to or below existing structures and utilities or other underground structures. Such observations serve a number of purposes and are carried out at different stages of the work viz., (i) site investigation for help in surveying and initial planning; (ii) during construction, for collection of data for design verification and also for construction control by modifying method and rate of progress as required and (iii) for monitoring the performance of the structure in service and ensuring its stability.

11.1.2 Stages of Instrumentation

During the investigation stage, investigations would cover determination of the characteristics of the soil along the alignment pore pressure, and permeability, especially in case of subway tunnels and fault zones. Para 2.5 gives details of type of geotechnical investigations.

In some cases, deep benchmarks are established as reference points by fixing the same at a depth in rock not likely to be disturbed by the excavation, since BMs existing on surface nearby are likely to be disturbed due to settlement of surrounding ground during tunnelling. Levels at these points are taken periodically with Precise Levelling instruments.

During construction stage the type of instrumentation cover for determining:

- Ground movement on alignment, adjacent to and away from tunnel,
- Building movement, stability and structural changes on structures within the zone of influence
- Movement within the tunnel under construction and adjacent tunnels/ utilities
- Stresses in rock and the supports/ lining provided in tunnel and dynamic effects on ground during drill and blast
- Ground water movement, pore pressure and percolation pattern

11.2 GROUND MOVEMENT STUDIES

The types of instrumentation used for study of ground movement are (FHWA Technical Manual)[1]:

- Extensometers which can be Probe type or Fixed bore hole type.
- Inclinometers- Conventional and in-place ones
- Heave gauges
- Convergence gauges.

11.2.1 Extensometers

Probe Extensometers

Extensometer is used for measurement of 'deformation in rock mass and surrounding soil with respect to a deep anchor'. It can be used for assessing settlement of ground at different levels with respect to the deep anchor. These can be in form of a portable Probe such as a Reed Switch Transducer. These measure distances between two or more points within a bore hole drilled in soft ground. A bore hole is drilled into the ground upto well below the potential influence zone of the tunnel and a Flush coupled rigid PVC Access pipe of smaller diameter is inserted into the same upto base. A spider magnet each is positioned at desired depths and held tight against the pipe by spring loaded anchors from the borehole wall, as shown in Figure 11.1. The annular space in the hole is filled with grout. The bottom reference point at base of this pipe would become the reference point from which the movement of the other points above can be judged. A Probe containing Reed Switch and connected to the reading unit by a survey tape is lowered into the access pipe. Probe is lowered into the bore to the bottom and as it is slowly raised, observations are made. This transducer can detect the position of the magnets at different levels by emitting a beep and operator notes down the position by listening to the beep. Any difference in position of any of the points with respect to the initial set would indicate the amount of settlement. He may repeat the observations to recheck and confirm. Thus results noted are operator sensitive. The observations at each location would take about 45 minutes.

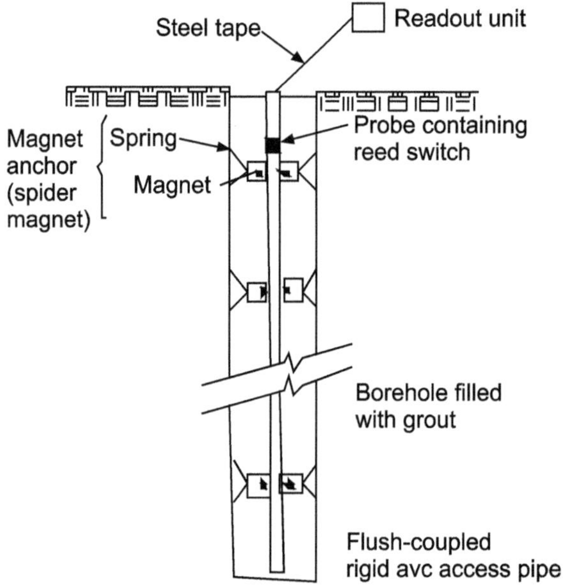

Figure 11.1 Probe Extensometer[1]

Fixed Borehole Extensometers:

These can be used in soft ground as well as in rock. They are installed at different depths from ground level. They can be Single point ones (SPBX) or MPBX, Multi Point Borehole Extensometers. Latter is generally preferred, in which upto six anchors can be used, placed at different depths in one bore. The anchors are grouted into the ground. This device is generally used by installing them at different distances above the crown of tunnel being bored. They are connected to the ground level by steel or fibre glass rods (of small diameter). The movement of the tops of rods at ground level would indicate how much each anchor has moved thus indicating the amount of movement of the soil or rock at that elevation. Readings can be taken by manual observations or with use of electrical data logger or transducer. These data loggers / transducers can be removed and reused at new locations when the need for observation at one location is over as the tunnel advances. These observations will help the engineer to take any precautionary or corrective measures in case of unanticipated settlements.

In case there is any problem in locating such bore holes on the alignment, falling on a carriageway, such probes can be fixed from sides/ pavements kept at an angle towards the crown of the advancing tunnel and results suitably interpreted. A schematic of the arrangement may be seen of the borehole in vertical position, in Figure 11.2.

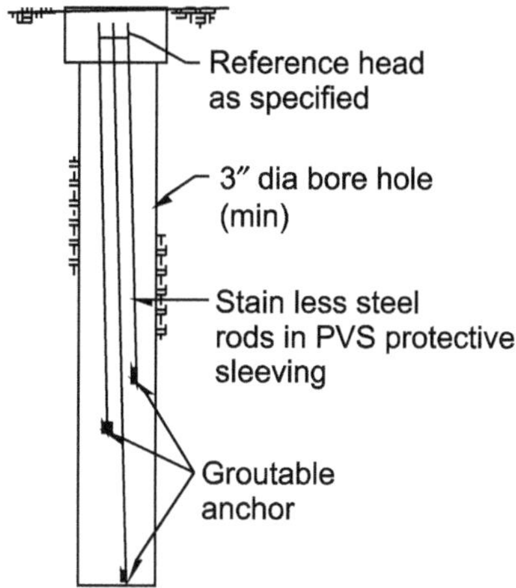

Figure 11.2 Multiple Position Borehole Extensometer1

Horizontal Borehole Extensometers

These are similar to the borehole extensometers mentioned above but installed in horizontal bore extending from the cut tunnel wall for measuring the movements in the transverse direction after the tunnel is excavated. The arrangement is shown in Figure 11.3. Installation of such instrument, protection of same and doing periodical readings is difficult as it will interfere with the ongoing work, as the stretch will be busy with follow up activities of tunnel work after excavations and is subject to frequent movement of miners , who may not understand their existence or importance.

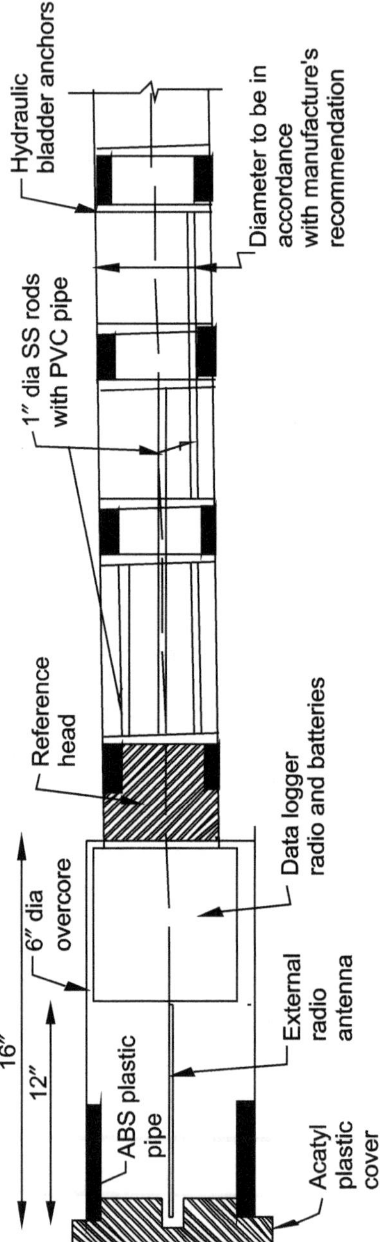

Figure 11.3 Horizontal Borehole Extensometer1

But they are helpful in design modifications and protection measures for structures on surface above such zone of influence.

11.2.2 Inclinometers

Conventional Inclinometers

Inclinometers are used for observing the movement of ground around an underground excavation. They are installed in a plastic casing drilled vertically on the side of the proposed excavation till it reaches a level well below the bottom of proposed excavation so that the bottom is at a stable ground, with reference to which movements of rock or ground at different levels above in transverse or longitudinal direction can be observed. The guide casing has tracking grooves for guiding the probe parallel to and perpendicular to the direction of excavation. It has universal joints at intervals for letting it tilt with the movement of the surrounding rock/ ground, (as shown in Figure 11.4). A probe containing tilt sensors and with guide wheels is lowered into the guide casing on a graduated cable to the bottom of the hole. While winching it up, collection of inclination data is done on a read out unit at surface, at fixed interval (about 600mm). By an iterative process the tilt of the pipe in that direction is determined with respect the position of the stable point at bottom and plotted. These would indicate the profile of the pipe in that direction, as shown in the inset in the figure. The probe is turned by 90 degrees and lowered again and winched up and tilts at different levels determined in a similar manner. Its path now will indicate the profile in the (longitudinal) direction. Such readings are taken once a day during construction, which will help in taking any preventive or remedial measure. This is found helpful specially for monitoring ground movement on the sides of diaphragm walls in cut and cover construction works.

Presently in-place inclinometers are also available. In this, 'gravity driven transducer sensors' are installed equidistantly at different levels in the casing and wired to a computer on top. Sensors are computer driven and readings can be taken automatically by a data logger on surface and tilts observed. More the number of sensors proposed to be installed in a bore, larger will be the diameter of such bore. The equipment is comparatively expensive and more complicated.

11.2.3 Heave Gauge

Heave gauges are required to determine the heave of the base of an open cut in cut and cover construction due to removal of overburden. No reliable instrument has been found to work for direct measurement of same. One of

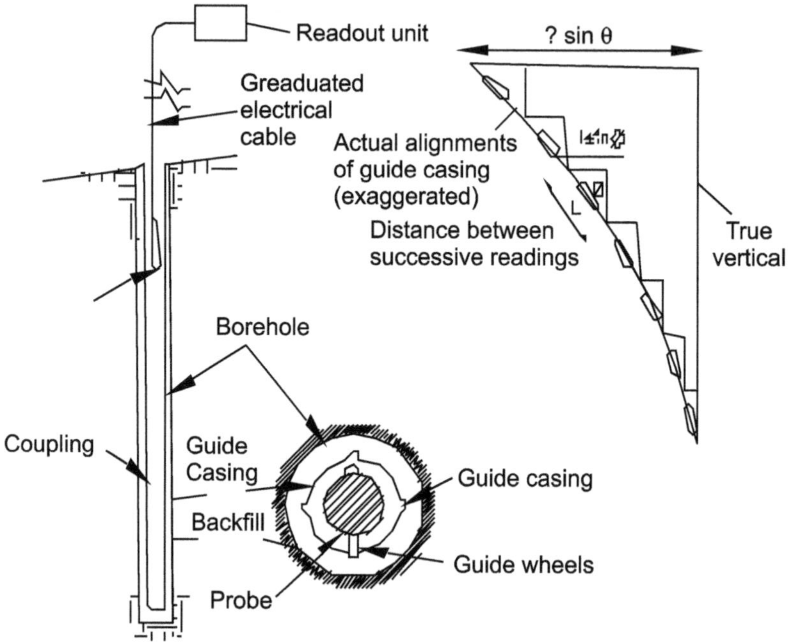

Source: FHWA Manual

Figure 11.4 Conventional Inclinometer[1]

the best alternatives found good is the use of a 'the magnet-reed switch gauge' as a probe extensometer for measuring the increasing distance instead of measuring decreasing distance of spider magnet **from a fixed target anchor** placed at the bottom of a pipe drilled into the ground, the pipe being located within the side walls of the proposed enclosure/ trench. Before start of the work, a pipe is bored in and an anchor /target placed at the bottom well below the influence zone. After initial readings, the pipe is sealed with an expanding plug about 1.5 to 3.0 metres below surface placed with an insertion tool, and the portion above plug, cut using an internal cutting tool. The position of the pipe is noted. When the excavation reaches near that, its level is noted by a reading. Care is taken not to disturb it and once excavation reaches that level the process is repeated till excavation reaches the final level. The changes in its level will give an idea of the heave of the excavation at that stage.

11.2.4 Convergence Gauges and Roof Anchors

It is necessary to note the quantum of movement of exposed tunnel walls and roof periodically, for which target anchors have to be fixed as quickly as possible before the tunnel starts to 'work'. Anchors are fixed on side walls and roof of arch and distance between them is measured periodically using a tape extensometer, as shown in Figure 11.5.

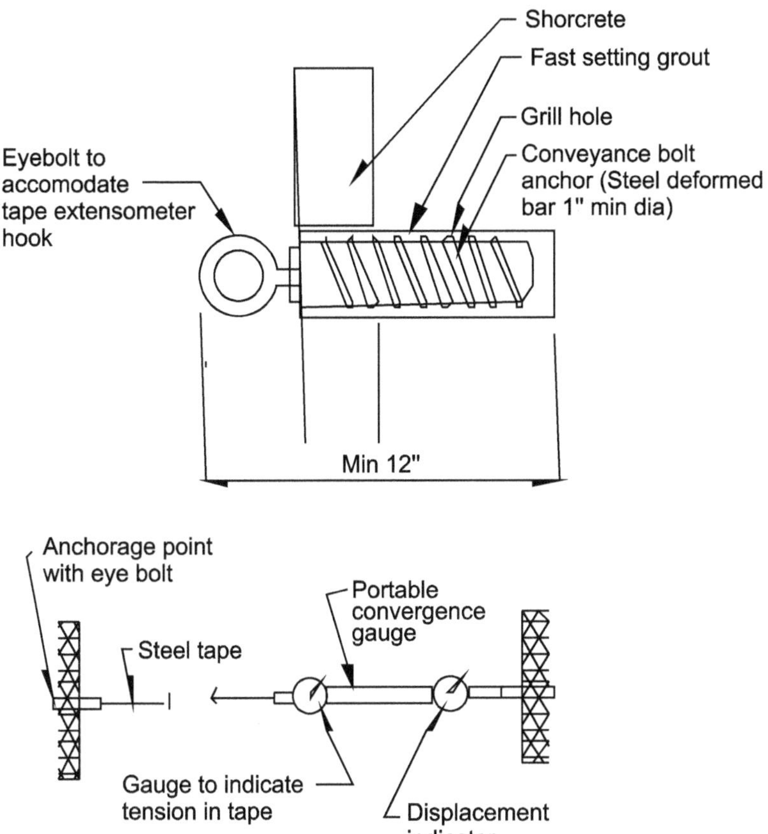

Figure 11.5 Tape Extensometer and Eye bolt Anchor[1].

But such measurements are not possible immediately behind the advancing face where TBM is used due to the presence of the trailing equipment and movement of people. In such cases, alternative equipment in form a distomat (or Total station) can be used by sighting from a distance in the rear. In such cases, the instrument determines distance by emitting a laser or infra red beam to a target fixed at the anchor location, instead of a bolt with an eyelet used for extensometer measurement. Even in this case, sight line may be obstructed by presence of some trailing equipment of TBM till it clears. Hence, the locations of targets will have to be judiciously chosen. Distomat is capable of measuring distances in x, y and z directions and give the three co-ordinates with respect to position of the instrument. Knowing the co-ordinates, the distance between pairs of them can be computed and changes can be determined.

In fact, the settlement of the ground around the tunnel have to be determined using such measurements only, as drilling deep holes and use of extensometers mentioned earlier will not be possible over depths normally involved in such construction.

11.3 MEASUREMENT AND MONITORING IMPACT ON BUILDINGS/ STRUCTURES

11.3.1 Settlement of ground above the tunnel and on sides may have little impact on structure on the surface in case of railway and highway tunnels through hilly terrain due to nature of soil and depth of excavation helping in distribution of load, as discussed in Chapter 4. Even in such cases, in case the ground is interspersed with soft soil some localized settlement at surface is possible. In case of shallow tunnels, done by boring through soil such settlement is countered by stabiiising the soil just ahead of the tunnelling face and on sides by freezing or grouting, as mentioned in Para 6.7. In case of cut and cover construction, as excavation proceeds in between the cofferdams, depending on properties of soil there will be settlement of soil for some distance on either side of excavation. In all these cases, if there are any structures on the surface they may also be subjected to some readjustment of load with likely consequence of being subjected to some deformation or damage in form of cracks. Suitable remedial measures will have to be taken in time, to ensure safety and also prevent claims for such damages. Instrumentation is done on vulnerable buildings, pavements and structures and for watching such developments on a continuous basis till the excavation is completed and the tunnel/ cut are provided with permanent supports and any pit/ cavity is filled up.

11.3.2 Measurement of Deformation

In case of roads and pavements, at intervals monitoring points are selected and a bolt like device embedded with the top exposed and position noted. It is desirable to fix such points away from area subjected to vehicle traffic. Their levels and relative positions are measured to an accuracy of 5 to 6 mm at periodic intervals. On buildings/ structures, such targets can be fixed on the vertical face of a wall forming integral part of the building. They can be provided with hexagonal nut plugged in a hole, the top of which can be taken as the reference point for measurement of level. Figure 11.6.shows an arrangement as given in FHWA Manual.

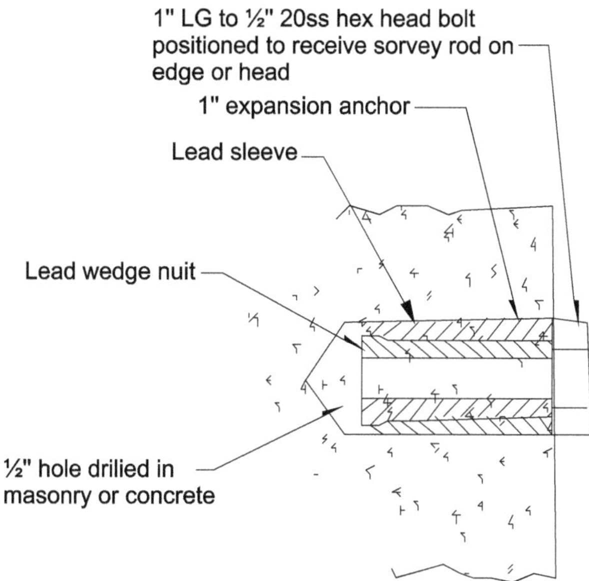

Figure 11.6 Deformation Measuring Point on Structures[1].

Similar reference points can be fixed for measuring lateral movements and settlement on pavements and hard surfaces by providing the bolts, with hemispherical heads, vertically.

Robotic Total Station

If real time data of movement of structures in an area is required, a Robotic Total station can be used. They can record the position of the targets in three dimensions. One station can be fixed to sight and take measurements of a number of targets fixed in its range. The instrument is installed 'atop small electric motors' in such a way that they rotate about their axes. They can be timed to automatically get actuated at pre-determined times and make the measurements. Such sophisticated instrumentation is quite expensive and may be called for only in case of very sensitive structures coming in the area of influence of the tunnel.

Tilt meters

It is important to monitor tilt in the floors, walls, columns etc. of structures also. Conventionally tilt is measured manually using hand held instruments on reference points fixed on plates fixed to the structures. More sophisticated electronic powered tilt meters with electrolytic level transducers are available. The transducers are housed in casings, which can be attached to the element of structure to be monitored. They are available for measurement in uni-axial or bi-axial mode. One such with bi-axial movement is shown in Figure 11.7.

11.4 STRUCTURAL MONITORING

11.4.1 Due to base settlement, deformation or shocks and vibration caused during excavation underground, the structure may develop some structural damage. This will be mostly in forms of cracks. If any such crack develops, likely cause has to be studied and if necessary method or rate of excavation modified to prevent further damage. The structure may have already had some structural or other superficial crack. Hence a detailed inspection of structures before start of work is done; and tell tales will have to be fixed for monitoring. Also, as soon as any such crack is noticed during construction, tell tales have to be fixed and their further behaviour monitored by periodical inspection. The tell tales can be in the form of glass strips, plastic strips etc.. There are crack gauges available which span across the crack, which should be fixed as soon as the crack is noticed. It consists of two overlapping plastic flat pieces. One is fixed on one either side of the crack on the wall with epoxy (or any other fixing mechanism provided). There will be crossed lines on upper plate and a graduated grid scale on lower one. The position of the cross over grid line is noted at installation and amount of movement periodically measured over the grid line, the difference giving amount of movement. Simultaneously, the ends of crack line can be marked on the wall and distance between noted initially. During monitoring, any extension of the ends of the crack is measured. The two sets of readings will give an idea of the damage being caused for taking remedial action.

11.4.2 Measurement of Loads and Stresses[2]

When a cavity is made inside the mountain or soil, there will be some release in load and readjustment of forces around the cavity. They will cause some displacement and movement around the surface of opening. Such displacement will be time dependant. The forces of gravity in ground around

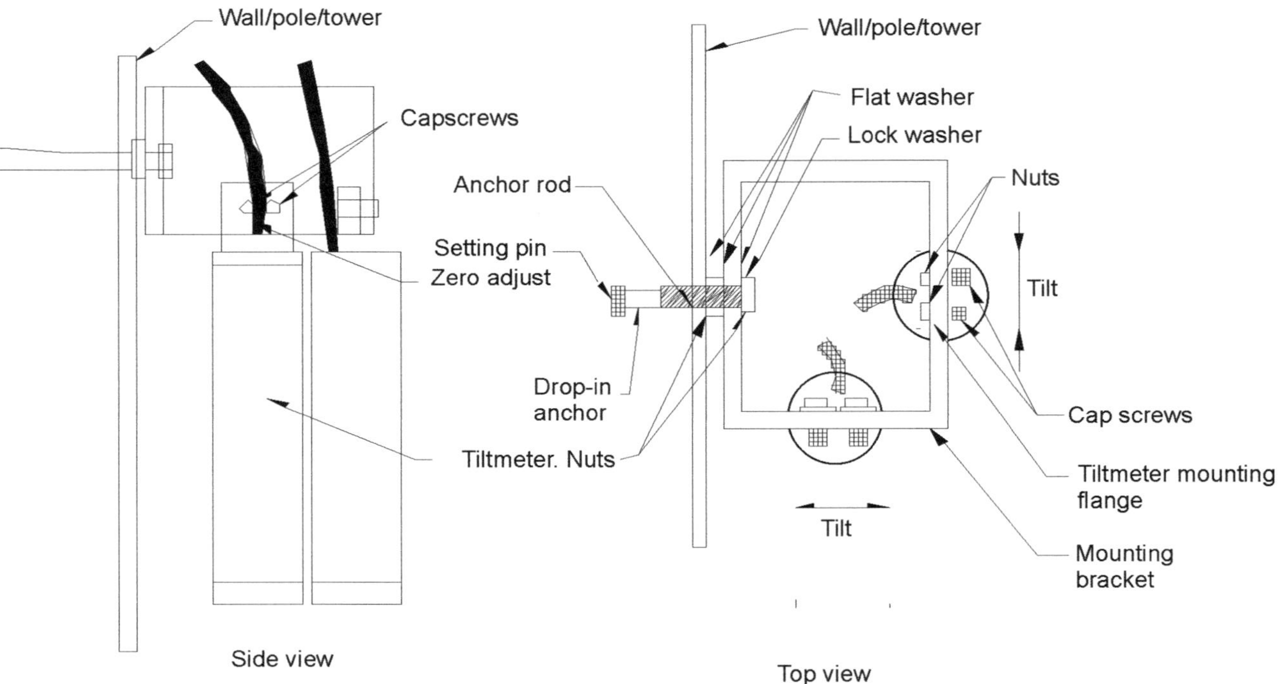

Figure 11.7 Bi-axial Electronic Tilt Meter[1]

on the excavated surface can result in instability, calling for provision of supports. In order to properly design such supports it is necessary to know the pressures/ forces, soon after excavation, and instrumentation is used for obtaining rock pressures including pore water pressure. They will help in verifying assumptions made in design of supports and lining and make necessary changes. Such instrumentation is done on the temporary steel support ribs and struts provided; rock bolts fixed deep into surrounding rock; and shotcrete and in case of NATM it is done to find loads /stresses in them so as to check their adequacy and also for design of final support/lining.

In olden days, they used to embed some form of load cells in the rock at some depth from excavated surface and measure loads. In rock tunnels, such measurements used to be done in the drifts to help in preliminary design and around. Presently many forms of electronic and hydraulic measuring devices are available for such measurements on a continuous basis. The force/ load on rock bolts and tie back anchors provided from supports are measured on a continuous basis using special load cells. The cylindrical load cell comes with a central hole so that it can be fixed at the surface end of the bolts wrapping round the bolt. They are fixed to the bolt on the exposed surface with a pair of packing plates on either end of the load cell. As the rock moves and the bolt or anchor is stressed, the pressure will come on the load cell and the same can be read on the load cell. Using electronic type of cells, the reading can be recorded at predefined intervals on a continuous basis.

In case of steel supports (arch ribs) provided to support the roof load and on sides, the effect of load coming on them can be measured periodically by fixing load cells or strain gauges welding small flanges between which the cell can be mounted on the members. Generally this is done at the springing of arch of the rib and at the crown. In some cases, additional load cells can be fixed at 45° position on arch ribs. Alternatively wire type strain gauges can be fixed with ends fixed by welding or epoxy on to the members to measure strain and co-relate for arriving at stresses. The latter are less expensive. Strain gauges have gauge length around 140mm and shorter ones of 50 mm are also available for use on struts, between diaphragm walls in cut and cover work.

Similarly in the NATM, strain gauges are embedded in the shotcrete for measuring stresses in concrete, both in radial (along the circle) direction and tangentially. They are fixed on short steel rods with ends fixed to rod by welding and embedded in shotcrete.

11.4.3 Pore Pressure Measurements

Vibrating wire type pore pressure meters are available for installation by drilling a hole through the lining or exposed rock surface for periodically measuring and monitoring any increase in pore pressure in the surrounding rock. Such increase can weaken the surrounding rock or soil. Cassagrande piezometer and standpipes can be used and measurements taken using dip meters and noting changes in levels. These are more often used in subway tunnels.

11.5 LOCATION OF MEASURING INSTRUMENTS

Figures 11.8 to 11.10 would give an idea of where such instruments and devices are fixed for monitoring deformation and forces/ stresses in the tunnels or in the zone of influence.

In a cavity or tunnel through rock, instruments for measurement of movement of the exposed surface and their convergence; load cells for measuring changes in rock pressures/ forces are most important. Additionally, the behaviour of rock beyond for some distance can be measured with use of probe extensometers fixed horizontally from the walls and vertically upward in the roof. Figure 11.8 shows one typical arrangement.

If the rock is such that it needs supports during excavation and /or needing support lining, load cells are provided in addition between joints in supporting ribs at crown and springing or alternatively strain gauges may be fixed on ribs at crown, near springing and near base of vertical support for measuring loads/ forces. This will help in design of the permanent lining.

Figure 11. 9 shows the arrangement of instrumentation done in case of tunnels bored using NATM, (as done in Pir Panjal Railway Tunnel).In this case strain gauges are embedded in the shotcrete for measuring the forces for which permanent lining is to be designed, in addition to load cells on rock bolts for understanding rock stresses and movement to be provided for in the design of permanent lining.

In case of cut and cover method, more elaborate instrumentation is called for, being in soft soil and high sub soil water levels causing high pressures and ground settlements. Soil behaviour and settlement are studied by use of a pair or more extensometer cum inclination meters on sides and PSP (Precise Levelling Points at suitable intervals on the surface Inclinometer gauge wells are provided in tunnel walls at intervals for monitoring lateral wall movements. Stand pipe or electronic type piezometers are provided for

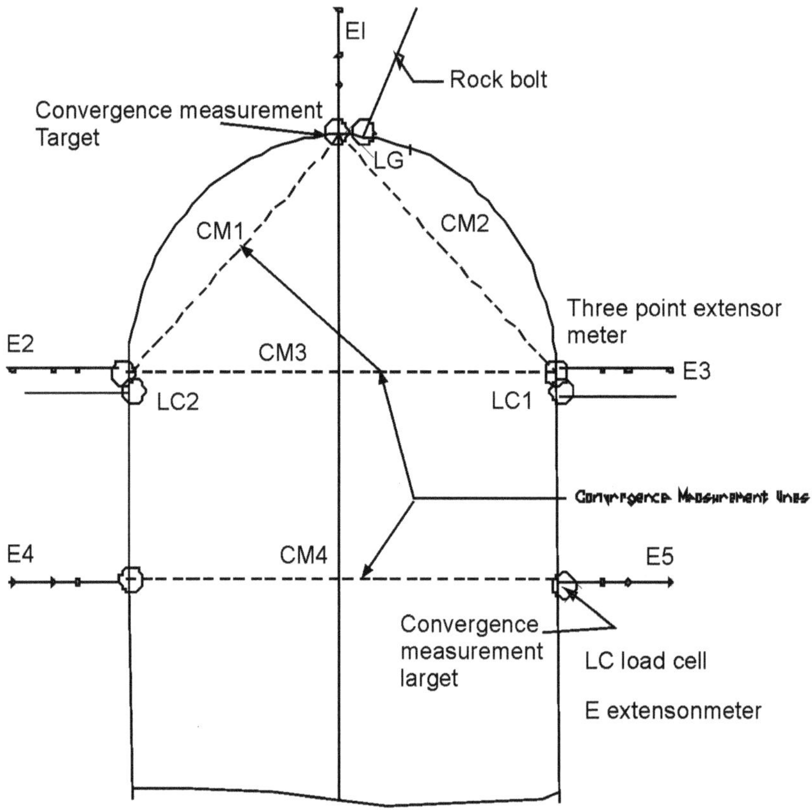

Adapted from reference 2

Figure 11.8 Typical Arrangement of Instruments in a Tunnel in Rock

ascertaining pore pressure. Heave extensometers (electronic or magnetic) are provided at intervals between the diaphragms and earth Pressure cells fixed below the floor to measure heave pressure. Figure 11.10 shows a typical arrangement.

For more details references quoted at end of the chapter may be seen. For monitoring circular subway tunnels bored with TBM, undermentioned instrumentation is done:

(i) Settlement markers on the surface to note levels perpendicular to alignment at 5 m centres
(ii) Inclinometer at a distance of 1.5z outside from centre of each tunnel bore where z is depth of axis of bore from surface.

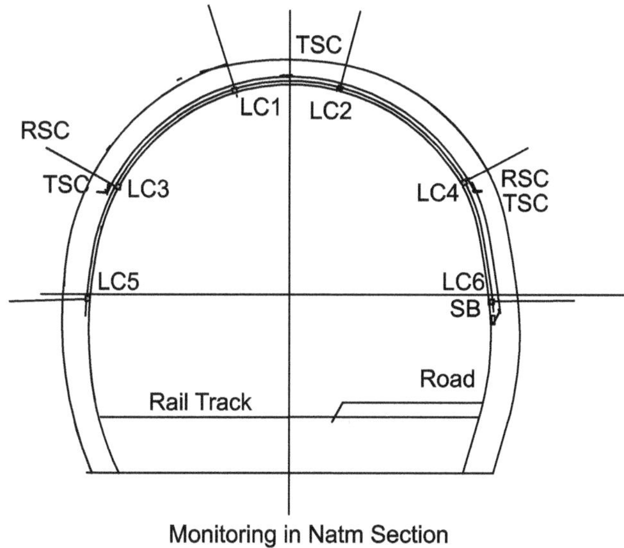

Figure 11.9 Typical Instrumentation in NATM Tunnel[2]

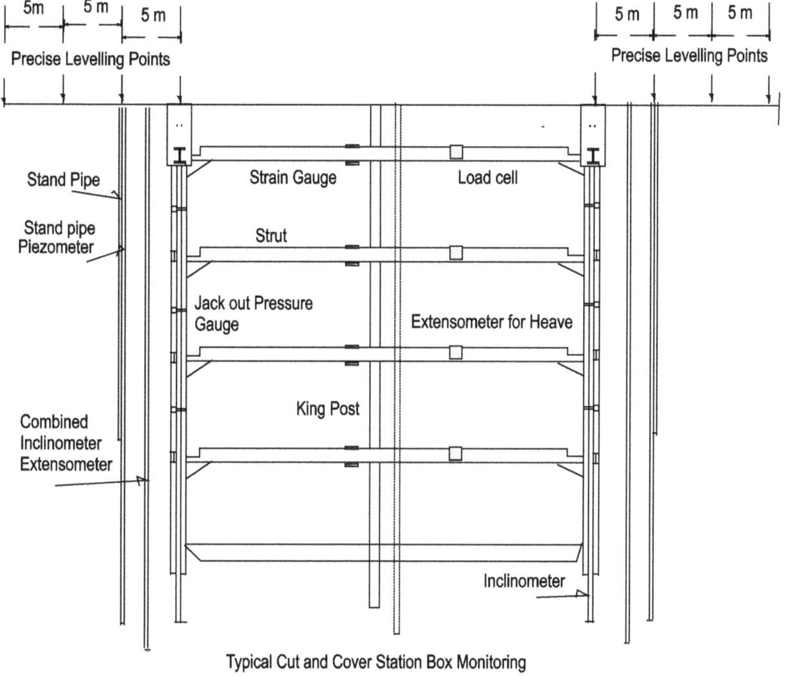

Figure 11.10 Cut and Cover Box Monitoring Arrangement of Instruments[2]

(iii) Inclinometer cum extensometer at the centre line between the bores, and one each on outer side of centre line at 0.5 D, D being the spacing between bores.

(iv) Stand pipe Piezometer and stand pipe at 0.75 D on the outside, on either side.

A view of the some of the instruments, mentioned above, can be seen in Figure 8.15

11.6 REFERENCES

1. FHWA: Technical Manual for Design and Construction of Road Tunnels- 'Chapter 15, Geotechnical and Structural Instrumentation' - USDOT, Washington, D.C.
2. Rastogi, V.K., 'Instrumentation and monitoring of underground structures and metro railway', World Tunnel Congress 2008 on Underground Facilities for Better Environment and Safety- India.

CHAPTER

12

Inspection and Maintenance of Tunnels

12.1 GENERAL

Tunnels, like any other structure require proper maintenance, which calls for periodical structured inspection and follow up action. Unlike any other over ground structure, tunnels are subjected to the stresses caused by the passing loads, have to withstand the forces exerted by the surrounding ground and overburden. While the older tunnels were generally bored through rocks and geologically more stable and settled soils, many of the newly constructed and under construction tunnels are through challenging geological conditions. Thus their maintenance throws up more challenges.

12.1.1 Basic Difference with Other Structures[1]

Physical and structural assessment of tunnels is much more complex than that of other structures like bridges, buildings, road pavements and rail track due to following differences and hence tunnel inspection procedures have to be necessarily different from those of the latter .

(i) The design of the tunnel including its profile depends greatly on the characteristics of surrounding ground and ground-structure interaction. It is very difficult to assess them precisely and our understanding of the forces they exert on and their interaction with the tunnel is limited, rendering it difficult to make proper design assumptios. The sub soil water condition of the ground also cannot be precisely predicted before design and construction as they may undergo changes over time with changes in surface water flow conditions, resulting in unexpected forces and seepage on the structures, causing deterioration of the tunnel structure.

(ii) Understanding of structural behaviour of tunnels is comparatively limited e.g., the effect of a rock bolt getting loose on the structure it supports.

(iii) Visual inspection of the tunnel intrados, the time honoured method of inspection, can only reveal surface defects like cracks, spalling, leakage of water etc. Using a small hammer to tap the surface can indicate hollowness in the structure. But such inspection does not give an idea of the health of the components above the intrados like the tunnel lining, other internal fixtures like waterproof membrane, the condition of rock bolts, and distress in the layers of overburden. On the other hand, every component of a bridge structure can be visually examined and/ or non- destructively tested.

(iv) Each tunnel located in varying types of soils and climatic condition is designed to suit local conditions and even in the same tunnel, the supporting arrangement may differ in different lengths. Each tunnel therefore becomes a unique structure by itself, while it is not so in the case of other structures like bridges for instance. Thus inspection of each tunnel offers a new challenge to understand the 'unknown'.

(v) Another factor which renders Inspection of the existing tunnels more difficult, is the non-availability of documentation on the design and construction of the old tunnels constructed in the earlier centuries. This emphasizes the need for proper documentation and preservation of permanent records of design and construction details of tunnels in recently executed or ongoing and future works.

The distinctive features of tunnel inspection described above underscore the need for the inspecting personnel to be aware of these facts as well as the previous history of the tunnel. As the normal maintenance staff may not be knowledgeable in these special aspects, they have to be suitably briefed or trained before inspecting tunnels. The Indian Railways are taking steps in this direction. Also, there may be occasions when some defects and conditions occur, the impact of which can be understood only by specialists. Hence special inspections by such personnel may become necessary periodically, say once in 5 or 6 years and at closer intervals in the case of tunnels known to be problematic

12.2 TYPES OF INSPECTION

Practices regarding tunnel inspections vary from country to country. The Indian Railways have laid down the periodicity and requirement of such inspections in Chapter 10 of the Indian Railways Bridge Manual (IRBM)[1], according to which, tunnels have to be inspected once a year in a specified month after the monsoon rains by the Section Engineer (P.Way Engineer generally and Works Engineer as required). He enters his observations in a register containing a standard proforma in two or three sheets allotted for each tunnel. It lists 8 components of the tunnel and approaches, one below other, and is tabulated in such a way that remarks of inspection for 15 years can be recorded serially in juxtaposition on each such sheet. Under each year, there are three columns to record (i) date of inspection, (ii) condition of the component and (iii) Action taken.

The section engineer, while sending the register along with a certificate listing important defects, to the Assistant Engineer of the sub-division, takes prompt action for repairs within his competence. The Assistant Engineer peruses the Register and gives explicit instructions as deemed necessary to the Section Engineer. The Assistant Engineer himself is required to inspect all the tunnels in his jurisdiction before the monsoon, accompanied by the Section Engineer. He records his observation in the Register and forwards it to the Divisional Engineer with his certificate and asking for the latter's orders. Divisional Engineer would inspect the tunnels he deems necessary and give further instructions. The Register is then sent to the Territoral Head of the Engineering Department of the Railway for his scrutiny and instructions, if any. Tunnels found to be in distress need to be inspected more frequently. These inspections at different levels have been mostly visual and carried out by normal maintenance personnel, not necessarily with special knowledge on tunnels. More rational procedures are now being adopted.

Recently (in August 2012) RDSO of the Indian Railways has supplemented these instructions with ‘Guide Lines on Inspection and maintenance of Tunnels No GE: G- 0015’. These guidelines provide for three types of inspections (as is being adopted in the case of bridges and also, tunnels in other countries). The guide lines also lay down procedures for better documentation of the inspection. (They can be followed for Highway Tunnels also). According to these guidelines, the inspection is of three kinds, viz.,

(a) Regular inspection (annually or more frequently) by maintenance engineers,
(b) Detailed Inspection by a tunnel expert once in 1 to 6 years depending on the geology and tunnel condition and

(c) When unexpected events like train derailments or occurrence of fire inside the tunnel, natural disasters like earthquakes, landslides, flooding etc. happen, special inspections may have to be conducted with the assistance of experts and special equipment to assess the damage caused to the structural components of the tunnel. There are a few other inspection procedures and record keeping measures proposed[2] in the RDSO Guidelines

(i) In order that the inspection is carried out effectively, it is necessary for the inspecting official to familiarize himself with the drawings of the tunnel if available, the individual components of the tunnel and their past history. He should study the, previous inspection reports, construction phase geological records and data obtained from instrumentation installed, if any.

(ii) There should be a referencing system established, which can be used for recording the precise location of a defect, so that its deterioration or otherwise can be monitored during subsequent inspections and remedial action taken as necessary.

(iii) The recording of the defects noticed during inspection should be detailed and specific, supported by sketches. Where the defects are considered to be of a serious nature, It will be preferable to photograph or even videograph them to have a permanent visual record. Similar action taken during subsequent inspections will help in monitoring the rate of deterioration over time. Such details should be included in a separate report accompanying the register.

(iv) Present status of the component on which a defect had been made in previous inspection should be mentioned, supported by sketches if any.

(v) Observations should be made of any visual signs of movement or deformation of the tunnel structure/its overburden by recording systematic convergence measurement. Such measurements are taken by installing permanent stainless steel reference points embedded in the tunnel wall by grout along the periphery of the tunnel section at suitable intervals along the length of the tunnel. Usually five markers are installed at each cross section, two at rail level, two at springing level and one at the crown. This will enable recording of inter distances along six diagonals and five straights. Any variation in the inter distances recorded on subsequent occasions would indicate deformation of the tunnel structure.* It is preferable to use a Tape extensometer for distance measurement to facilitate ease of use and accuracy.

(vi) A consistent abbreviation system should be adopted in recording defects. List of abbreviations should be prominently included in the Register/ Report.

(vii) A standard grading system should be adopted for indicating severity of defect and their priority for attention.

* While taking such measurements, it is assumed that the reference points themselves do not shift in position, which may not be true when the tunnel itself suffers rotational deformation. Rigorous mathematical methods have been developed for calculating the revised coordinates of the reference points due to deformation. Reference may be made to the paper on 'Absolute deformation profile measurement in Tunnels using relative convergence Measurements by Mahdi Moosavi and Saeid Khazaei, Mining Engineering Department, The University of Tehran, Iran - Proceedings, 11th FIG Symposium on Deformation Measurements, Santorini, Greece, 2003'

12.3 INSPECTION COVERAGE AND DOCUMENTATION

12.3.1 Apart from the Tunnel Inspection Register, in which the defects are entered using condition codes and comments and short sketches as mentioned above, the RDSO guidelines stipulate that a 'Supplementary Tunnel Information Register' should be maintained containing more detailed information on the defects noted, along with detailed sketches and photographs of these defects. Similar detailed information on the defects noticed during the previous and subsequent inspections should also be incorporated in the Supplementary Register, so that any improvement or deterioration in the condition of the defect over a period of time can be ascertained. As far as possible, such information should be entered in the Supplementary Register during the inspection itself. Where it is not possible because of the need to photograph or videograph the defects, such information should be entered in the Register not later than two weeks from the date of inspection The Supplementary Register should also contain reference to any special report that might have been prepared on the tunnel / defects noticed/ repairs carried out etc., but not included in the register.

12.3.2 Inspection Coverage

The elements of the tunnel and approaches to be inspected and what to look for are tabulated below (based on provisions in the IRBM 1986):

Table 12.1 Coverage of Annual Inspection of Tunnels.

Sl. No.	Segment to be inspected	Details
1	Tunnel approaches and cuttings	Any change in cutting slope, slips, condition of slope pitching, vegetation (to be cleared/ controlled), drainage at toe, catch water drains on top, any loose boulders
Ii	Portals at either end	Any signs of slips in the slopes above the portals, condition of masonry, catch water drains above the portal, any signs of percolation into the overburden, growth of any tress or hanging on slopes (to be cleared)
iii	Tunnel walls, roof, invert, lining	Any noticeable change in profile Condition of rock bolts; condition of weep holes; Any fresh seepage; Condition of seepage at joints, bolt holes, previously noticed; Condition of lining, any damage caused by any moving part. Any rust streaks, specially at cracks and rock bolt/ target insertion points
	-do-Structural condition	Any noticeable crack, condition of tell-tales on cracks noted in earlier inspections; any flaking or breakage in masonry, concrete; Condition of pointing in masonry; Any hollowness in concrete, specially at locations showing bulging
iv	Drainage	Condition of side drains inside and on approaches upto outfall, adequacy,
v	Refuges/ shelters/ cross passages	General condition, cleanliness, freedom from any vegetation, freedom from any obstructions
vi	Ventilation shafts, Adits	Their condition of the openings, whether they are free of any blockages and their adequacy. Condition of peripheral surfaces, freedom from any vegetation, slips,
vii	Integrity of profile/ clearances	Internal dimensions should be recorded at specific section and compared with previous records to detect any movement / infringement to Moving dimensions and minimum clearance requirement, Any

		signs of bulging, leaning- (Reference points should be fixed for the purpose at completion stage of new tunnels). Use of tape extensometers recommended.
viii	Track	Line, Level, condition of rail, sleepers and fittings, cleanliness of ballast; condition of base slab, plinths and fittings in the case of paved track; general drainage condition
	Road pavement	Surface evenness, soundness of the pavement, presence of any ruts, pot holes; condition of walkways and drainage
ix	Lighting, Fire fighting arrangements, C communication lines.	Their working conditions, adequacy as noticed during working should individually be inspected and defects noted and rectified.
	General	For tunnels over 200 m length, the level of pollution and temperature condition should be checked by enquiry from the patrolling keymen / gang mate on railway tunnels and other maintenance personnel in other than railway tunnels

While inspecting and recording structural defects in tunnel the component of the tunnel structure inspected should be clearly described viz. concrete, steel (structural / reinforcement), masonry, shotcrete etc and the defects classified as 'minor', 'moderate' and 'severe'[3]. Locations of any joint or any insert should be carefully examined to see if there is any widening and ingress of water or soil through it. Any sign of spalling of concrete or widening of cracks and their severity should be noted, supporting it with sketches/ photographs. The locations sounding 'hollow' on knocking with a light hammer should be tested non-destructively normally. In case of doubts, cores can be taken, and tested, taking suitable precaution to properly seal the locations where core is taken immediately, in consultation with the designer/ expert. In the case of spalls, the effect of external causative factors like leakage, ingress of toxic water, frost damage, and likely effect of pollution / smoke should be examined. The manual gives guidance for classifying defects as 'minor', 'moderate' or severe'. For example cracks upto 0.80 mm deep are 'minor', those between 0.8 mm and 3.2 mm are 'moderate' defects, and any crack deeper than 3.2 mm is a 'severe' one. Description of type of defect is also simplified by using abbreviations like B- Buckle; CR-Crack; SC- scaling; C- corrosion E- Efflorescence; LK- leaking etc..

Examination of the upper parts of tunnel wall and roof cannot be done effectively standing at the road or rail level. Stagings erected will have to be clear of moving dimensions and will have to be well held against disturbance by the moving vehicle. Hence it is desirable to use stagings erected on a trolley/ dip- lorry and use them for the purpose of inspection under 'traffic block'. Alternatively they can be erected on a wagon attached to a slow moving train or vehicle.

12.3.3 Defect Grading and Condition Rating

The RDSO Guidelines mentioned above recommends that the defects under the broad headings of inspection (listed in Table 12.1, e.g., 'Tunnel portals. Including approaches', 'Tunnel walls, roof, and invert' etc.,) should be numerically rated as given in Table 12.2 below. Rating will be based on the location, type, amount / size and extent to which the structure retains its original character and capacity despite the defect. It will therefore be necessary for the recording/ rating official to be aware of how the structure had been designed, so that he can determine to what extent the defect would affect the objectives of the design and safety margins including clearance between structure and moving vehicles.

Table 12.2 Numerical Grading of Defects

Rating	Description
5	Excellent condition- No defects found
4	Good condition- No repairs necessary. Isolated defects found
3	Fair condition - Minor repairs required but element is functioning as Original designed.
	Minor, moderate, and isolated severe defects are present but no significant loss.
2	Poor condition- Major repairs are required and element is not functioning as originally designed. Severe defects are present.
1	Serious condition- major repairs required immediately to keep structure open to traffic.

Source: Reference 2 and RDSO Guidelines GE: G- 0015.

While summarizing the inspection details and making recommendations, the inspecting official categorizes the defects as 'Severe', 'Moderate', and 'Minor'. He should also mention the priority of repairs to be carried out as 'Critical', 'Priority' and 'Routine'.

'Critical' would refer to those defects that are likely to cause danger to the traffic and maintenance personnel. They need immediate remedial action: the defects need to be under continuous observation. Vehicular traffic may require observing appropriate speed restriction while negotiating the defective section of the tunnel. In worst cases, the tunnel may have to be closed to traffic until rectification of the defects, followed by permanent remedial measures.

'Priority' attention refers to those that call for further investigations, study, design and carrying out temporary and / or permanent repairs on priority basis. It may be kept under close watch till action is taken.

'Routine' type of repairs refers to those that can be carried out as part of scheduled maintenance repairs.

A 'Tunnel Inspection Report' has to be submitted along with the Inspection Register, which inter alia should contain a summary of findings of inspection, catgorised list of repairs required, grouping them under 'minor', 'major', and 'severe' tunnel wise. It should also contain the inspecting official's recommendation regarding repairs and rehabilitation measures where called for. While giving such recommendations, he should classify repair/ rehabilitation measures under the three categories; 'Critical,' ' Priority' and 'Routine'.

12.4 MAINTENANCE, REPAIRS AND REHABILITATION

1. Cuttings on Approaches and Adits

Any wild growth and trees, likely to obstruct the view of the loco pilot or likely to endanger the stability of the slopes or cause damage to the portal or other masonry on approaches and adits should be cleared and the roots dug out. Any loose boulder on the cutting slopes and slopes behind Portals likely to drop on the track or roadway should be removed by jacking and levering. Those that cannot be removed should be broken, if necessary by blasting and cleared. Side drains on the approaches and catch water drains on the top of the cuttings and behind portals should be cleaned. Likewise, any cracks in the masonry/ pitching provided at such locations should be repaired and rendered leak proof so that there is no danger of water seeping down through them and causing slips. The ground surface in the zone of influence should be examined for presence of any cracks which may let water seep down and cause slips or seepage in the tunnel. Where necessary, retaining walls should be provided at the toe of slopes.

2. Portals

Any cracks or peeling of masonry or damage to the portal walls, arch etc., should be promptly repaired with polymer or other suitable mortar, epoxy resin etc.. Any lean in the walls should be corrected with additional strengthening arrangements. In the worst case, it will be better to replace the affected portions of the wall under traffic block. If there is low overburden on the portal, it may cause seepage / leakage in the tunnel in that the vicinity. Proper catch water drains, through insertion of perforated pipes may have to be provided at such locations to prevent accumulation of water and seepage into the tunnel.

3. Tunnel walls, roof etc.

Cracks in the tunnel wall and roof which are wide should be epoxy grouted and their behavior kept under watch. If such cracks are accompanied by any bulge in the masonry, reasons there for should be explored by checking the design and studying whether there has been of any change in the local geology. Remedial measures should be decided under expert advice. If there are any loose rock bolts found, reasons for their looseness, including possible corrosion should be explored. Spalling of concrete should be attended to, as any major repair to concrete, by proper recessing, cleaning by jetting or blowing air and finishing with polymer based concrete or mortar to suit the location. Heavy spalling may be attended to with application of concrete after providing a few dowels into base concrete and shotcreting preferably with wire mesh reinforcement. At locations with hollow concrete, the bad concrete should be chipped out and replaced in the same manner, if it is considered local. Major cases of hollow concrete need investigation to see if there is any external cause for such hollowness.

Leakage: Seepage and leakage are among the major maintenance problems in tunnels, especially in old tunnels. They can aggravate and lead to difficult problems like damage to/ deterioration in linings. It may lead to muddy track also.[1] They occur generally where the overburden depth is low and is made of porous material. Presence of underground springs and streams can also cause such seepage/ leakage. Leakages generally occur at joints in masonry lining and construction joints in concrete. Seepage over a wide area should be carefully examined to locate the cause. Functioning of weep holes should be checked and blocked ones cleared of obstructions, talus etc.. If any of them cannot be cleaned, another weep hole should be drilled close by. Another source of leakage is old shafts. The surface of the rock/ lining should be kept clear of any soot and deposits and it may be preferable to clean and whitewash the shaft periodically for providing better illumination.

In case the seepage is heavy, as an immediate relief, holes at about 1m intervals, staggered if more than one row is required, should be drilled through the lining at a level as near to the side drain as possible and the water led into the drain by providing proper grooving. The cause of heavy seepage should be examined and if possible, accumulation of water above the lining can be drained away by driving horizontal perforated pipes through the side soil. Construction joints or cracks with seepage should be sealed by cutting a vee-shaped groove along the crack/joint and sealing it with proper grouting/ guniting and caulking. Where stopping leakage by this method is not successful at joints or cracks, it will be preferable to attend to the joint by sealing/ shotcreting but with a recess along, and to let the seeping water out by drilling a series of holes through in the recessed drain channel upto the side drain below.

4. Changes in Profile

If any signs of closing in on the walls or settlement in roof is noticed, it should first be checked whether there is any infringement to moving dimensions: if so, suitable speed restrictions should be imposed and the same location monitored by frequent checking. If it is noticed that movement is not stabilised, but continues, more detailed investigations will need to be done to study the external movements of the soil/ ground and remedial action taken with expert advice.

5. Shafts/ Adits

The main task is to keep there the shafts/adits clean and clear and any structural defect must be attended to, so that they serve the purpose they were intended to serve.

6. Track and Road pavement

In addition to the normal day to day or periodical attention they receive, immediate remedial action should be taken to rectify any defect found during these inspections on a priority basis. Drains should be cleaned periodically and it is preferable to cover the drains for workmen to walk freely.

Similarly the other utilities like Ventilation, lighting, fire fighting facilities, water pipe line and communication lines and equipment will need to be kept in working order and any defects found during these Inspections should be attended to on priority.

12.5 PRACTICE ELSEWHERE

The Inspection and Systematic maintenance of tunnels and Safety aspects is now being given significant importance in other countries like USA, Japan etc.. They have brought out Manuals, devoted for the purpose and/ or Safety Aspects[4]. Examples are:

(i) **USA:** Tunnel Management System - brought out jointly by the Federal Highway Administration and Federal Transit Administration: containing five chapters viz., Introduction; Tunnel construction & Systems; Fundamentals of Tunnel Inspection; Inspection Procedures; General Discussion and Inspection Documentation. (Reference 4 gives more details of the contents.)

(ii) **Japan:** Japanese Railways had been adopting the traditional manual methods of inspection in the past. But 'Smart technology' tunnel inspection methods are increasingly being used, such as Image scanning system of tunnel lining using line-sensor CCD camera, Image scanning system of tunnel lining using laser beam, Tunnel lining inspection system Using Multipath array radar, as described in a Paper on 'Maintenance of railway tunnels with smart technology' by KIWAMU TSUNO Railway Technical Research Institute submitted in CSIC/JSPS International Symposium 13 Nov 2012

(iii) **German Railways:** Guidelines ER1 (1)

(iv) **Swiss Railways:** Guidelines SIA 197 (11), (12)

(v) **UK-** One chapter in their Manual for Bridges and Tunnels.

The FHWA Manual is very exhaustive and it covers a chapter on Design aspects also. It is like 'hands on' guide book for Tunnel Inspection. Chapter 3 in the manual covers exhaustively what to inspect in each major component, discipline wise (civil/ structural; mechanical; and electrical components) instructions and formats to be used. Chapter 4 covers frequency of inspection, what to look for in each component, how to classify defects and numerical grading of the component concerned and action to be taken. Two tables have been suggested for grading. Table 4.01 is a General Condition Code. Table 4.02 is in more detail for covering each type of lining like cut and cover lining, linings in tunnels in soft ground, linings in tunnels in rock and timber lined tunnels. They have adopted a 10 point grading o to 9 in both cases. - Grade 9 refers new condition; 8 - Excellent with no defect; 7-Good - with isolated defects; 6 - Shade in between 5 and 7; 5- Fair- minor repairs required, element functioning as intended; 4- Shade between 3 and 5; 3-Poor - Major repairs required to keep it open; 2 Serious- Major repairs required to keep it open for traffic; Grade I is for 'Critical condition when the tunnel is to be closed for traffic and study of feasibility of repairs; Rate 0 refers to a tunnel closed and beyond repairs.

Frequency of detailed inspection specified, apart from daily, weekly, monthly routine inspections, is once in 5 years for new tunnels and upto once in two years for older tunnels, depending on the age and condition of tunnel; for mechanical equipment like fans, pumps, motors, etc. Table 4.03 of the Manual indicates a frequency of inspection varying from daily, fortnightly, monthly, quarterly and annual, and going upto ten years. The percentages of tunnel owners who enforce these inspections are also indicated.

While describing individual defects, they are to be classified as ''Minor', 'Moderate', or 'Severe' according to criteria laid down in the manual for each component of the tunnel. Though the method of inspection suggested is visual and non- destructive methods, it gives the inspector freedom to call for collection of samples (cores for determination of strength, freeze/ thaw characteristics etc.)[4].

For 'track' in Railroad tunnels, it lays down instructions regarding tolerances in various elements of track geometry and condition of individual components like 'rail', 'sleeper' etc. For more details, References 4 and 5 may be referred to.

12.6 RECENT DEVELOPMENTS

Modern practices in many countries abroad are to do non invasive study of the condition of tunnels and overburden.

12.6.1 Geophysical Techniques

Modern geophysical techniques can be harnessed to determine ground properties and geometry for heights varying from a few meters to more than one kilometer above the tunnel. Multichannel seismic reflection methods, developed for Oil and Gas exploration, can be used to determine the structural discontinuities in the overburden such as faults, folds or cavities in the bed rock strata. This will be of great use both in new tunnel projects and in maintenance of existing tunnels. In the past two decades, Ground penetrating radar (GPR) is being increasingly used as a non invasive technique, particularly in masonry and concrete lined tunnels to assess the condition of the lining, the faults in the overburden or hidden construction shafts. GPR uses electromagnetic radiation in the microwave band (UHF/ VHF frequencies) of the radio spectrum, and detects the reflected signals from subsurface structures. The radar survey is conducted by repeatedly profiling an antenna along the tunnel intrados generally using a hydraulic access platform which can be road or rail mounted.

12.6.2 MOBILE INSPECTION SYSTEM[6]

A new mobile inspection system for an accurate assessment of the tunnels has been developed by Euroconsult Group in Spain. Known by the name "Tunnellings", it is capable of taking 3D and 2D photographs of the visible tunnel structure as it moves in the tunnel. It basically consists of sensors comprising a laser light source and a digital camera fixed at an angle to the light source. See Figure 12.1.The sensors are pointed towards the interior surface of the tunnel, the light switched on and the digital camera operated. The distances to each point are calculated by trigonometry and used to prepare 3-D and 2-D photographs. A maximum of six sensors and cameras are placed in a frame with adjustable arms to suit the profile of the tunnel and mounted on a rail or road vehicle. These six sensors are capable of covering one half of the tunnel perimeter during the forward movement of the vehicle, the other half being covered during the return run of the vehicle The camera has a depth accuracy of 0.5 mm and a longitudinal accuracy of 1 mm (adjustable).

Source: Reference 6

Figure 12.1 A road mounted vehicle carrying the sensors

The monitoring and control system consists of an industrial computer for the sensors' synchronization and a monitor for control and monitoring tasks. A high resolution odometer is also provided to record the longitudinal distances. The vehicle is moved along the tunnel at speed of 30 kmph to record the data of the tunnel surface. The 3-D and 2-D images are captured and stored in the computer and can be downloaded to a computer in the central office to have a permanent record of the inspection. Such automation of inspection is claimed to be very useful in maintenance of very long tunnels, greatly reducing manpower requirements for manual inspection of long tunnels. A typical image captured during inspection of a lining is presented in Figure 12.2. Even 2-D pictures are clear enough to show cracks and surface defects on the lining.

A comparison of a number of inspections of the same tunnel can be carried out quickly also. Thus structural defects and evolution of defects in the lining can be assessed with this facility. This comparison facility can make it possible progress to assess of any geometric variations like convergence, segment displacements etc., on the section.

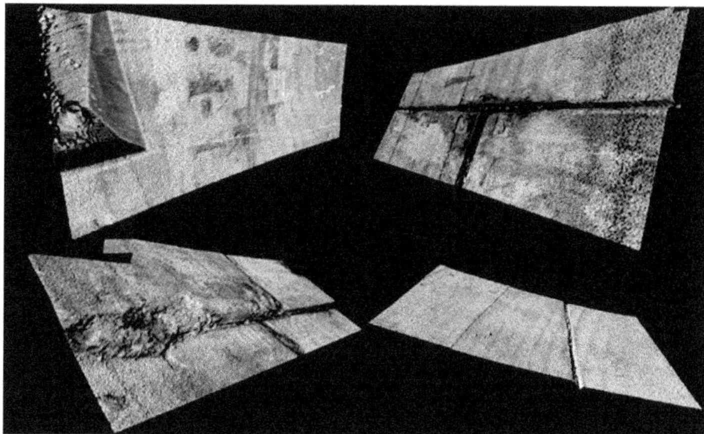

Source: Reference 6.

Figure 12.2 Photograph showing a 3-D reconstruction of the data logged, giving a clear picture of the health of the tunnel lining, including cracks, leakage and spalls

The system is also capable of assessing the condition of the permanent way and other track structure in the tunnel. It includes an assessment of transverse section (line and level) by means of '3-D geometry,' and detection flaws like corrosion, defects in fixtures, cracks in sleepers/ plinth of slab tracks etc..

12.7 REFERENCES

1. IRBM (1986) - Indian Railways Bridge Manual - Chapter 10; Railway Board, Ministry of Railways, New Delhi
2. Brijesh Kumar and Sanjay Mishra (2014), 'Tunnel Inspection - A Practical Approach: Key Provisions of RDSO' National Technical Seminar on 'Management of P.Way Works through Need Based Outsourcing & Design, Construction and Maintenance of Railway tunnels' Jaipur 2014. (Volume II)-pp 357-369
3. RDSO, (2012), Guidelines No GE: G-0015. 'Guidelines for Civil Engg. Inspection, Maintenance and Safety in existing Tunnels' 41P
4. Agarwal, M. and Miglani, K.K. (2014) -'Global Experience of Design, Construction and Maintenance of Railway Tunnels with Special Reference to The Indian Railways', National Technical Seminar on 'Management of P.Way works through Need Based Outsourcing & Design, Construction and Maintenance of Railway tunnels' Jaipur 2014. (Volume II), Institution Permanent Way Engineers (India), New Delhi.
5. 'Highway and Rail Transit Tunnel Inspection Manual', US Department of Transportation, Federal Highway Administration.
6. Gavilán M., Sánchez M, F., Ramos and J.A., Marcos, O., (2013) 'Mobile inspection system for High-resolution assessment of tunnels by EUROCONSULT GROUP', 6th International Conference on Structural Health Monitoring of Intelligent Infrastructure, Hong Kong, December 2013

Abbreviations

BLT	Ballastless Track
CL	Centre Line
CMRL	Chennai Metro Rail Ltd.
DBK	Dandakaranya Bolangir Kiriburu Project
DMRC	Delhi Metro rail Corporation
ETBM	Earth Balancing type Tunnel Boring Machine
GPR	Ground Peneteration Radar
ICJ	Indian Concrete Journal
IPWE (I)	Institute of Permanent Way Engineers (India)
IR	Indian Railways
IRICEN	Indian Railways Institute of Civil Engineering
ITA	International Tunnelling Association
KRCL	Konkan Railway Corporation Ltd..
LEL	Lower Explosive Limit
MPBX	Multiple Point Borehole Extensometer
NATM	New Austrian Tunnelling Method
NGI	Nolwegian Geotechnices Institute
NMT	Norwegian method of Tunnelling
PSC	Prestrssed Concrete
RC, RCC	Reinforced Concrete
RDSO	Railway Research and Designs Organisation
RMR	Rock Mass Rating
RQD	Rock Quality Designation
SDA	Self Drilling Anchor

SOD	Schedule of Dimensions
SEM	Sequential Excavation
SPBX	Single Point Borehole Extensometer
SPL	Springing Level
SVJNL	Sutluj Vidhyut Jal Nigam Ltd..
TBM	Tunnel Boring Machine
tlm	Tunnel lane Metre
UCS	Ultimate Compressive Strength
UIC	International Union of Railways
USBRL	Udhampur Srinagar Baramulla Rail Link

Bibliography

'Closed Face Tunnelling Machines and Ground Stability', A Guideline for Best Practice. Institution of Civil Engineers, London

Advances in Tunnelling Technology and Subsurface Use, 2:3 (1982), International Tunnelling Association

Bickel, John O., and Kuesel, TAR., (Eds.) 1982, `Tunnel Engineering Handbook', Van Nostrand Reinhold Company, New York, 670 P.

Bindra, S.P. and Bindra, K.1976. 'Elements of Bridge, Tunnel and Railway Engineering', Dahnpat Rai & Sons, New Delhi

Bowles, S.K., 1985, 'Foundations Analysis and Design', McGraw-Hill Kokagusha Limited, Tokyo, Japan (2nd Edn.), 816 P.

'Code of Practice for Construction of Tunnels– IS: 5878-

- Part I-1971: 'Precision Survey and Setting Out',
- Part II-1970-Section 1: 'Underground Excavation in Rock-Drilling and Blasting'
- Part II-1971-Section 2: 'Underground Excavation in Rock-Ventilation, Lighting, Mucking and Dewatering.
- Part III- 'Underground Excavation in Soft Strata' Bureau of Indian Standards, New Delhi.

IS:4756-1968, "Safety Code in Tunnelling", Bureau of Indian Standards, New Delhi.

IS:4089-1967, "Safety Codes in Blasting and Related Drilling Operations" Bureau of Indian Standards, New Delhi.

Code of Practice 2006 Tunnels under Construction-Work Cover NSW, Australia.

Daw, A.W. and Daw, E.W., 1909, 'The Blasting of Rock mines, quarries, tunnels', Second Edn. Spon, London

Finn,B. and Michelson, P.G., 2002, 'Evaluation of Wide offset angle and Pre-stack depths Migration for Application to Seismic Imaging of Complex Geology.

Gallagher, V.P., 1985, 'Tunnel Lighting Design Procedure, Report No. FHWA-IP-85-9', US Department of Transportation, Washington,

Gallagher, Vincent P. (1985) 'Tunnel Lighting Design Procedures', Report No. FHWA-IP-85-9, Federal Highway Administration, U.S. Dept. of Transportation, Washington.

Highway and Rail Transit Tunnel Inspection Manual, US Department of Transportation, Washington 2003

Hooper, S.K., 1970, 'Foundation Interaction Analysis', in: "Developments in Soil Mechanics," C.R. Scot (Ed.), Applied Science Publishers, London. IS: 4089-1967,

Kani, Gasper, 'Analysis of Multi-storey Frames', Frederic Ungar Publishing Company, New York, 664 P.

Lesser, N., Horowitz, F., and King, K., 1987, 'Transverse Ventilation System of the Holland Tunnel Evaluated and Operated in Semi-Transverse Mode', Transportation Research Record 1150, USDOT, Washington, D.C.

Limaye, G.K., 1979, 'Tunnels-Lecture Notes', Institution of Advanced Track Technology, Pune (Unpublished).

Megaw, T.M. and Bartlett, J.V., 1982, 'Tunnels: Planning, Design and Construction', Vol. 2, Ellis Harwood, 321 P.

Morton, D.J., 1982, 'Subway Tunnels', In: 'Tunnel Engineering Handbook', Bickel, John 0. and T.R. Kuesel, (Eds.), Van Nostrand Reinhold Company" New York, pp. 417-444.

Narayanan, G.A., 1988, Air-Conditioning and Ventilation of Calcutta Metro', International Seminar on Transit Technology of Tomorrow, Bombay.

Narayanan, G., 2009, 'Aerodynamic Problems in Tunnels in High Speed lines', International Technical Seminar on High Speed Corridors and

Higher Speed on Existing Network', Mumbai, Institute of Permanent Way Engineers (India).

National Technical Seminar on 'Management of P.Way Works through Need Based Outsourcing & Design, Construction and Maintenance of Railway tunnels' Jaipur 2014. (Volume II)

Padmanabhan, V.C.A., 1965, 'Notes on D.B.K. Railway Project', Unpublished Indian Railways Report, 143 P.

Parker, A.D., (1970)' Planning and Estimating Underground Construction', McGraw-Hill Publications, New York, 300

Pequignot, C.A., 1963, 'Tunnels and Tunnelling', Hutchinson Scientific and Technical, London, 555 P.

Pokrovsky, N.M., 1977, 'Driving Horizontal Workings and Tunnels', Translation by Savic, I.V., M.I.R. Publishers, Moscow, 421 P.

Proceedings of International Seminar on 'Metro Railways-Profile and Prospects' Kolkata 1985. Metro Railway, Kolkata.

Proctor and White, 1946, "Rock Tunnelling with Steel Supports', Commercial Shearing and Stamping Co, Youngstown, Ohio

Ramamurthy, T (2011), 'Engineering in Rocks for Slopes, Foundations and Tunnels',

RDSO, Ministry of Railways, Lucknow, 'Important Tunnels on Indian Railways',Research, Design and Standards Organisation.

Reynolds, H.R. 1961, 'Rock Mechanics', Crossbey- Lockwood, London

RITES, 1993-Project Report for Metro Rail, Delhi (Unpublished).

'Safety Code for Tunnelling', Bureau of Indian Standards, New Delhi. IS: 5878 (Part 1)-1971.

'Safety Code for Working Compressed Air', Bureau of Indian Standards, New Delhi. IS: 4756-1968,

'Safety Codes for Blasting and Related Drilling Operations', Bureau of Indian Standards, New Delhi. IS: 4137-1967,

Seetharaman, R., Tunnel and Airport Engineering, Umesh Publications,

Sen, P.K., 1985, 'Computerised Analysis of Subway Boxes for Calcutta' Metro'. Proceedings of international Seminar on Metro Railway-Problems and Prospects, Calcutta-Paper 11.

Sharma, S.C., 'Tunnel Engineering, Jain Book Company, New Delhi

Silas H. Woodard 1943, 'Dams, Aqueducts, Canals, Shafts, Tunnels' in Ed Thadeus Merriman,T and Wiggin H.Wiggin, 'Civil Engineers Handbook, John Wiley and Sons Ltd., New York, 5th Edn 1943.pp 1614-1652,

Srinivasan, R "Harbour, Dock and Tunnel Engineering', Charotar Publications, New Delhi.

Szechy, K 1973, ' The Art of Tunnelling', Akademiai Kiado, Budapest, Hungary (2nd Edn.). Thadani, B.N., 1964, 'Modern Methods in Structural Mechanics', Asia Publishing House, London

Technical Manual for Design and Construction of Road tunnels', (in 15 Chapters) - Bridges and Structures, US Department of Transportation, Washington web site.

Turner, J., 1959, 'Newest Trends in the Design of Underground Railways', Journal of Institution of Civil Engineers. Vos, Charles J., (1993),

Vinod Kumar (2008), 'Construction of India's latest Underground Railway Project', Underground Facilities for Better Environment and Safety, India-World Tunnel Congress, 2008, pp 1301-1312

Vos, Charles (1993) 'Submerged Tunnels Examples of Maxima Structures', Seminar Papers of International Association of Bridge and Structural Engineers Conference, New Delhi.

Index

D

E

F

G

H

I

J

K

L

M

N

O

P

R

S

T

U

V

About the Authors

Dr S. Ponnuswamy is former Additional General Manager, Southern Railway, Chennai. He took his B E (Civil) from Madras University, M S (Transportation) from IIT Madras and PhD from Anna University. He is a Fellow of the Institution of Engineers (India), Fellow of the Chartered Institute of Logistics and Transportation, London and Member, Institution of Structural Engineers, London. He gained wide experience on planning, design, construction and maintenance of Rail and Road way and structures serving in different parts of the country and in Nigeria. Since his retirement from Railways, he had been a Guest Faculty in Indian Institute of Technology Madras and College of Engineering, Guindy teaching Transportation System and Transport Structures related subjects. He has been associated as a Consultant on several Transportation and Bridge projects for RITES, New Delhi. He has authored over 40 papers published in India and abroad, and authored a book each on 'Bridge Engineering' and 'Railway Transportation' and co-authored a book on 'Urban Transportation'.

Late Dr David Johnson Victor, former Professor of Civil Engineering and Head of Transportation Division, Indian Institute of Technology, Madras had earned his BE (Civil) from Madras University, M.Tech (Structures) from IIT, Kharagpur and Ph.D. from University of Texas, Austin Texas, USA. He was a Fellow of the Institution of Engineers (India). He founded and headed the Transportation Division for over 18 years. He served on invitation as Professor at Toyohashi University of Technology, Japan. He had long and wide experience in Teaching, Research and Consultancy in fields of Transportation Planning and Structures and authored over 75 papers published in India and abroad. He had been awarded medals and prizes for three of his research papers from the Institution of Engineers (India) including Railway Board Gold medal and E. P. Nicholaidas prize and commendation certificate for one from the Indian Roads Congress. He had participated in a number of Indian and international conferences and chaired some sessions on Transportation. His book on 'Essentials of Bridge Engineering', now in sixth edition is adopted as a textbook in most universities in India and he is the lead author of the book on 'Urban Transportation'.

www.ingramcontent.com/pod-product-compliance
Ingram Content Group UK Ltd.
Pitfield, Milton Keynes, MK11 3LW, UK
UKHW011454240425
457818UK00021B/825